Use R!

Series Editors
Robert Gentleman Kurt Hornik Giovanni Parmigiani

More information about this series at http://www.springer.com/series/6991

Jaroslaw Harezlak • David Ruppert • Matt P. Wand

Semiparametric Regression with R

Springer

Jaroslaw Harezlak
School of Public Health
Indiana University Bloomington
Bloomington, Indiana, USA

David Ruppert
Department of Statistical Science
Cornell University
Ithaca, New York, USA

Matt P. Wand
School of Mathematical
and Physical Sciences
University of Technology Sydney
Ultimo, New South Wales, Australia

ISSN 2197-5736 ISSN 2197-5744 (electronic)
Use R!
ISBN 978-1-4939-8851-8 ISBN 978-1-4939-8853-2 (eBook)
https://doi.org/10.1007/978-1-4939-8853-2

Library of Congress Control Number: 2018953727

This Springer imprint is published by the registered company Springer Science+Business Media, LLC part of Springer Nature.
The registered company address is: 233 Spring Street, New York, NY 10013, U.S.A.

To Xiaochun, Anna, and Sofia

—Jarek

To Susan, Joan, and Dale

—David

To Handan, Declan, and Jaida

—Matt

Preface

Our goal is to provide an easy-to-follow applied book on semiparametric regression methods using R. Semiparametric regression has a large literature, but much of it is geared towards data analysts with advanced knowledge of statistical methods. This book is intended for applied statistical analysts who have some familiarity with R. It is accompanied by the website:

semiparametric-regression-with-r.net

and contains pointers to datasets and R code relevant to this book. Given the ever-changing nature of software, we recommend regular checking of this website for news, updates, and errata.

Semiparametric regression builds on parametric regression models by allowing for more flexible relationships between the predictors and response variables. For example, the impact of alcohol consumption on heart disease risk may be U-shaped. Semiparametric regression accommodates this type of relationship. Examples of semiparametric regression include generalized additive models, additive mixed models, and spatial smoothing. R now has a great deal of semiparametric regression functionality. However, many of these developments have not trickled down to rank-and-file statistical analysts. This book helps to close the gap between the available methodology and their use in practice.

Semiparametric regression research continues to progress at a rapid pace, with R as the dominant computing environment. Many semiparametric regression papers now include R code or a reference to publicly available R packages. Mixed model and hierarchical Bayesian representations of semiparametric regression models mean that packages for general mixed model analyses and Bayesian inference also play an important role in contemporary semiparametric regression.

We have assembled a broad range of semiparametric regression R analyses and put them in a form that is useful for applied researchers. Where feasible, we provide R code in the text. Full code for all examples is contained in scripts within the R package HRW that accompanies this book.

For their detailed checking and generous feedback during the book's final production phase, we thank Alex Asher, Ray Carroll, Eli Kravitz, and Tianying Wang. In addition, we are grateful for the help received from Brian Caffo, Bob

Carpenter, Ciprian Crainiceanu, Andrew Gelman, Trevor Hastie, Giles Hooker, Hon Hwang, Chris Jones, Cathy Lee, Yang Liu, Thomas Lumley, Marianne Menictas, Fabian Scheipl, Kunlaya Soiaporn, Yi Su, Elizabeth Sweeney, Julie Vercelloni, Naisyin Wang, Simon Wood, Luo Xiao, Thomas Yee, James Yu, and Vadim Zipunnikov.

Bloomington, Indiana, USA — Jaroslaw Harezlak
Ithaca, New York, USA — David Ruppert
Ultimo, New South Wales, Australia — Matt P. Wand
June 2018

Contents

Chapter 1
Introduction

1.1 Semiparametric Regression

Regression is used to understand the relationships between *predictor variables* and *response variables* and for predicting the latter using the former. In *parametric regression*, the effect of each predictor has a simple form, for example, is a linear or exponential function, so that its overall shape is dictated by the model, not the data. In contrast, with *nonparametric regression* the model is flexible enough to allow any smooth trend in the data; see Fig. 1.1 for an example. *Semiparametric regression* combines parametrically modeled effects for some predictors with nonparametric modeling of the effects of the other variables.

Because of its flexibility, semiparametric regression has proven to be of great value in many applications in fields as diverse as astronomy, biology, medicine, economics, and finance. Using semiparametric regression models, one can extract important information from often messy datasets. An introduction to the field can be found in the book *Semiparametric Regression* by Ruppert et al. (2003) and its follow-up survey article, Ruppert et al. (2009).

1.2 The R Language

R (R Core Team 2016) is a major computing programming language for statistical methodology. The emergence of R around the start of mainstream Internet usage in the mid-1990s leads to a revolution of sorts and has allowed statistical methodologists from around the world to share their code much more easily than ever before, using the so-called *packages*. The primary website for R is the *Comprehensive R Archive Network* (`cran.r-project.org`) and contains the latest version of R and thousands of packages. Familiarity with R is assumed throughout this book. A reader with no such familiarity should first consult some of the numerous R

J. Harezlak et al., *Semiparametric Regression with R*, Use R!,
https://doi.org/10.1007/978-1-4939-8853-2_1

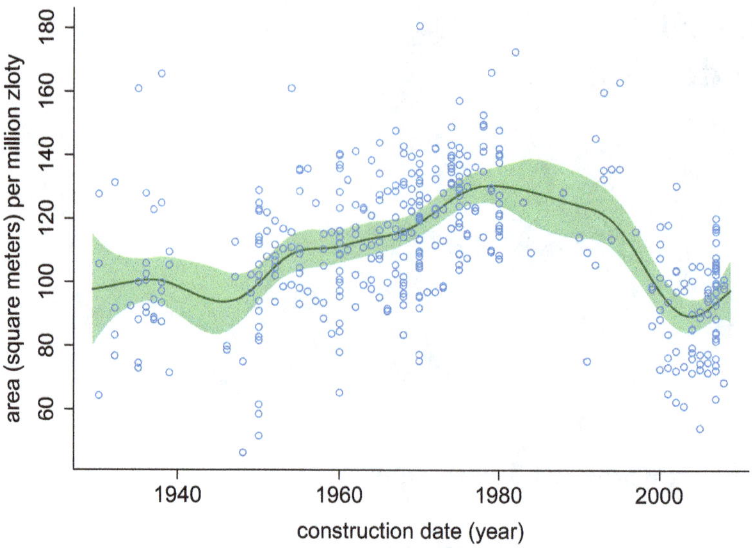

Fig. 1.1 Area/price ratio versus construction date for the Warsaw apartment data in the data frame `WarsawApts` within the R package `HRW`. The curve is an estimate of the mean area/price ratio given the construction date. The shaded region indicates approximately 95% pointwise confidence intervals.

tutorials and notes that are available on the Internet, including on the Comprehensive R Archive Network.

1.3 Some Examples

To illustrate features of semiparametric regression, in this chapter we discuss one dataset taken from each of the subsequent chapters. Table 1.2 at the end of the chapter describes all datasets used in this book.

1.3.1 Warsaw Apartments

The Warsaw apartments dataset is used throughout this book's early chapters to illustrate the most fundamental semiparametric regression models. It contains data on several variables for 409 apartments sold in the city of Warsaw, Poland, during 2007–2009. The data are stored in the data frame `WarsawApts` within the R package, `HRW`, that accompanies this book. This data frame is a subset of one named `apartments` in the R package `PBImisc` (Biecek 2014). The full description of `apartments` can be found in the `PBImisc` package's help files.

A question of interest is how the ratio of floor area to price depends on the construction date. The basic unit of currency in Poland is the *złoty*. Figure 1.1 contains a plot of area per million złoty versus construction date with a nonparametric regression function estimate and *variability bands* which have approximately 95% pointwise confidence interval validity. "Pointwise" means that there is a 95% coverage probability at each value of the predictor. We see from Fig. 1.1 that there is an interesting nonlinear relationship between area/price ratio and construction date. The first three turning points in the mean function correspond to major events in Warsaw's history: (1) the German invasion of 1939, (2) the end of World War II and beginning of communist rule in 1945, and (3) the start of martial law in 1981. During communist rule building quality declined. Hence buildings constructed in 1975 have a larger mean area/price ratio compared with those constructed before 1940. Poland became a democracy in 1989 and around 2000 pre-war building quality was restored. In Chap. 2, we use the `WarsawApts` dataset to illustrate the basic concepts of semiparametric regression modeling.

Another question of possible interest is "Are there differences between districts of Warsaw in terms of how construction date impacts the area/price ratio?" Figure 1.2

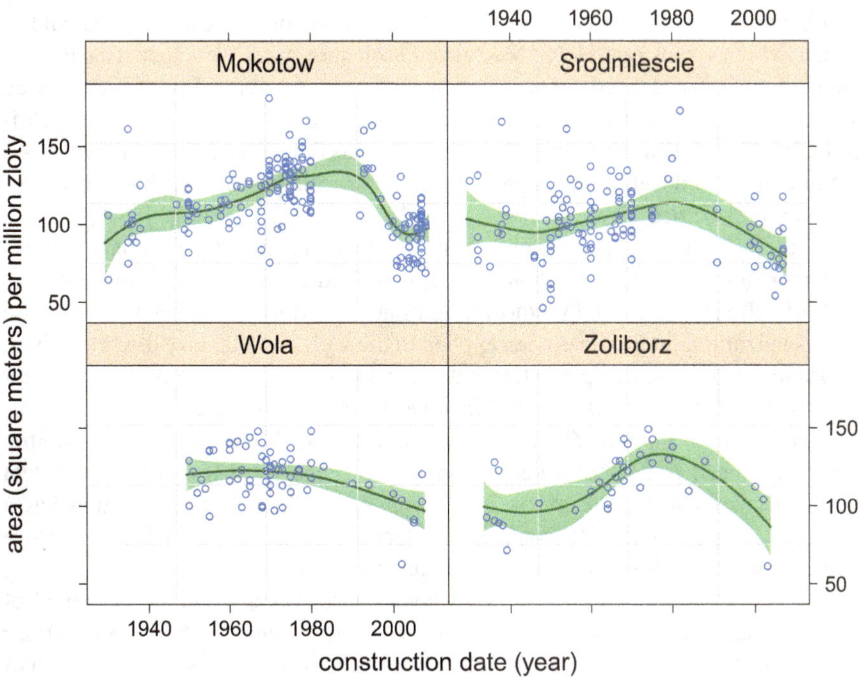

Fig. 1.2 The same data as shown in Fig. 1.1 but broken down according to the district in Warsaw in which each apartment is located. The curve in each panel is an estimate of the mean area/price ratio given the construction date for that district treated separately. The shaded regions indicate approximately 95% pointwise confidence intervals.

plots the data of Fig. 1.1 broken down according to district. This plot uses graphics supported by the R package `lattice` (Sarkar 2017). The regression function estimates and approximate pointwise confidence intervals are obtained individually for each district. Some differences among the districts are apparent. For example, the Mokotow curve is higher than that for Srodmiescie—the latter being the central business district of Warsaw. This suggests that buyer's get more floor space for their money in Mokotow than Srodmiescie for apartments built around the same time.

1.3.2 Boston Mortgage Applications

Generalized additive models (GAMs) are useful when there are several predictors each having a nonlinear effect. In GAMs, the linear predictor is a sum of nonparametrically modeled functions of univariate predictors. GAMs are covered in Chap. 3.

We illustrate GAMs using a dataset concerning mortgage applications in Boston, USA, during the years 1997–1998. The data frame `BostonMortgages` in the `HRW` package contains data on several variables concerning 2380 applications. `BostonMortgages` is a subset of the `Hdma` data frame in the package `Ecdat` (Croissant 2016). This name "Hdma" is an apparent typographic error and should be Hmda, which stands for "Home Mortgage Disclosure Act." We selected a subset of the predictors and deleted cases with missing values to create this smaller dataset. The response of interest is `deny`, the status of the mortgage application which is coded as "yes" when the mortgage application was denied and "no" otherwise. We are interested in developing a regression model for the probability that a mortgage application is denied.

Figure 1.3 is a visual display of the data in which the variable of primary interest, *indicator of mortgage application denied*, is plotted against the 12 other variables in `BostonMortgages`. The yes/no variables are coded as $0 =$ no and $1 =$ yes. To aid visualization, jittering has been applied to the variables that take discrete values.

There are 12 possible predictors but, for now, we concentrate on the predictor *ratio of the debt payments to total income* which is shortened to *debt payments to income ratio* in Fig. 1.3. The curve in Fig. 1.4 shows that the probability that a mortgage is denied is decreasing in the range from 0 to 0.3 of the debt payments to income ratio and is increasing after 0.3. The shaded region has a pointwise approximate 95% confidence interval interpretation. In Sect. 3.3.3, we will incorporate additional predictors that feature in Fig. 1.3.

Munnell et al. (1996) investigated whether race was a factor in the denial of mortgage applications after adjustment for the other variables. The variable `black` is the indicator of Black or Hispanic ethnicity. In Chap. 3 we investigate the effect of `black` using semiparametric regression to adjust for possible confounding variables.

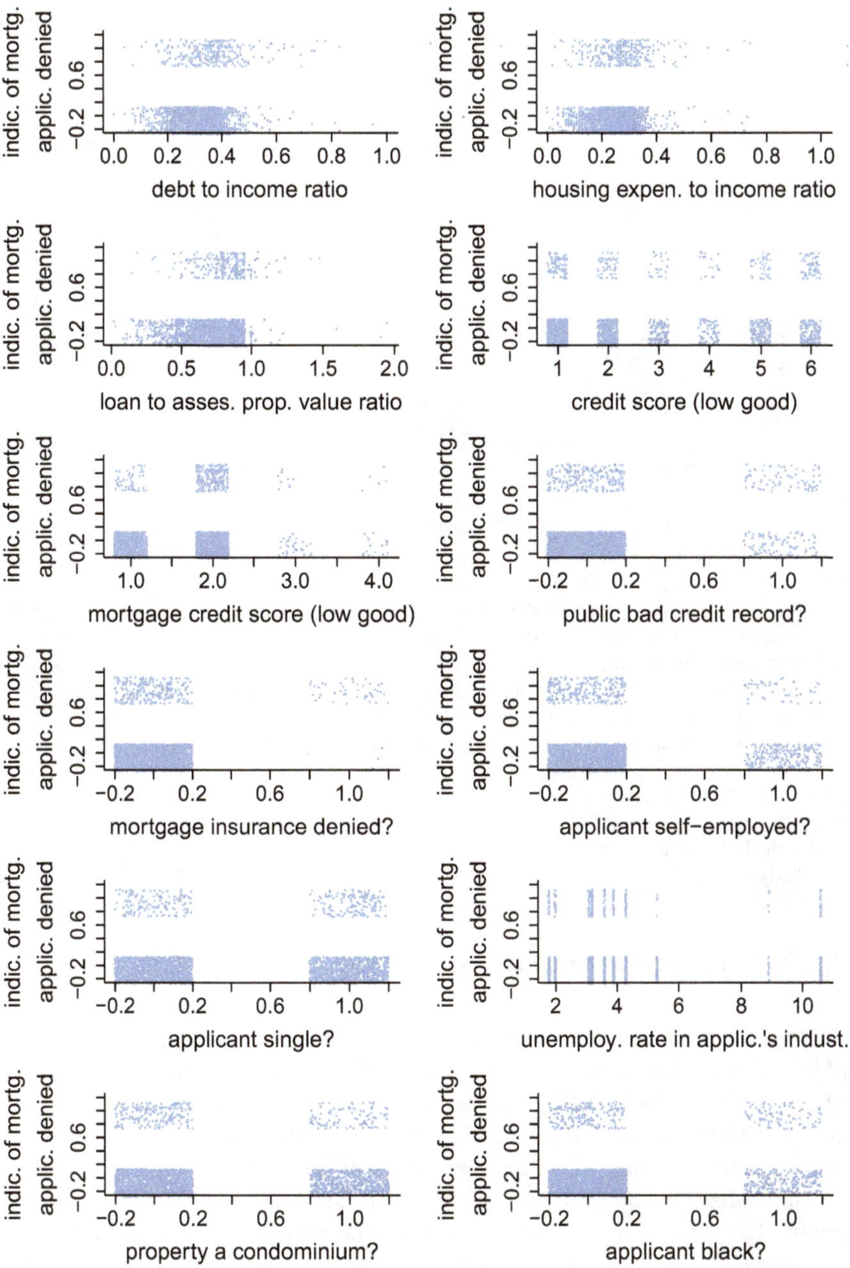

Fig. 1.3 Plots of indicator of a mortgage application denied against the other variables in the data frame `BostonMortgages` within the R package `HRW`. The yes/no variables are coded: $0 = \text{no}$ and $1 = \text{yes}$. To aid visualization, jittering has been applied to the discrete variables data.

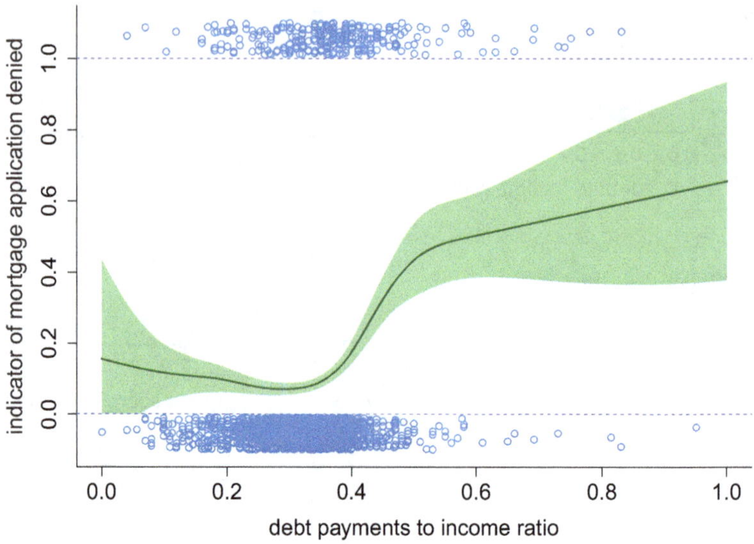

Fig. 1.4 Estimated probability of mortgage denial as a function of the debt payments to income ratio based on the data shown in the top-left panel of Fig. 1.3. The blue circles show the data with jittering of the response values to aid visualization. The shaded region is an approximate 95% confidence band. This fit was obtained using the `gam()` function in the R package `mgcv`; see Chap. 3. Of the 2380 mortgage applications, 5 have debt payments to income ratios between 1.16 and 1.42 and one has a debt payments to income ratio of 3. These cases were used during estimation but, to focus attention on the majority of the cases, they are not shown in the plot.

Table 1.1 Cross-tabulation of adolescents by gender and race in the Indiana adolescent growth dataset.

	Black	White
Female	30	70
Male	28	88

1.3.3 Indiana Adolescent Growth Data

The *Indiana adolescent growth data* were obtained from a study of the mechanisms of human hypertension development conducted at the Indiana University School of Medicine, Indianapolis, USA, that started in the 1980s and is still continuing. Pratt et al. (1989) contains a full description of the study. The data are from a *longitudinal* study and are a special case of *grouped* data, which is the topic of Chap. 4.

The Indiana adolescent growth dataset is stored in the data frame named `growthIndiana` in the `HRW` package. Note that `growthIndiana` is restricted to the subset of 216 adolescents in the original study who had at least nine height measurements. Table 1.1 is a cross-tabulation of the adolescents by race and gender.

Figure 1.5 shows the entire dataset using `lattice` graphics in R. The panels in Fig. 1.5 plot height against age for each of the 216 adolescents, with color-coding according to gender/race status. Such data are often referred to as *growth curves*. It is

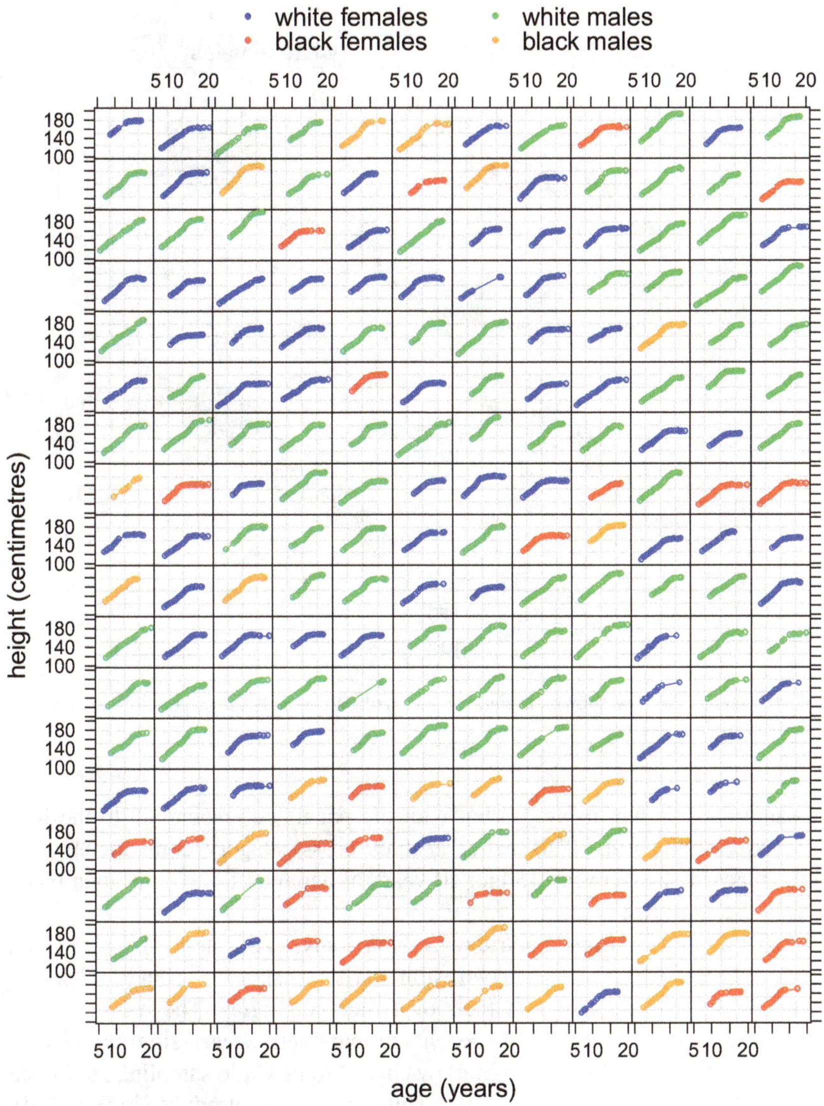

Fig. 1.5 The Indiana adolescent growth data stored in the data frame `growthIndiana` in the R package `HRW`. Each panel plots height (centimeters) against age (years) for each of 216 adolescents. Color-coding is used to indicate combined gender/race status.

not easy to fit these data using common parametric models. An additional challenge arises from proper accounting for dependencies between measurements on the same adolescent.

Comparison of growth between the gender and race categories is often of interest and will be studied in Chap. 4. Figure 1.6 is a different `lattice` graphics plot

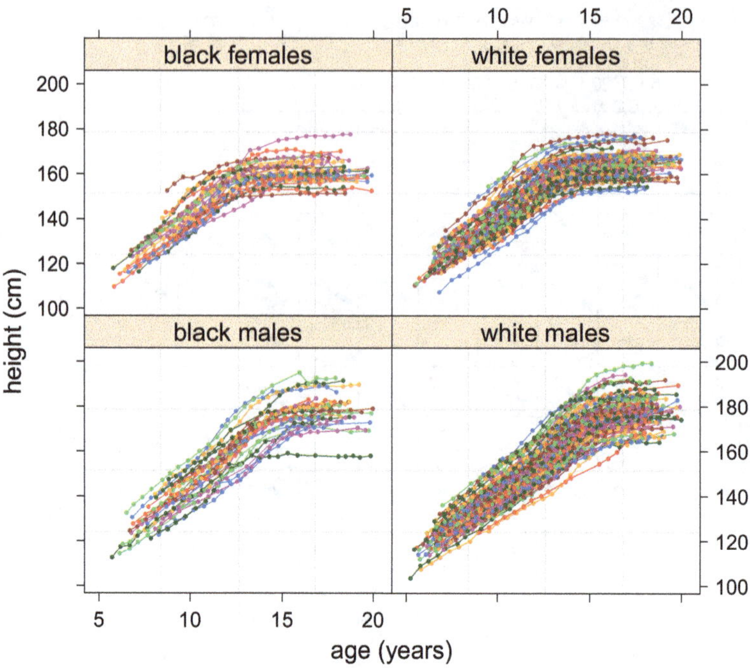

Fig. 1.6 The same data as shown in Fig. 1.5 but with the panels corresponding to the four gender/race combinations.

of the same data shown in Fig. 1.5 but with the panels corresponding to the four gender/race combinations. This better enables cross-category comparisons. For example, black males between 15 and 20 years of age tend to be taller than black females in the same age bracket.

To give a flavor of semiparametric regression analyses of interest for such data, described in Chap. 4, Fig. 1.7 shows two estimated *contrast functions* in which males and females are compared within their own race categories. The estimates and variability bands are based on a *Bayesian* semiparametric regression model with approximate inference achieved via Markov chain Monte Carlo sampling facilitated by the R package `rstan` (Guo et al. 2017). This approach is introduced in Sect. 2.10.

From Fig. 1.7 we see that there is little difference, statistically, between males and females up to the age of 12. After that males are significantly taller, with the gap bigger for the black race than it is for the white race. There is more variability in the black race contrast function since it is based on fewer observations—only about a quarter of the subjects in the study are black.

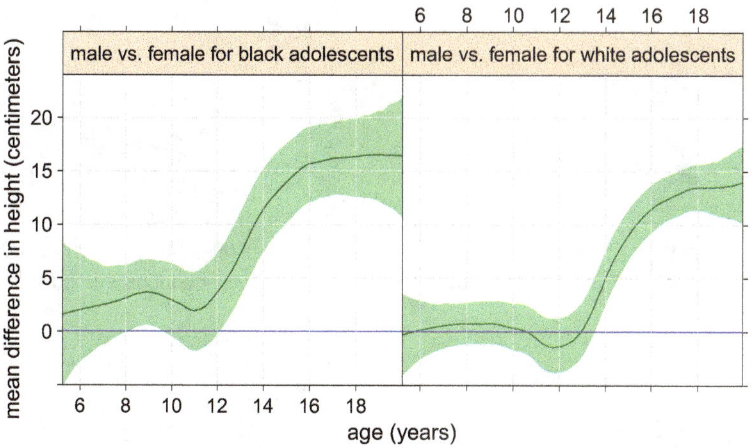

Fig. 1.7 Estimated contrast functions and approximate pointwise 95% credible sets based on a Bayesian semiparametric regression model fitted to the data shown in Figs. 1.5 and 1.6. Approximate Bayesian inference, based on Markov chain Monte Carlo, was performed using the R package `rstan`.

1.3.4 Sydney Real Estate Data

The *Sydney real estate data* were collected as a part of an unpublished study by A. Chernih and M. Sherris at the University of New South Wales, Australia. The data consist of 39 variables on 37,676 houses sold in Sydney, Australia, during the year 2001 and are stored in the data frame `SydneyRealEstate` in the `HRW` package.

Of central interest is the nature of the dependence of house prices on the other variables. Figure 1.8 depicts some of the individual dependencies through scatterplots of the logarithm of sale price against 8 of the potential predictors. For example, the top-left panel in Fig. 1.8 shows the intuitively obvious positive correlation between price and lot size. Underneath that, distance to the coastline is seen to have a negative impact on price.

Figure 1.9 shows the average log-prices on a 50×50 equal-sized geographical mesh. A strong spatial effect is apparent. The higher-priced areas tend to be near Sydney's waterways and ocean front. Rather than estimating univariate regression functions a *bivariate* function of longitude and latitude seems to be appropriate to model the behavior exhibited in Fig. 1.9. The bivariate extension of semiparametric regression analysis is dealt with in Chap. 5.

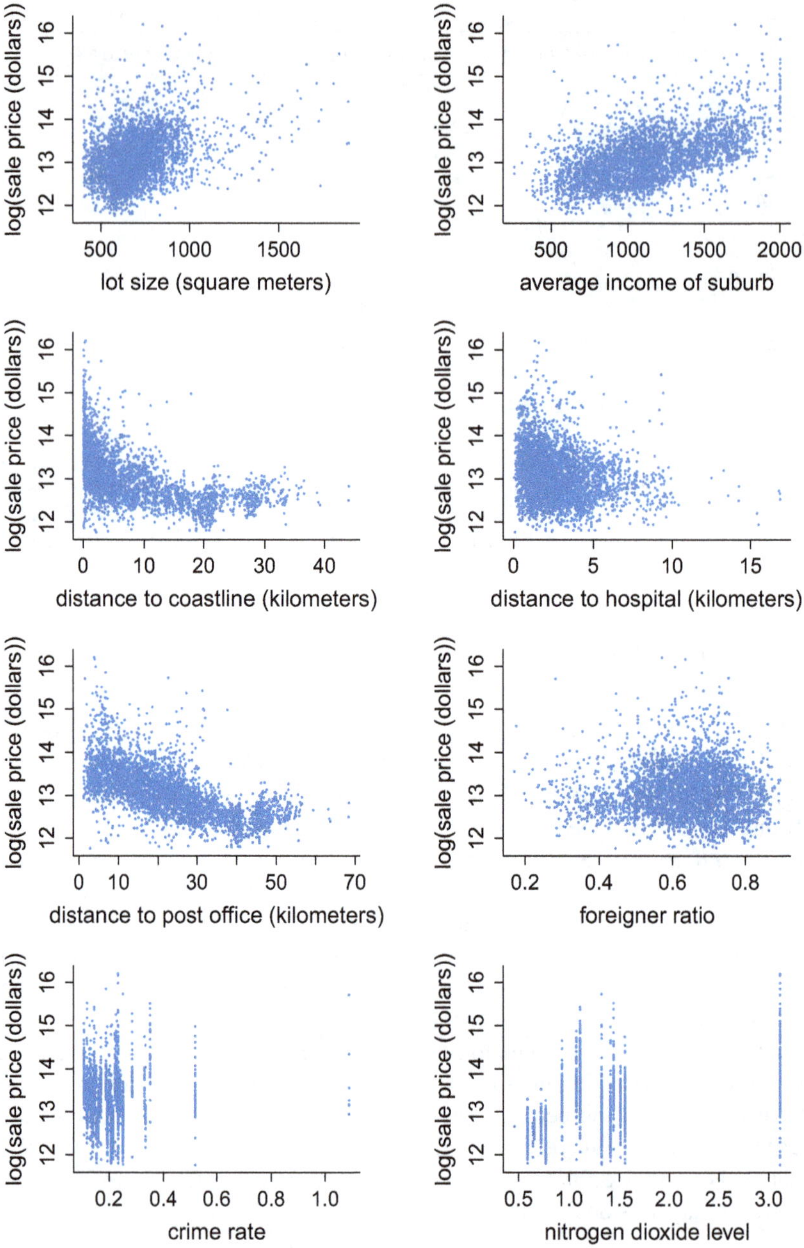

Fig. 1.8 Plots of logarithm of sale price (dollars) against some of the other variables in the data frame `SydneyRealEstate` within the R package `HRW`. To aid visualization, a 10% random subset of the data is used in the plots.

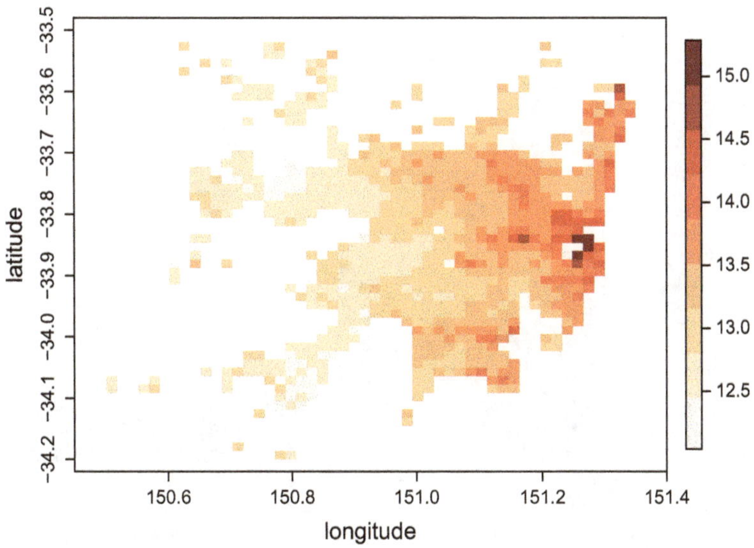

Fig. 1.9 The spatial variation in log price of the houses sold in Sydney, Australia, during 2001 based on the dataset `SydneyRealEstate` within the R package `HRW`. The averaging was done on a 50 by 50 rectangular longitude by latitude pixel mesh. The pixels where no data were recorded are left blank. Data are present in only 836 out of 2500 pixels.

1.3.5 Michigan Panel Study of Income Dynamics Data

The scatterplot on the left panel of Fig. 1.10 is household income excluding income from the wife's work versus the wife's age for 3382 households in the year 1987. These data are part of a much larger dataset from the Michigan Panel Study of Income Dynamics (e.g. Lee 1995). The 1987 cross-section is in the data frame `Workinghours` in the R package `Ecdat`.

A question of interest is the impact of wife's age on other household income but, unlike the situation in Fig. 1.1, the response variable here is highly skewed and includes some strong positive outliers. The methodology used to fit the mean response curve to the Fig. 1.1 scatterplot is not appropriate for the Fig. 1.10 scatterplot and the conditional mean function is not necessarily a good way of summarizing the response/predictor relationship. Instead we use conditional *quantile* functions. The right panel shows the 1, 5, 25, 50, 75, 95, and 99% estimated quantiles of other household income conditional on the wife's age. This plot allows appreciation for the effect of the predictor on the response in a different way than Fig. 1.1 and is more appropriate for such a skewed and outlier-ridden response variable.

The `Workinghours` data frame has data on several other variables such as education level of the wife, occupation of the husband, and number of children in the household. In Chap. 6 we explore semiparametric quantile regression models that incorporate multiple predictor effects.

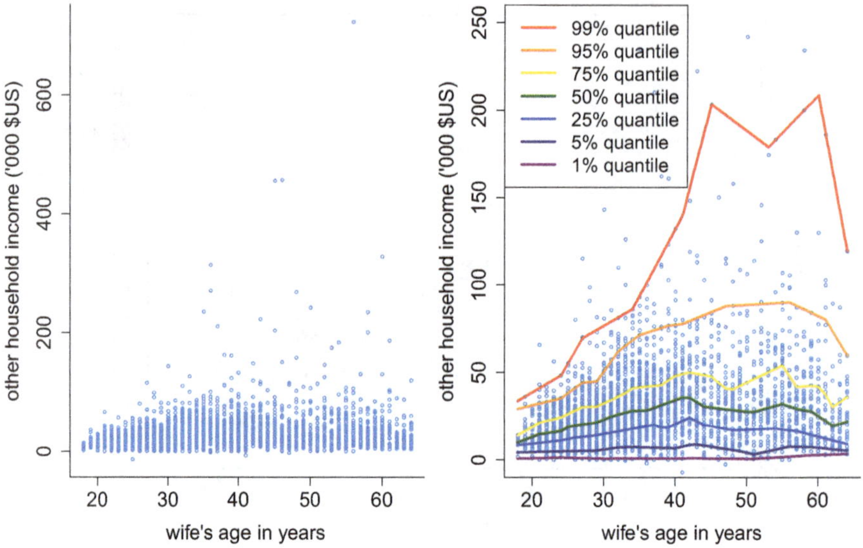

Fig. 1.10 Left panel: Household income from sources other than the wife's work (thousands of U.S. dollars) versus wife's age (years). Right panel: Zoomed view of left panel plot with restriction to households for which other income does not exceed $250,000. The curves correspond to nonparametric quantile function estimates with color-coding for the level of the quantile. The estimates were obtained using the function `rqss()` in the R package `quantreg`.

1.3.6 All of the Datasets Used in This Book

Table 1.2 lists and briefly describes each of the datasets used in this book, and the sections in which they are analyzed.

1.4 Aim of This Book

Semiparametric regression is a major area of methodological development and is being used widely in applications. See, for example, Ruppert et al. (2009) for a summary of the state of affairs in the late 2000s. Nevertheless, we believe that semiparametric regression should be used even more widely by applied researchers and that ongoing contributions to the R computing environment make this increasingly easier. The aim of this book is to demonstrate how semiparametric regression analyses can be carried out with only minimal knowledge of R. We do not get into the intricacies of semiparametric regression methodology and its underlying theory. Instead, our focus is implementation in R.

Relevant R packages include `gamlss` (Stasinopoulos and Rigby 2017), `nlme` (Pinheiro et al. 2017), `mgcv` (Wood 2017), `quantreg` (Koenker 2017), `refund`

Table 1.2 Datasets used in this book and sections where they are analyzed.

R data frame (package)	Brief description	Sections
WarsawApts (HRW)	Apartments sold in Warsaw, Poland, during 2007–2009	1.3, 2.2–2.10 2.12, 3.6
BostonMortgages (HRW)	Mortgage applications of residents of Boston, USA	1.3, 3.2 3.3, 3.6
growthIndiana (HRW)	Longitudinal heights of adolescents in Indiana, USA	1.3, 4.3
SydneyRealEstate (HRW)	Real estate sold in Sydney, Australia, during 2001	1.3, 5.3
Workinghours (Ecdat)	Income and attributes of households in Michigan, USA	1.3, 6.2
OFP (Ecdat)	Physician visits and attributes of elderly USA residents	3.2, 3.3
Caschool (Ecdat)	School test scores and attributes in California, USA	3.3, 3.4 3.4.2, 3.5
femSBMD (HRW)	Longitudinal spinal bone mineral density in the USA adolescents	4.2, 5.7
protein (HRW)	Longitudinal protein intake from a USA nutrition study	4.4
indonRespir (HRW)	Longitudinal respiratory infection status of children in Indonesia	4.5
ozoneSub (HRW)	Ozone concentrations in the midwest region of the USA	5.2
capm (HRW)	Daily USA stock returns and indices during 1993–2003	5.4
gasoline(refund)	Near infrared spectra and octane numbers for gasoline samples	5.6
brainImage (HRW)	Brain image coronal slice	5.8
DTI (refund)	Diffusion tensor imaging data	6.3
tecator (fda.usc)	Content of meat samples	6.3, 6.5
yields (HRW)	Yield curves	6.6
carAuction (HRW)	Attributes of auction-bought cars	6.7
PimaIndiansDiabetes (mlbench)	Diabetes status and attributes of the USA study of Pima Indians	6.8
BCR (HRW)	Mental health scores from a drug/placebo clinical trial	6.8
CHD (HRW)	Coronary heart status and attributes from a U.S. study	6.8
coral (HRW)	Alive/death status of coral organisms in French Polynesia	6.9
Ozone (mlbench)	Daily ozone levels and weather Los Angeles area during 1976	6.9

Datasets used in exercises only are not listed here but can be found in the index

(Goldsmith et al. 2016), `rstan` (Guo et al. 2017), and `VGAM` (Yee 2017). The index has the full list of packages mentioned in the book. Our intention is to describe in a straightforward way the relevant steps needed to conduct semiparametric regression analyses using R packages such as these.

This book will be useful to anybody who has a basic knowledge of R and is interested in exploring and modeling data where simple parametric assumptions are not realistic. Biostatisticians, data analysts, econometricians, and social scientists should find this book of special interest. We expect that the material presented here will be accessible to any reader who has taken courses in linear regression and generalized linear models. To fully appreciate Bayesian model fitting, an introductory course in Bayesian inference will be helpful.

In Chap. 2 we give a detailed account of the main semiparametric regression building block: *penalized splines*. Chapter 3 covers the important family of semiparametric regression models known as *generalized additive models*. Then in Chap. 4 we deal with extensions to *grouped data*, which includes longitudinal, multilevel, panel, and small area data as special cases. Chapter 5 is concerned with *bivariate* extensions of penalized splines and *spatial* semiparametric regression models. The last chapter, Chap. 6, is a collection of additional topics such as building in robustness and accounting for missing observations in semiparametric regression analysis.

Chapter 2
Penalized Splines

2.1 Introduction

In this chapter, we study nonparametric regression with a single continuous predictor. This problem is often called *scatterplot smoothing*. Our emphasis is on the use of penalized splines. We also show that a penalized spline model can be represented as a linear mixed model, which allows us to fit penalized splines using linear mixed model software. In R the most relevant packages are `mgcv` (Wood 2017), which supports a wide range of semiparametric regression models, `nlme` (Pinheiro et al. 2017) which supports the mixed model approach and `rstan` (Guo et al. 2017) which supports the Bayesian approach. Detailed R scripts for the examples presented in this book are in the `HRW` package (see `semiparametric-regression-with-r.net`).

2.2 Penalized Spline Basics

Penalized spline smoothing is a simple way of fitting a curve to a scatterplot and is a major building block for semiparametric regression. For now we focus on the *nonparametric regression* model

$$y_i = f(x_i) + \varepsilon_i, \quad 1 \le i \le n, \tag{2.1}$$

for predictor/response pairs (x_i, y_i). Here f is a smooth but otherwise arbitrary function and the errors, ε_i, are independent random variables such that $E(\varepsilon_i) = 0$. A simple penalized spline model for f is

$$f(x) = \beta_0 + \beta_1 x + \sum_{k=1}^{K} u_k (x - \kappa_k)_+ \tag{2.2}$$

J. Harezlak et al., *Semiparametric Regression with R*, Use R!,
https://doi.org/10.1007/978-1-4939-8853-2_2

where $x_+ \equiv \max(x, 0)$ for any $x \in \mathbb{R}$ and the κ_ks, $1 \le k \le K$, are pre-set values, usually taken to be approximately equally spaced with respect to the quantiles of the x_is. We will discuss choice of K in Sect. 2.4 but it is usually set to a number between about 10 and 50. Functions of the form (2.2) are piecewise lines "tied together" at the κ_ks and, for this reason, the κ_ks are called *knots*. The coefficients β_0, β_1 and u_k, $1 \le k \le K$, are chosen according to the constrained optimization problem

$$\text{minimize} \sum_{i=1}^{n} \{y_i - f(x_i)\}^2 \text{ subject to } \sum_{k=1}^{K} u_k^2 \le C \tag{2.3}$$

for some $C > 0$. An equivalent optimization problem is

$$\text{minimize} \left[\sum_{i=1}^{n} \{y_i - f(x_i)\}^2 + \lambda \sum_{k=1}^{K} u_k^2 \right] \tag{2.4}$$

which is known as *penalized least-squares*, with $\lambda \ge 0$ labeled the *smoothing parameter*. The fitted function is

$$\widehat{f}(x; \lambda) = \widehat{\beta}_0 + \widehat{\beta}_1 x + \sum_{k=1}^{K} \widehat{u}_k (x - \kappa_k)_+$$

where the hatted coefficients are obtained from (2.4). The effect of λ on $\widehat{f}(\cdot; \lambda)$ and strategies for choosing its value are described in Sect. 2.3.

The upper panel of Fig. 2.1 shows the penalized spline fit to

$$x_i = \text{year of construction of } i\text{th apartment}$$
$$\text{and } y_i = \text{area (square meters) per million złoty of the } i\text{th apartment}$$

for $1 \le i \le 409$ from the Warsaw apartment data introduced in Sect. 1.3.1, stored as `WarsawApts` in the R package `HRW`. The smoothing parameter is $\lambda = 100$ and the knots are placed at

$$\kappa_k = \left(\frac{k+1}{K+2} \right) \text{th sample quantile of the unique } x_i\text{s}$$

with $K = 20$. The corresponding spline basis functions $(x - \kappa_k)_+$, $1 \le k \le 20$, are shown in the lower panel.

Even though the truncated line basis is simple and intuitive, it has some disadvantages: (a) regression function fits have artificial kinks due to the restriction that f is piecewise linear, and (b) the basis functions are unbounded and far from orthogonal, which can lead to numerical problems. These shortcomings have been long recognized in numerical analysis and have resulted in the development of smoother and more stable spline bases. Increased smoothness can be achieved quite

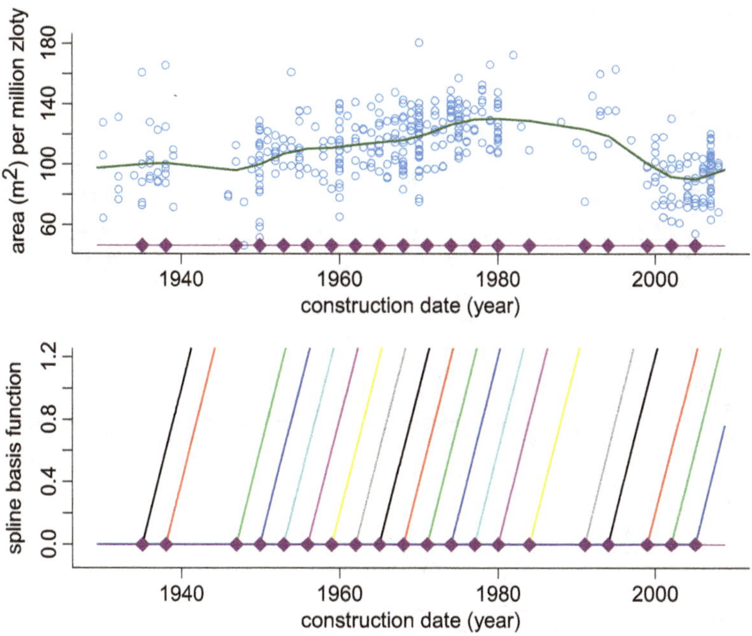

Fig. 2.1 Upper panel: Penalized spline fit to the Warsaw apartments running example regression dataset based on (2.2) with $\lambda = 100$ and $K = 20$. The locations of the knots, κ_k, are shown by the purple diamonds. Lower panel: the truncated line spline basis functions $(x - \kappa_k)_+$, $1 \leq k \leq 20$.

easily—by working with $\{(x - \kappa_k)_+\}^p$ instead of $(x - \kappa_k)_+$ for integers $p > 1$. The basis functions, and hence the fits, are such that their $(p - 1)$th derivatives are continuous. The most popular choice is $p = 3$, which leads to piecewise cubic fits with continuous second derivatives. Numerical stability is achieved via linear transformation of truncated polynomial bases with *B-splines* (e.g. de Boor 2001) being the standard choice (Exercise 1). As explained in the Chap. 5 appendix of Hastie et al. (2009), additional modifications can be made near the boundaries to achieve good linear extrapolation properties. Further linear transformation, detailed in Sect. 4 of Wand and Ormerod (2008), is required to ensure that the simple penalized least-squares form (2.4) is appropriate for the spline basis coefficients. The upshot of all of this is replacement of (2.2) by

$$f(x) = \beta_0 + \beta_1 x + \sum_{k=1}^{K} u_k z_k(x) \tag{2.5}$$

where the z_k, $1 \leq k \leq K$, are new and improved spline basis functions. For reasons given later in this section, we call the z_k cubic *O'Sullivan* splines. Figure 2.2 is analogous to Fig. 2.1 but with the $(x - \kappa_k)_+$ replaced by $z_k(x)$ with knots at the same locations. The bottom panel is a vertically zoomed version of the middle panel.

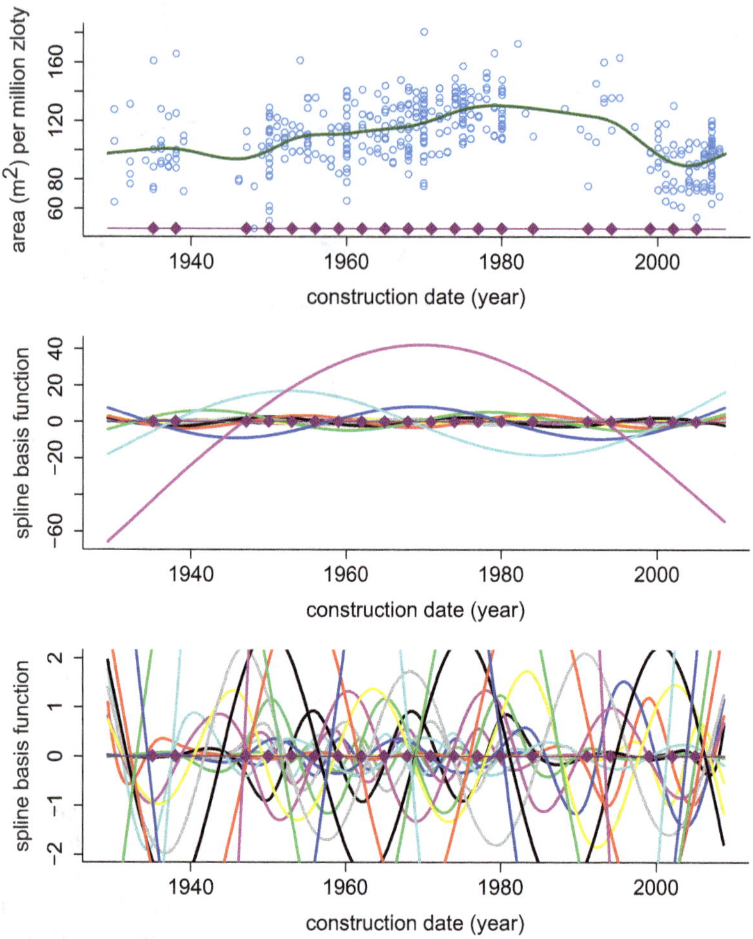

Fig. 2.2 Top panel: Penalized spline fit to the Warsaw apartments running example regression dataset based on (2.5) with $\lambda = 100$ and $K = 20$. The locations of the knots, κ_k, are shown by the purple diamonds. Middle panel: the cubic O'Sullivan spline basis functions $z_k(x)$, $1 \leq k \leq 22$. Bottom panel: A zoomed view of the middle panel plot with the vertical range restricted to -2 to 2.

Unlike truncated lines, the z_k functions do not have simple mathematical expressions. The function `ZOSull()` in the `R` package `HRW` supports the evaluation of $z_k(x)$ for arbitrary arguments x within a prespecified range. Specifically, the z_k, $1 \leq k \leq K$, are defined with respect to *range* values a and b and *interior knots* $\xi_1, \ldots, \xi_{K-2}$ for some positive integer K such that

$$a < \xi_1 < \cdots < \xi_{K-2} < b.$$

Once these are specified, then $z_k(x)$ is defined for all $a \leq x \leq b$, but not for x outside of the interval $[a, b]$. Usage of `ZOSull()` is illustrated by the following code, which produces the middle panel of Fig. 2.2:

```
> library(HRW); data(WarsawApts)
> x <- WarsawApts$construction.date
> a <- 1.01*min(x) - 0.01*max(x)
> b <- 1.01*max(x) - 0.01*min(x) ; numIntKnots <- 20
> intKnots <- quantile(unique(x),seq(0,1,length =
+                     (numIntKnots+2))[-c(1,(numIntKnots+2))])
> xg <- seq(a,b,length = 1001)
> Zg <- ZOSull(xg,range.x = c(a,b),intKnots = intKnots)
> plot(0,type = "n",xlim = range(xg),ylim = range(Zg),
+       bty = "l",xlab = "construction date (year)",
+       ylab = "spline basis function")
> for (k in 1:ncol(Zg)) lines(xg,Zg[,k],col = k,lwd = 2)
> lines(c(min(xg),max(xg)),rep(0,2),col = "darkmagenta")
> for (k in 1:numIntKnots)
+    points(intKnots[k],0,pch = 18,cex = 2,col = "darkmagenta")
```

The object Zg is a 1001×22 matrix with kth column containing z_k evaluated over an equally spaced grid of length 1001 between $a = 1929.22$ and $b = 2008.78$.

O'Sullivan penalized splines also have the attraction of being a natural generalization of *smoothing splines* (e.g. Green and Silverman 1994). Smoothing splines, parameterized by $\lambda > 0$, arise as the solution to the optimization problems

$$\text{minimize}\left[\sum_{i=1}^{n}\{y_i - f(x_i)\}^2 + \lambda\int_{-\infty}^{\infty} f^{(m)}(x)^2\,dx\right] \tag{2.6}$$

over all functions f. The integral in (2.6) is known as a *roughness penalty*. For $m = 2$, the solution to (2.6) is a linear combination of approximately n cubic basis functions with knots at the x_is. O'Sullivan (1986) provides a representation of the solution in terms of cubic B-splines. Since the number of basis functions grows with the sample size, so does the computational overhead of fitting a smoothing spline. On the other hand, penalized splines with an O'Sullivan spline basis are barely affected by high sample sizes and, for reasonable choices of K, are very good approximations to smoothing splines. Because of the several attractive features of O'Sullivan penalized splines, we will use them as the default for the remainder of this book. From now on, a penalized spline estimate is assumed to use a cubic O'Sullivan spline basis unless we say otherwise.

The R function `smooth.spline()` is so-named because it fits a smoothing spline to the input data. However, its default is to use the O'Sullivan penalized spline approximation when the sample size exceeds 50. The R code

```
> library(HRW); data(WarsawApts)
> x <- WarsawApts$construction.date
> y <- WarsawApts$areaPerMzloty
> fitSS <- smooth.spline(x,y,spar = 0.56)
```

leads to a fit that is similar to that shown in the top panel of Fig. 2.2. The `spar` in `smooth.spline()` is a monotonically increasing transformation of λ that is described in this function's help pages.

The choice of K and positioning of the interior knots have a relatively minor effect on the fit. For most signals that arise in practice, including all examples in this book, $K = 35$ O'Sullivan spline basis functions is sufficient. In keeping with smoothing splines, a good default for the interior knots is

$$\xi_k = \left(\frac{k}{K-1}\right) \text{th sample quantile of the unique } x_i\text{s}, \quad 1 \leq k \leq K-2.$$

An in-depth discussion on choosing K and formally checking whether a particular K is adequate is given in Sect. 2.4.

In contrast to K, the value of λ has a very big influence on the penalized spline fit. This is illustrated in Fig. 2.3 which shows four fits to the same data as in Figs. 2.1 and 2.2 with different values of λ.

The upper left panel of Fig. 2.3 has a very small λ and is close to an ordinary least-squares fit. Since the basis functions are so numerous and wiggly, a high degree of overfitting is observed. Setting $\lambda = 100$ leads to a more pleasing fit. If λ is made very large, as in the lower two panels of Fig. 2.3, then the spline basis functions have too little an influence and the fit becomes close to the ordinary least-squares line. We next explain how a good λ value can be selected using the data.

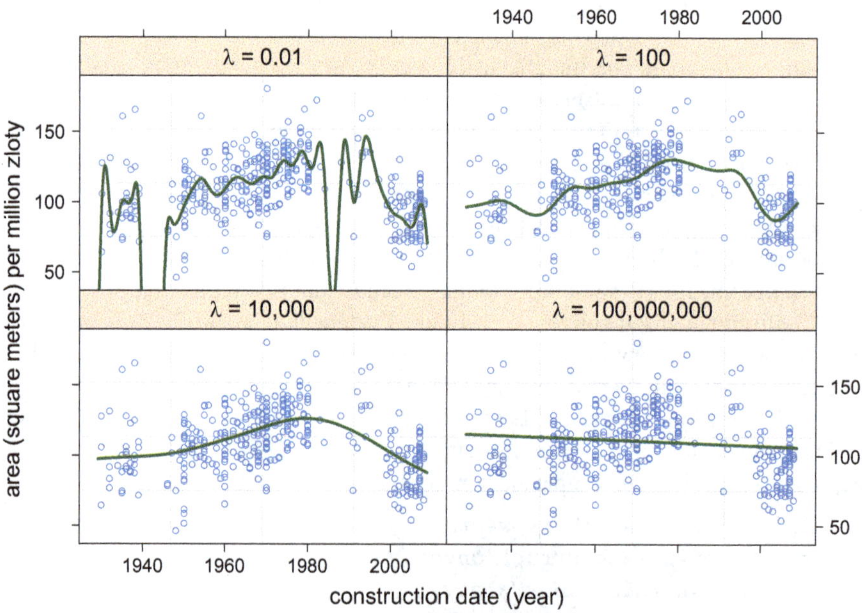

Fig. 2.3 Four penalized spline fits to Warsaw apartments running example regression dataset based on (2.5) with the smoothing parameter λ set to a different value in each panel. The number of spline basis functions is $K = 35$.

2.3 Choosing the Smoothing Parameter

Figure 2.3 shows that λ has a profound impact on the resultant penalized spline. Choosing λ by eye is one possibility and, as discussed in the previous section, $\lambda = 100$ looks about right. However, an objective means of choosing the smoothing parameter is desirable for a number of reasons. These include reproducibility, users with varying degrees of experience and high throughput settings where very many nonparametric regressions need to be performed and human involvement is not feasible. There now exist several data-based algorithms that aim to select λ so that the fit is optimal in some sense. We will not get into the technicalities of what it means for a nonparametric regression estimate to be optimal here. Chapter 5 of Ruppert et al. (2003), for example, provides details on this issue.

Two of the main penalized spline functions in R, `smooth.spline()` and `gam()` in the `mgcv` package, select λ from the data using a method known as *generalized cross-validation (GCV)* (Craven and Wahba 1979). The mathematical details of GCV are given in Sect. 2.6 but the salient feature to note here is that it produces an estimate of a theoretically optimal λ that depends only on the regression data (x_i, y_i), $1 \leq i \leq n$. We denote this estimate by $\widehat{\lambda}_{\text{GCV}}$.

The following code illustrates penalized spline smoothing with λ chosen automatically via GCV:

```
> library(HRW); data(WarsawApts)
> x <- WarsawApts$construction.date
> y <- WarsawApts$areaPerMzloty
> fitSSauto <- smooth.spline(x,y)  ; print(fitSSauto)
```

This leads to the output:

```
Call:
smooth.spline(x = x, y = y)

Smoothing Parameter  spar= 0.4977603  lambda= 8.835914e-05
(11 iterations)
Equivalent Degrees of Freedom (Df): 15.56644
Penalized Criterion (RSS): 18509.75
GCV: 322.2859
```

showing that the amount of smoothing has been chosen automatically via GCV. The `spar= 0.4977603` and `lambda= 8.835914e-05` represent monotone transformations of the λ parameter used in Sect. 2.2 and explanations are produced by `help(smooth.spline)`. The line `GCV: 322.2859` indicates the minimum of the GCV criterion function, which is given by (2.9) in Sect. 2.6.

GCV is also used for default smoothing parameter choice by the function `gam()` in the R package `mgcv` (Wood 2017). Even though the primary aim of `gam()` is to support sophisticated multi-predictor models, as discussed in Chap. 3, it supports nonparametric regression via penalized splines as a special case. The code:

```
> library(mgcv) ; fitGAMauto <- gam(y~s(x,bs = "cr",k = 52))
> print(fitGAMauto)
```

mimics the default call to `smooth.spline()` given above by using a similar type of basis (`bs = "cr"`) and number of basis functions (`k = 52`). Note that the argument `k` in the function `s()` corresponds to $K + 2$ in our notation since `k` is the total number of basis functions including those for the linear component. The resultant output is:

```
Family: gaussian
Link function: identity


Formula:
y ~ s(x, bs = "cr", k = 52)


Estimated degrees of freedom:
14.6  total = 15.58


GCV score: 322.2575
```

Notice that `GCV score: 322.2575` is relatively close to that obtained from `smooth.spline()` above. The total `Estimated degrees of freedom` is 15.58 which is also close to the `Equivalent Degrees of Freedom` value of 15.56644 in the `smooth.spline()` output. These are λ-dependent measures of the model complexity that this book calls *effective degrees of freedom* and are explained in Sect. 2.6.

The two GCV-based penalized spline fits are plotted in Fig. 2.4. As one might expect, the curves are virtually indistinguishable.

Choosing λ equal to $\widehat{\lambda}_{\text{GCV}}$ leads to a pleasing nonparametric regression fit in this example. Note, however, that $\widehat{\lambda}_{\text{GCV}}$ is an estimator of the theoretically optimal λ and, hence, is susceptible to estimation error. Therefore, $\widehat{\lambda}_{\text{GCV}}$ may not always lead to such a good fit, especially in high noise situations (e.g. Härdle et al. 1988).

2.4 Choosing the Basis Size

The number of spline basis functions, K, is not as crucial as the smoothing parameter. However, a choice still has to be made. Assuming uniqueness of the predictor values, smoothing splines defined by (2.6) choose $K \approx n$. This has the advantage of catering for arbitrarily wiggly signals and, for this reason, much of the classical smoothing spline literature (e.g. Wahba 1990) is confined to this $K \approx n$ situation. In R the call:

```
> fitClassicSS <- smooth.spline(x,y,all.knots = TRUE)
```

achieves such a regression fit with GCV smoothing parameter selection. However, what has become known as the *low-rank* alternative (e.g. Hastie 1996) where $K \ll$

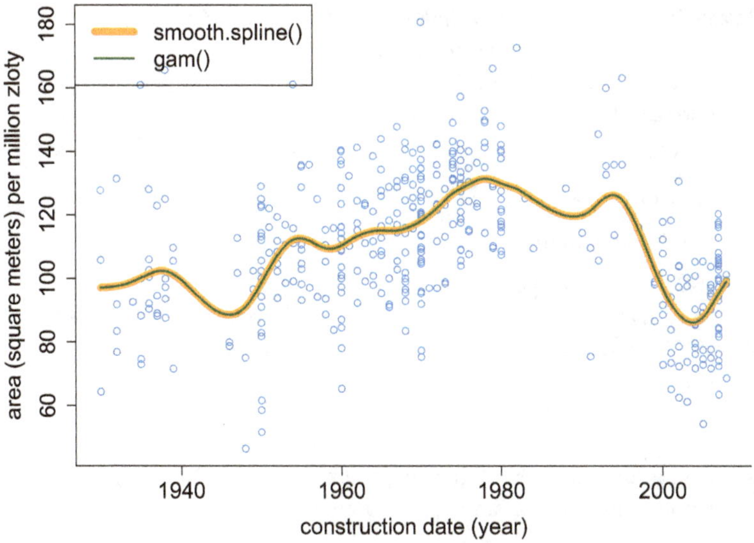

Fig. 2.4 Penalized spline fit to the Warsaw apartments running example regression dataset obtained from the default calls to the function `smooth.spline()` and `gam()`. In each case, the smoothing parameter is chosen according to generalized cross-validation.

n for large n, has a number of practical advantages: (a) the computations do not become too burdensome with very large sample sizes, (b) it is more amenable to mixed model and Bayesian representations, and (c) extension to complicated models such as those involving multiple predictors is more straightforward. In addition, if the signal is not overly wiggly, then low-rank O'Sullivan penalized splines are very close to classical smoothing splines for values of K that are much smaller than n (e.g. Li and Ruppert 2008; Wand and Ormerod 2008; Kauermann et al. 2009). For signals that typically arise in applications $K \approx 35$ will often be more than enough. The R commands:

```
> library(mgcv) ; help(choose.k)
```

give more discussion on this topic (Wood 2017).

There is a small literature on automatic choice of K from the data for penalized spline regression and includes, for example, Ruppert (2002) and Kauermann and Opsomer (2011). While such procedures are reasonably simple to implement in R, they are not part of any of the main semiparametric regression packages. Instead we will describe choice of K via the function `gam.check()` in the `mgcv` package (Wood 2017), which performs a hypothesis test for the adequacy of the number of spline basis functions in a `gam()` fit. The test, based on the so-called *k-index*, is explained in the `gam.check()` help page. A fuller description is in Pya and Wood (2016). For example, if we obtain the default `gam()` fit with cubic O'Sullivan splines using:

```
> library(mgcv) ; library(HRW) ; data(WarsawApts)
> x <- WarsawApts$construction.date
> y <- WarsawApts$areaPerMzloty
> fitGAMdefault <- gam(y~s(x,bs = "cr"))
```

then the command:

```
> gam.check(fitGAMdefault)
```

leads to the following output:

```
Method: GCV   Optimizer: magic
Smoothing parameter selection converged after 6 iterations.
The RMS GCV score gradient at convergence was 0.0002739156.
The Hessian was positive definite.
Model rank =  10 / 10

Basis dimension (k) checking results. Low p-value (k-index<1)
may indicate that k is too low, especially if edf is close
to k'.

        k'  edf k-index p-value
s(x) 9.00 8.05    0.95    0.17
```

Note that ten basis functions are used in the fit, including two basis functions for the linear component. Hence, the fit uses $K = 8$ spline basis functions. The combination of low `p-value`, `edf` being close to `k'` and `k-index` less than 1 is a cause for concern and indicates that a higher number of spline basis functions is needed. A `gam.check()` of the $K = 50$ fit demonstrated in Sect. 2.3:

```
> fitGAM50splines <- gam(y~s(x,bs = "cr",k = 52))
> gam.check(fitGAM50splines)
```

leads to the output:

```
Method: GCV   Optimizer: magic
Smoothing parameter selection converged after 6 iterations.
The RMS GCV score gradient at convergence was 0.004910069.
The Hessian was positive definite.
Model rank =  52 / 52

Basis dimension (k) checking results. Low p-value (k-index<1)
may indicate that k is too low, especially if edf is close
to k'.

        k'  edf k-index p-value
s(x) 51.0 14.6    1.01    0.53
```

We see that $K = 50$ passes the k-index test with flying colors. Finally, a check that $K = 25$ is also sufficient:

```
> fitGAM25splines <- gam(y~s(x,bs = "cr",k = 27))
> gam.check(fitGAM25splines)

Method: GCV   Optimizer: magic
Smoothing parameter selection converged after 4 iterations.
The RMS GCV score gradient at convergence was 0.005054014.
The Hessian was positive definite.
Model rank =  27 / 27

Basis dimension (k) checking results. Low p-value (k-index<1)
may indicate that k is too low, especially if edf is close
to k'.

      k' edf k-index p-value
s(x) 26  14    1.01    0.53
```

shows a similar `k-index` score and p-value.

Figure 2.5 shows each of the three fits. The lessons are: (a) the default $K = 8$ for `gam()` is too low and does not allow for the finer structure in the signal to be recovered, (b) $K = 25$ is a much better choice and passes the k-index test, and (c) for this particular signal there is no advantage in having more than 25 O'Sullivan

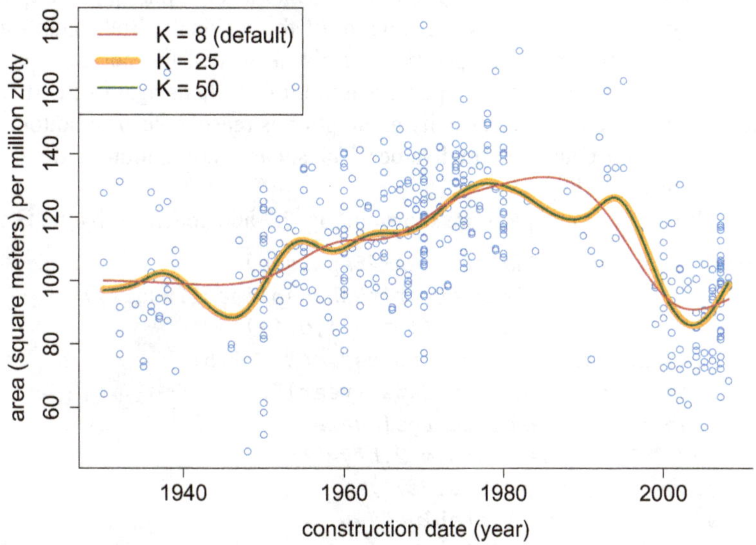

Fig. 2.5 Illustration of the effect of K (the number of spline basis functions) on penalized splines to the Warsaw apartments running example regression dataset.

spline basis functions since the $K = 25$ low-rank approximation to the classical smoothing spline is excellent and has less computational baggage than penalized splines having many more basis functions.

2.5 Checking the Residuals

The examination of residuals is an important part of any regression analysis. Residuals are used to check model assumptions and to locate outliers and other potential problems with the data. Assuming model (2.1), the *residual* for the ith case is $y_i - \widehat{f}(x_i)$.

In the previous two sections we have established that the choice

$$(K, \lambda) = (25, \widehat{\lambda}_{\text{GCV}})$$

is a good one for penalized spline fitting of the Warsaw apartments running example regression dataset. As with ordinary least-squares regression, checks of the residuals are in order for reasonableness of assumptions such as homoscedasticity and normality. If a penalized spline fit is performed using the `gam()` function in the `mgcv` package, then the `gam.check()` function applied to the `gam()` fit object (used in the previous section to guide choice of K) also provides residual plots. The code:

```
> print(gam.check(fitGAM25splines))
```

leads to the plot shown in Fig. 2.6. The top left panel is a Normal quantile-quantile plot and the bottom left panel is a histogram of the residuals. Both plots show a mild departure from normality, specifically right skewness, but not enough to be of a serious concern. The top right panel is a plot of residuals against fitted values and indicates that the homoscedasticity assumption is reasonable. The bottom right plot is of the response and the fitted values. The squared correlation between these variables is known as R^2.

Two additional residual plots are shown in Fig. 2.7 and generated by the R code:

```
> stdResids <- (residuals(fitGAM25splines)
+                /sqrt(summary(fitGAM25splines)$scale))
> par(mfrow = c(1,2),mai = c(1,1,0.1,0.1))
> plot(x,stdResids,col="dodgerblue",cex = 1.5,
+      xlab = "construction date (year)",
+      ylab = "standardized residuals",
+      cex.lab = 2,cex.axis = 2,bty="l",
+      ylim=c(-3,max(stdResids)))
> abline(h = 0,col = "slateblue")
> for (hVal in c(-3,-2,2,3))
+    abline(h=hVal,lty = 2,col = "slateblue")
> acf(stdResids[order(x)],col="darkgreen",cex.lab = 2,
+     cex.axis = 2,main="")
```

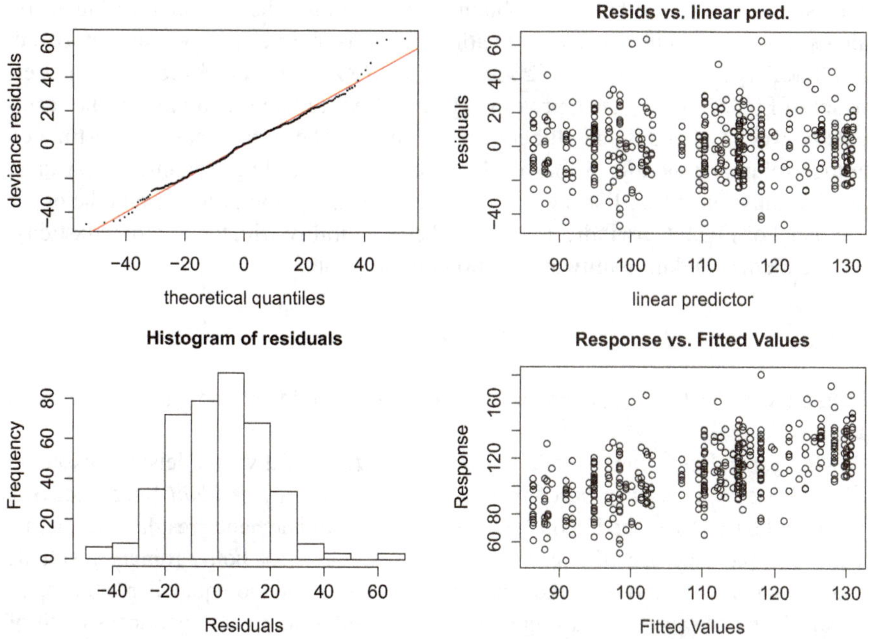

Fig. 2.6 Residual plots for the `gam(y ~ s(x,bs = "cr",k = 27))` penalized spline fit to the Warsaw apartments running example regression dataset. These plots are obtained from the call to the function `gam.check()` applied to the `gam()` fit object.

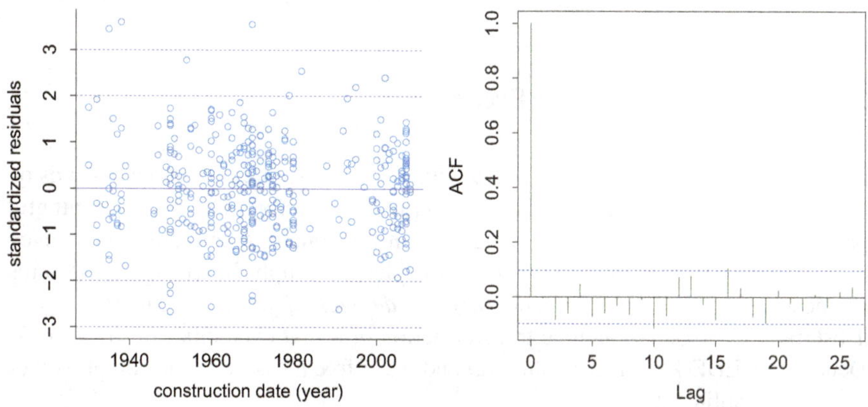

Fig. 2.7 Left panel: standardized residuals versus predictor values for the `gam(y ~ s(x,bs = "cr",k = 27))` penalized spline fit to the Warsaw apartments running example regression dataset. The dashed horizontal lines at −3, −2, 2, and 3 aid assessment of the standard normality. Right panel: estimated autocorrelation function of the standardized residuals with ordering with respect to the predictor values.

The standardized residuals are obtained by dividing the ordinary residuals by an estimate of their standard deviation. This is extracted from the fit object `fitGAM25splines` via `sqrt(summary(fitGAM25splines)$scale)`. The left panel of Fig. 2.7 shows an approximately equal-width band with most of the residuals between −3 and 3 and all between −4 and 4. The right panel is the estimated autocorrelation function of the residuals, ordered according to construction date, and has only one borderline significant spike. Taken together, Fig. 2.7 and the upper left panel of Fig. 2.6 are indicative of good compatibility with the homoscedasticity, independence, and normality assumption on the errors:

$$\varepsilon_i \stackrel{\text{ind.}}{\sim} N(0, \sigma_\varepsilon^2)$$

for the $(K, \lambda) = (25, \widehat{\lambda}_{\text{GCV}})$ penalized spline fit obtained from the `gam()` function in `mgcv`.

The generalized linear models and generalized additive models discussed in Chap. 3 use a generalization of residuals called *deviance residuals*; see Sect. 3.2. With model (2.1) assumed in this chapter, residuals and deviance residuals coincide. The `mgcv` package uses the term "deviance residuals" for both ordinary residuals and their generalization. In the same vein, the sum of the squared residuals, or *residual sum of squares*, is a standard statistic used to assess the goodness-of-fit of linear models as well as the penalized spline models in this chapter. As discussed in Chap. 3, for generalized linear models and generalized additive models, the *deviance* is the generalization of residual sum of squares. Note that the `mgcv` package uses the term "deviance" for both residual sum of squares and its generalization.

2.6 Effective Degrees of Freedom

The smoothing parameter λ associated with penalized spline smoothing depends on the units of measurement of the regression data. Therefore λ is not a meaningful parameter in terms of the effect that it has on the penalized spline fit. For a given fit, λ can be made as small or big as one pleases simply by changing the units of measurement of the data. The *effective degrees of freedom* (e.g. Hastie and Tibshirani 1990) is a monotone transformation of λ that overcomes this drawback. Denoted by EDF(λ), it is a meaningful and scale-free measure of the complexity of a penalized spline fit.

We need some algebra to derive the appropriate form of EDF(λ). First, define the matrices

$$\boldsymbol{C} \equiv \begin{bmatrix} 1 & x_1 & z_1(x_1) & \cdots & z_K(x_1) \\ \vdots & \vdots & \vdots & \ddots & \vdots \\ 1 & x_n & z_1(x_n) & \cdots & z_K(x_n) \end{bmatrix} \quad \text{and} \quad \boldsymbol{D} \equiv \begin{bmatrix} \boldsymbol{0}_{2\times 2} & \boldsymbol{0}_{2\times K} \\ \boldsymbol{0}_{K\times 2} & \boldsymbol{I}_K \end{bmatrix} \qquad (2.7)$$

where, for example, $\mathbf{0}_{2\times 2}$ is the 2×2 matrix of zeroes and $\boldsymbol{I}_K$ is the $K \times K$ identity matrix. Straightforward algebra (Exercise 3) can be used to show that the vector of fitted values of a penalized spline fit is

$$\begin{bmatrix} \widehat{f}(x_1;\lambda) \\ \vdots \\ \widehat{f}(x_n;\lambda) \end{bmatrix} = \boldsymbol{C}(\boldsymbol{C}^T\boldsymbol{C} + \lambda\,\boldsymbol{D})^{-1}\boldsymbol{C}^T\boldsymbol{y}.$$

Under the nonparametric regression model (2.1) with the homoscedasticity condition $\text{Var}(\varepsilon_i) = \sigma_\varepsilon^2$, $1 \le i \le n$, one can show that

$$\sum_{i=1}^{n} \text{Cov}\big(\widehat{f}(x_i;\lambda), y_i\big) = \sigma_\varepsilon^2\,\text{tr}\{(\boldsymbol{C}^T\boldsymbol{C} + \lambda\,\boldsymbol{D})^{-1}\boldsymbol{C}^T\boldsymbol{C}\}.$$

This suggests the definition

$$\text{EDF}(\lambda) \equiv \frac{1}{\sigma_\varepsilon^2}\sum_{i=1}^{n} \text{Cov}\big(\widehat{f}(x_i;\lambda), y_i\big) = \text{tr}\{(\boldsymbol{C}^T\boldsymbol{C} + \lambda\,\boldsymbol{D})^{-1}\boldsymbol{C}^T\boldsymbol{C}\} \tag{2.8}$$

as a scale-invariant measure of the effective degrees of freedom possessed by $\widehat{f}(\cdot;\lambda)$. Note that

$$\text{EDF}(0) = \text{number of columns in } \boldsymbol{C}$$

which shows that the effective degrees of freedom reduces to the number of parameters being fit when there is no penalization.

Definition (2.8) is a widely used standard in semiparametric regression, including R software. For example, the fit object from a call to `smooth.spline()` includes the effective degrees of freedom as the `$df` list component. An illustration is:

```
> library(HRW); data(WarsawApts)
> x <- WarsawApts$construction.date
> y <- WarsawApts$areaPerMzloty
> fitSS <- smooth.spline(x,y,spar = 0.56)
> print(fitSS$df)

[1] 12.42063
```

showing that EDF(λ) is about 12.4 for this fit. A rough interpretation is that the fit has the same complexity as an 11th degree polynomial ordinary least-squares fit—the latter having exactly 12 (effective) degrees of freedom.

Figure 2.8 is a reproduction of Fig. 2.3 with λ replaced by EDF(λ). The fit in the top left panel is close to the ordinary least-squares fit with 37 basis functions (since $K = 35$) and this is reflected in an EDF value being close to this maximal value. At

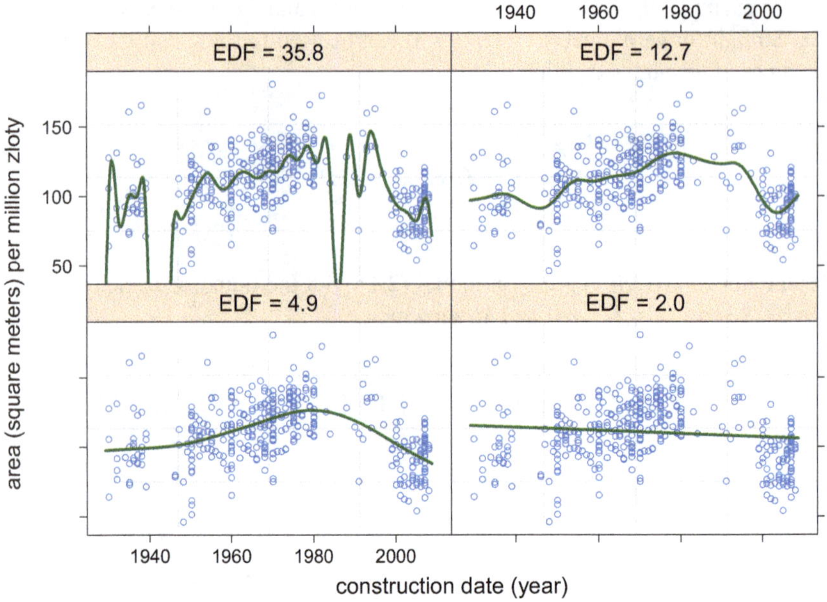

Fig. 2.8 The four penalized spline fits of Fig. 2.3 but with effective degrees of freedom (EDF) values instead of smoothing parameter values.

the other end of the spectrum, the bottom right panel is close to the minimum EDF value of 2, corresponding to the least-squares line.

Common automatic smoothing parameter selectors are simple functions of EDF(λ) and the *residual sum of squares (RSS)*:

$$\mathrm{RSS}(\lambda) \equiv \sum_{i=1}^{n} \{y_i - \widehat{f}(x_i;\lambda)\}^2.$$

For example, $\widehat{\lambda}_{\mathrm{GCV}}$ discussed in Sect. 2.3 is the minimizer of

$$\mathrm{GCV}(\lambda) \equiv \frac{\mathrm{RSS}(\lambda)}{\{1 - \mathrm{EDF}(\lambda)/n\}^2}. \tag{2.9}$$

The denominator on the right-hand side of (2.9) guards against the RSS-minimizing choice of $\lambda = 0$. Details are given in, for example, Sect. 5.3 of Ruppert et al. (2003).

There are several other ways by which EDF(λ) and RSS(λ) can be combined to produce sensible smoothing parameter selection criteria. For example,

$$\mathrm{AIC}(\lambda) \equiv \log\{\mathrm{RSS}(\lambda)\} + 2\,\mathrm{EDF}(\lambda)/n \tag{2.10}$$

is known as *Akaike's information criterion (AIC)* (Akaike 1973). Yet another criterion is *corrected AIC* (Hurvich et al. 1998) given by

$$\text{AIC}_C(\lambda) \equiv \log\{\text{RSS}(\lambda)\} + \frac{2\{\text{EDF}(\lambda)+1\}}{n-\text{EDF}(\lambda)-2}.$$

The semiparametric regression literature is not in total agreement regarding the definition of AIC for Gaussian response models. For example, Hastie et al. (2009) use a definition of AIC that differs from (2.10). However, note that (2.10) matches Akaike (1973) and the majority of the literature.

Finally, we note that criteria such as AIC and GCV have wider usage outside of smoothing parameter selection for penalized splines. They are commonly used to choose between competing models regardless of whether nonparametric regression is involved. Hence, they are known as *model selection* criteria.

2.7 Mixed Model-Based Penalized Splines

An alternative penalized spline model is

$$y_i = \beta_0 + \beta_1 x_i + \sum_{k=1}^{K} u_k\, z_k(x_i) + \varepsilon_i$$

where the spline coefficients, u_k, $1 \le k \le K$, are independent zero mean random variables distributed independently of the ε_is. This is a special case of the linear mixed model (e.g. Robinson 1991). If it is also assumed that

$$\text{Var}(u_k) = \sigma_u^2,\ 1 \le k \le K, \quad \text{and} \quad \text{Var}(\varepsilon_i) = \sigma_\varepsilon^2,\ 1 \le i \le n, \tag{2.11}$$

then the *best linear unbiased predictor* of $f(x)$ exactly matches the penalized spline estimator with smoothing parameter

$$\lambda = \sigma_\varepsilon^2/\sigma_u^2.$$

Details of this equivalence are given in Sect. 4.9 of Ruppert et al. (2003). We call this construction *mixed model-based penalized splines*. As we will see throughout this book, this representation of penalized splines is extremely useful.

If (2.11) is strengthened to

$$u_k \overset{\text{ind.}}{\sim} N(0,\sigma_u^2) \quad \text{and} \quad \varepsilon_i \overset{\text{ind.}}{\sim} N(0,\sigma_\varepsilon^2) \tag{2.12}$$

then *restricted maximum likelihood (REML)* estimation of the variance parameters, σ_u^2 and σ_ε^2, is supported by contemporary mixed model software and yields an alternative approach to automatic penalized spline smoothing (e.g., Sect. 4.5.4 of Ruppert et al. 2003). Let

$$\widehat{\lambda}_{\text{REML}} \equiv \widehat{\sigma}_\varepsilon^2 / \widehat{\sigma}_u^2$$

denote the estimator of the optimal λ based on the REML estimators $\widehat{\sigma}_\varepsilon^2$ and $\widehat{\sigma}_u^2$.

A matrix representation of the mixed model-based penalized spline with Gaussian assumptions (2.12) is

$$\boldsymbol{y}|\boldsymbol{u} \sim N(\boldsymbol{X\beta} + \boldsymbol{Zu}, \sigma_\varepsilon^2 \boldsymbol{I}), \qquad \boldsymbol{u} \sim N(\boldsymbol{0}, \sigma_u^2 \boldsymbol{I}) \tag{2.13}$$

where $\boldsymbol{y}$ is the $n \times 1$ vector containing the y_is,

$$\boldsymbol{X} \equiv \begin{bmatrix} 1 & x_1 \\ \vdots & \vdots \\ 1 & x_n \end{bmatrix} \quad \text{and} \quad \boldsymbol{Z} \equiv \begin{bmatrix} z_1(x_1) & \cdots & z_K(x_1) \\ \vdots & \ddots & \vdots \\ z_1(x_n) & \cdots & z_K(x_n) \end{bmatrix} \tag{2.14}$$

are design matrices containing, respectively, the linear and spline basis functions of the predictors, and

$$\boldsymbol{\beta} \equiv \begin{bmatrix} \beta_0 \\ \beta_1 \end{bmatrix} \quad \text{and} \quad \boldsymbol{u} \equiv \begin{bmatrix} u_1 \\ \vdots \\ u_K \end{bmatrix}$$

are the corresponding coefficient vectors. In mixed model parlance, the entries of $\boldsymbol{\beta}$ are called *fixed effects* and those of $\boldsymbol{u}$ are called *random effects.*

The R function `lme()` in the package `nlme` (Pinheiro et al. 2017) supports mixed model-based penalized splines with restricted maximum likelihood smoothing parameter selection. The following code provides illustration:

```
> library(HRW);   data(WarsawApts)
> x <- WarsawApts$construction.date
> y <- WarsawApts$areaPerMzloty
> numIntKnots <- 23
> intKnots <- quantile(unique(x),seq(0,1,length=
+                  (numIntKnots+2))[-c(1,(numIntKnots+2))])
> a <- 1.01*min(x) - 0.01*max(x)
> b <- 1.01*max(x) - 0.01*min(x)
> Z <- ZOSull(x,range.x = c(a,b),intKnots = intKnots)
> library(nlme)
> dummyID <- factor(rep(1,length(x)))
> fit <- lme(y ~ x,random = list(dummyID = pdIdent(~-1+Z)))
> betaHat <- fit$coef$fixed ; uHat <- unlist(fit$coef$random)
> sigsqepsHat <- fit$sigma^2
> sigsquHat <- as.numeric(VarCorr(fit)[1,1])
```

The key command in this code chunk is the call to `lme()`. This function supports best linear unbiased prediction and restricted maximum likelihood estimation for a variety of linear mixed models, but geared towards grouped data settings. To fit (2.13) for a general random effects design matrix as in (2.14) requires some coercion. The object `dummyID` contains the full index set and `random = list(dummyID = pdIdent(~-1+Z))` forces all of the data to be in a single group with a multiple of the identity matrix covariance matrix imposed the random effects. The specification `-1+Z` ensures that a column of ones is not added to $\boldsymbol{Z}$ matrix—which happens by default. Also, `unlist(fit$coef$random)` converts the object `fit$coef$random` from an R list to an R array.

Note that the function `gam()` in the package `mgcv` uses `lme()` in this way when `method` is set to `"REML"`. Nevertheless it useful to know how `lme()` can be used for fitting penalized splines since such knowledge aids extension to more elaborate models, such as those described in Sect. 4.3, with user-specified design matrices.

The estimated variance parameters and their ratio are obtained as follows:

```
> outVec <- c(sigsqepsHat,sigsquHat,sigsqepsHat/sigsquHat)
> print(round(as.numeric(outVec),2))
```

```
[1] 317.09   1.46 217.75
```

The last of these numbers is the smoothing parameter selected by REML.

Given the best linear unbiased predictors of $\boldsymbol{\beta}$ and $\boldsymbol{u}$, which we denote by $\widehat{\boldsymbol{\beta}}$ and $\widehat{\boldsymbol{u}}$, the estimate of $f(x)$ for any $x \in \mathbb{R}$ is

$$\widehat{f}(x) = \boldsymbol{X}_x \widehat{\boldsymbol{\beta}} + \boldsymbol{Z}_x \widehat{\boldsymbol{u}}$$

where

$$\boldsymbol{X}_x \equiv [1\ x] \quad \text{and} \quad \boldsymbol{Z}_x \equiv [z_1(x)\ \cdots\ z_K(x)].$$

Plotting an estimate over a grid simply involves stacking versions of $\boldsymbol{X}_x$ and $\boldsymbol{Z}_x$, with x taking the value of each grid-point, into matrices. The following code does this for an array, `xg`, of `ng` grid-points. The R matrices `Xg` and `Zg` are the grid-wise versions of $\boldsymbol{X}$ and $\boldsymbol{Z}$:

```
> ng <- 1001 ; xg <- seq(a,b,length=ng)
> Xg <- cbind(rep(1,ng),xg)
> Zg <- ZOSull(xg,range.x = c(a,b),intKnots = intKnots)
> fHatg <- as.vector(Xg%*%betaHat + Zg%*%uHat)
> plot(x,y,bty = "l",xlab = "construction date (year)",
+      ylab = "area (square meters) per million zloty",
+      col = "dodgerblue",cex.lab = 1.5,cex.axis = 1.5)
> lines(xg,fHatg,col = "darkgreen",lwd = 2)
```

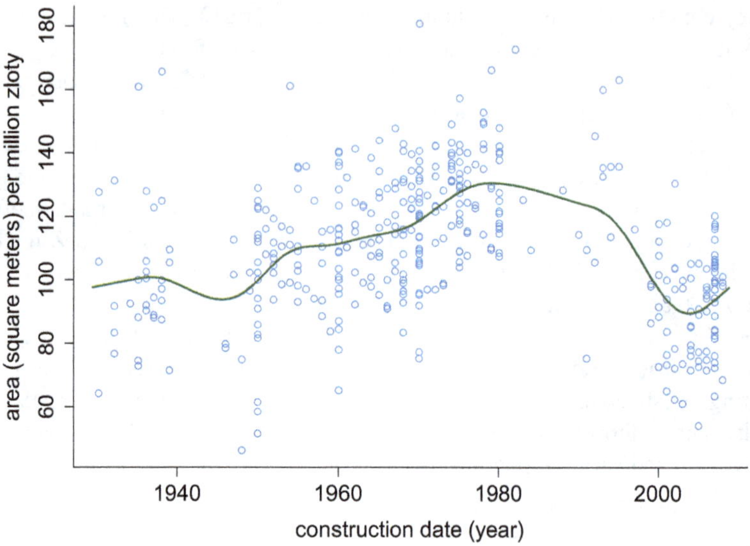

Fig. 2.9 Mixed model-based penalized spline fit to the Warsaw apartments running example regression dataset obtained using the function `lme()`. The smoothing parameter is chosen according to restricted maximum likelihood.

The fit shown in Fig. 2.9 results.

Further calculations lead to

$$\widehat{\lambda}_{\text{REML}} = 217.8 \quad \text{and} \quad \text{EDF}(\widehat{\lambda}_{\text{REML}}) = 10.72.$$

This second value can be compared with

$$\text{EDF}(\widehat{\lambda}_{\text{GCV}}) = 15.05,$$

indicating that the REML fits a penalized spline with less fine structure than GCV. This is in keeping with the simulation study in Sect. 5.4 of Ruppert et al. (2003) and, in particular, Fig. 5.7 given there. Theoretical insight on the behavior of $\widehat{\lambda}_{\text{REML}}$ compared with more traditional smoothing parameter selectors is given in Kou and Efron (2002), Reiss and Ogden (2009) and Krivobokova (2013). Exercise 8 explores this issue further.

Mixed model-based splines have profound implications for semiparametric regression. As we will see in upcoming chapters, they allow flexible function estimation to be incorporated into sophisticated models that deal with complications such as grouping and missingness using software for mixed models and hierarchical Bayesian models. The latter requires *Bayesian* mixed model representations of penalized splines, which is discussed in Sect. 2.10.

2.8 Variability Bands

Figure 2.10 is an embellishment of Fig. 2.9 that has a *variability band* added to the penalized spline fit. Valid inferential statements concerning the variability band is a delicate topic and is discussed in detail in Sects. 6.2–6.4 of Ruppert et al. (2003). An approximate interpretation is one of pointwise 95% confidence intervals for the mean response. For example, the dashed vertical line at 1980 on the horizontal axis cuts the variability band at 124.2 and 136.6. Therefore, 124.2 m^2/(million złoty) to 136.6 m^2/(million złoty) is an approximate 95% confidence interval for the mean area per million złoty of Warsaw apartments constructed in 1980. *Simultaneous* confidence statements require a wider band (e.g., Sect. 6.5 of Ruppert et al. 2003). Most R packages produce only pointwise confidence intervals. An exception is the `locfit()` function in the `locfit` package (Loader 2013). This package uses local polynomial regression (Loader 1999; Wand and Jones 1995) rather than penalized splines.

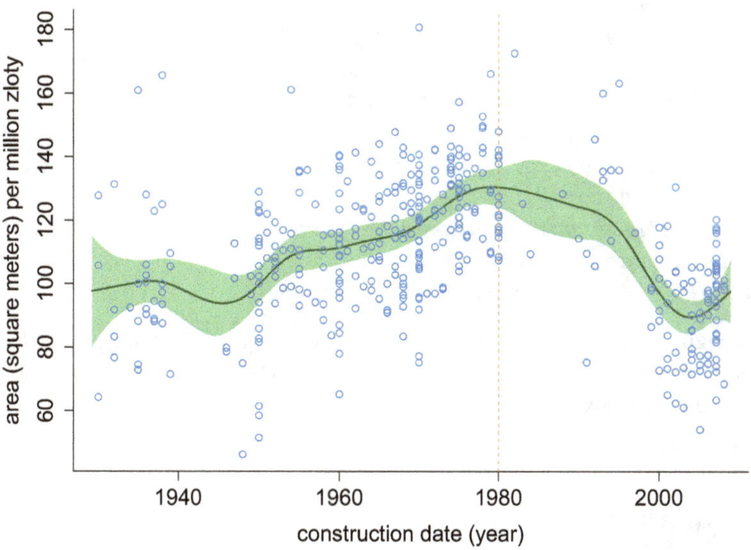

Fig. 2.10 The mixed model-based penalized spline fit of Fig. 2.9 with a variability band added according to (2.15) and (2.16). The dashed vertical line passes through 1980 on the horizontal axis and the points at which it crosses the boundaries of the variability band, 124.2 and 136.6 square meters per million złoty, are the limits of an approximate 95% confidence interval for mean area per million złoty for apartments constructed in 1980. Analogous pointwise confidence interval statements apply to the points where other vertical lines cross the boundaries of the variability band.

The formulae for the lower and upper limits of the variability band shown in Fig. 2.10 are

$$\widehat{f}(x) \pm 2\,\widehat{\text{st.dev.}}\{\widehat{f}(x) - f(x)\} \tag{2.15}$$

where

$$\widehat{\text{st.dev.}}\{\widehat{f}(x) - f(x)\} = \widehat{\sigma}_\varepsilon\, \boldsymbol{C}_x \left(\boldsymbol{C}^T\boldsymbol{C} + \frac{\widehat{\sigma}_\varepsilon^2}{\widehat{\sigma}_u^2}\boldsymbol{D}\right)^{-1} \boldsymbol{C}_x^T \tag{2.16}$$

with $\boldsymbol{C}$ and $\boldsymbol{D}$ as defined by (2.7) and $\boldsymbol{C}_x \equiv [\boldsymbol{X}_x\ \boldsymbol{Z}_x]$. The following R code produces the variability band shown in Fig. 2.10 by computing (2.16) over a grid:

```
> Cg <- cbind(rep(1,ng),xg,Zg)
> C <- cbind(rep(1,length(y)),x,Z)
> D <- diag(c(0,0,rep(1,ncol(Z))))
> sdg <- sqrt(sigsqepsHat)*sqrt(diag(Cg%*%solve(crossprod(C)
+          + (sigsqepsHat/sigsquHat)*D,t(Cg))))
> CIlowg <- fHatg - 2*sdg ; CIuppg <- fHatg + 2*sdg
> plot(x,y,bty = "l",xlab = "construction date (year)",
+        ylab = "area (square meters) per million zloty",
+        type = "n",cex.lab = 1.5,cex.axis = 1.5)
> polygon(c(xg,rev(xg)),c(CIlowg,rev(CIuppg)),col = "palegreen",
+           border = FALSE)
> lines(xg,fHatg,col = "darkgreen",lwd = 2)
> points(x,y,col = "dodgerblue")
> abline(v = 1980,lty=2,col = "darkorange")
```

Although variability bands are useful for assessing the uncertainty of $\widehat{f}$, hypothesis testing is needed to address questions such as whether f is linear or nonlinear. Testing will be discussed in Sect. 2.9.

2.9 Hypothesis Testing

In many applications, it is of interest to test whether a simple model is satisfactory or instead a more complex model is needed to describe the data adequately.

In a parametric setting, when the models are nested a likelihood ratio test is often used. The approximate distribution of twice the logarithm of the likelihood ratio (LR) is

$$2\log(\text{LR}) \sim \chi^2_{\text{DF1}-\text{DF0}} \tag{2.17}$$

where DF_1 and DF_0 are the degrees of freedom of the fit for the full and reduced (or null) models, respectively. For an F-test, $\mathrm{DF}_1 - \mathrm{DF}_0$ is the numerator degrees of freedom.

For non-nested models, no approximate null distribution is readily available, but a comparison can be made between the values of a model selection criterion such as AIC.

Although we have stressed a certain algebraic equivalence between penalized least-squares estimation and mixed models, they have led to rather different types of tests. This difference is due to a fundamental difference between the likelihood functions for fixed effects models and for mixed effects models. Let $\boldsymbol{y}$ be the response vector, let $\boldsymbol{\beta}$ be a vector of coefficients that we will always treat as fixed effects, and let $\boldsymbol{u}$ be another vector of coefficients. Assume that we wish to test that $\boldsymbol{u} = 0$.

If we treat the elements of $\boldsymbol{u}$ as fixed effects, then the likelihood is based on

$$p(\boldsymbol{y}; \beta, \boldsymbol{u}, \sigma_\varepsilon^2), \tag{2.18}$$

the density of $\boldsymbol{y}$ as a function of $(\boldsymbol{\beta}, \boldsymbol{u}, \sigma_\varepsilon^2)$. Here σ_ε^2 is the variance of the ε_i. On the other hand, if we treat $\boldsymbol{u}$ as a vector of random effects, then $\boldsymbol{u}$ must have a density which we will denote by $p(\boldsymbol{u}; \sigma_u^2)$. Let $p(\boldsymbol{y}|\boldsymbol{u}; \boldsymbol{\beta}, \sigma_\varepsilon^2)$ be the conditional density function of $\boldsymbol{y}$ given $\boldsymbol{u}$ as a function of $(\boldsymbol{\beta}, \sigma_\varepsilon^2)$. Then the likelihood is

$$p(\boldsymbol{y}; \beta, \sigma_\varepsilon^2, \sigma_u^2) = \int_{\mathbb{R}^K} p(\boldsymbol{y}|\boldsymbol{u}; \boldsymbol{\beta}, \sigma_\varepsilon^2) p(\boldsymbol{u}; \sigma_u^2)\, d\boldsymbol{u}, \tag{2.19}$$

the density of $\boldsymbol{y}$ as a function of $(\boldsymbol{\beta}, \sigma_\varepsilon^2, \sigma_u^2)$. Note that $\boldsymbol{u}$ has been integrated out and so does not appear on the left-hand side of (2.19).

One approach to testing is to use (2.18) as the likelihood with $\boldsymbol{u}$ estimated using a penalty. For Gaussian responses, this leads to approximate F-tests. This approach is used extensively in Hastie and Tibshirani (1990) and is implemented in the `mgcv` package. With this approach, an approximation to the null distribution of twice the logarithm of the LR is

$$2\log(\mathrm{LR}) \sim \chi^2_{\mathrm{EDF1}-\mathrm{EDF0}}.$$

For the approximate F-test, $\mathrm{EDF}_1 - \mathrm{EDF}_0$ is the numerator effective degrees of freedom, in analogy to (2.17).

Crainiceanu et al. (2005a) take a mixed model approach based on using (2.19) as the likelihood function. As described in Sect. 2.7, testing that $\boldsymbol{u} = 0$ reduces to the restricted likelihood ratio test (RLRT) that $\sigma_u^2 = 0$. (It is assumed that $E(\boldsymbol{u}) = 0$ so $\sigma_u^2 = 0$ implies that $\boldsymbol{u} = 0$.) Crainiceanu et al. (2005b) found that in this setting, approximation (2.17) can be very inaccurate and is not recommended. The problem is that the assumptions made by this approximation are violated in two ways: (a) the null value of the variance component being tested is 0 which is on the boundary of the parameter space, whereas the approximation assumes that it is in the interior; and (b) the random effects imply that *all* of the y_is are correlated under the alternative hypothesis, whereas the approximation assumes that they are independent, or at least can be divided into a large number of independent clusters. The approximation (2.17) would be asymptotically correct if the number of independent clusters is increased to infinity, provided the parameter is in the interior of the parameter space. However, typically all y_is are correlated.

In contrast, the penalized least-squares approach to testing uses the null hypothesis that $\boldsymbol{u} = 0$ where $\boldsymbol{u}$ is a nonrandom parameter. The null value of $\boldsymbol{u}$ is clearly in the interior of the parameter space since both positive and negative values of the components of $\boldsymbol{u}$ are possible. Moreover, in the penalized least-squares approach $\boldsymbol{u}$ cannot induce dependencies between the y_i since $\boldsymbol{u}$ is nonrandom. Therefore, the assumptions behind approximation (2.17) are not violated in the penalized least-squares approach to testing.

The mixed model approach to testing that the variance component σ_u^2 is 0 is implemented by the function `exactRLRT()` in the `RLRsim` package (Scheipl and Bolker 2016). In the case of testing that a single variance component is 0, the exact null distribution of the restricted likelihood ratio test statistic has been derived in Crainiceanu and Ruppert (2004). However, it does not have a simple form and its quantiles are not available in closed form. Instead, `exactRLRT()` simulates of the null distribution. The default number `nsim` of simulated draws is 10,000 giving a good approximation to the exact null distribution. Moreover, the computation is very fast so much larger values of `nsim` are feasible.

When treating $\boldsymbol{u}$ as nonrandom, that is, using likelihood (2.18), the F-test's type I error rate can be far from the nominal value when testing the null hypothesis of a parametric model versus a smooth regression. This problem occurs when the difference in EDF between the parametric and smooth fits is very close to zero because smooth fit is nearly the same as the parametric fit; this is likely when the null hypothesis holds. If this difference in EDF values is large, at least greater than 1, then there is less concern about a possibly erroneous p-value. A possible solution to this problem is to select the smoothing parameter so that the EDF of the smooth fit is, say, 3 more than the parametric fit. An easy way to do this is by letting the smoothing parameter be zero and fixing the dimension of the spline basis appropriately, e.g., 3 more than the dimension of the parametric model. See Exercise 6.

We now return to the running example involving the Warsaw apartments dataset. If we want to test existence of a relationship between the mean area/price ratio and the construction date, then the null hypothesis is $\beta_1 = 0$ and $\boldsymbol{u} = \boldsymbol{0}$, and then we can use the following code that implements an approximate F-test:

```
> library(HRW) ; data(WarsawApts) ; library(mgcv)
> fitPenSpl <- gam(areaPerMzloty ~
+                      s(construction.date,bs = "cr",k = 27),
+                      data = WarsawApts,method = "REML")
> anova(fitPenSpl)
```

and gives output:

```
Family: gaussian
Link function: identity

Formula:
areaPerMzloty ~ s(construction.date, bs = "cr", k = 27)

Approximate significance of smooth terms:
                          edf Ref.df     F p-value
s(construction.date)  9.703 11.878 18.58  <2e-16
```

Note that `method = "REML"` in the call to `gam()` leads to employment of mixed model-based penalized splines with REML smoothing parameter selection, instead of the default GCV smoothing parameter selection. Nonetheless, the F-test is based on (2.18).

The value of the F-statistic is 18.58 with the associated p-value less than 2×10^{-16}, so we strongly reject the null model of no relationship between mean area/price ratio and construction date. The EDF of the fit is 9.7, so there is little concern about an approximation error in the calculation of the p-value.

We might instead wish to test the null hypothesis of a parametric model versus a nonparametric fit. If the parametric model is a linear, quadratic, or cubic polynomial, then the F-test is performed with the following code:

```
> WarsawApts$const.date.sq <- WarsawApts$construction.date^2
> WarsawApts$const.date.cu <- WarsawApts$construction.date^3
> fitLinear <- lm(areaPerMzloty ~ construction.date,
+                     data = WarsawApts)
> fitQuadratic <- update(fitLinear, .~. + const.date.sq)
> fitCubic <- update(fitQuadratic, .~. + const.date.cu)
> anova(fitLinear,fitQuadratic,fitCubic,fitPenSpl,test = "F")
```

which gives:

```
Analysis of Variance Table

Model 1: areaPerMzloty ~ construction.date
Model 2: areaPerMzloty ~ construction.date + const.date.sq
Model 3: areaPerMzloty ~ construction.date + const.date.sq
                                             + const.date.cu
Model 4: areaPerMzloty ~ s(construction.date, bs = "cr", k = 27)
  Res.Df    RSS     Df Sum of Sq        F    Pr(>F)
1  407.0 197242
2  406.0 151979 1.0000     45262 142.6687 < 2.2e-16 ***
3  405.0 140195 1.0000     11784  37.1444 2.592e-09 ***
4  398.3 126362 6.7026     13833   6.5052 4.568e-07 ***
---
Signif. codes:  0 '***' 0.001 '**' 0.01 '*' 0.05 '.' 0.1 ' ' 1
```

The small p-values lead to rejection of all three parametric models. Moreover, `fitPenSpl` uses 9.7 effective degrees of freedom compared to 4 or fewer (effective) degrees of freedom for the polynomial fits which is in keeping with these low p-values.

We now use the `exactRLRT` in the `RLRsim` package to apply the mixed model approach to the `WarsawApts` dataset and test for no effect for construction date. Here we use 50,000 simulations. The computation is very fast, even with this many simulations.

```
> library(RLRsim) ; library(mgcv)
> fitVIAgamm <- gamm(areaPerMzloty ~
+                          s(construction.date,bs = "cr",k = 27),
+                          data = WarsawApts,method = "REML")
> print(exactRLRT(fitVIAgamm$lme,nsim = 50000))

        simulated finite sample distribution of RLRT.

        (p-value based on 50000 simulated values)

data:
RLRT = 144.87, p-value < 2.2e-16
```

The test results are similar to those obtained earlier in this section (see the output on page 38) using the approximate F-test in the package `mgcv`. Note that, although both `gam()` and `gamm()` can fit a mixed model-based penalized spline, only the output of `gamm()` is compatible with `exactRLRT()`.

2.10 Bayesian Penalized Splines

Mixed model-based penalized splines can also be fit by adopting a Bayesian approach. We call these *Bayesian penalized splines*. The advantages of a Bayesian approach compared to the frequentist mixed model approach include taking into account uncertainty associated with the variance components and the ability to deal with complications, such as heteroscedasticity and missing data, which cannot be handled using standard mixed model software. Such complications are tackled in Chap. 6.

We assume that the reader has some familiarity with Bayesian statistical analysis. Practical Bayesian inference for semiparametric regression, including that using R, usually involves *Markov chain Monte Carlo (MCMC)* samples, or *chains*, from the posterior distributions of parameters of interest. In the 1990s the Bayesian inference Using Gibbs Sampling (BUGS) project (Lunn et al. 2013) established itself as the most prominent software suite for MCMC-based Bayesian analyses.

The packages BRugs (Ligges et al. 2017) and R2WinBUGS (Gelman et al. 2015) provide access to BUGS from R. In the early 2010s the Stan project (Carpenter et al. 2017) emerged. It uses versions of MCMC known as *Hamiltonian Monte Carlo* (e.g., Sect. 12.4, (Gelman et al. 2014)) and the *no U-turn sampler* (Hoffman and Gelman 2014). Stan is accessible from R via the package rstan (Guo et al. 2017). For the majority of Bayesian semiparametric regression models that appear in this book we have found Stan to be the superior MCMC software platform. It is usually faster, supports more distributions, has a richer programming language, and runs on each of the operating systems supported by R. Therefore, rstan is the main package used in this book for Bayesian semiparametric regression. However, note that Stan is a long-term project that is still in a state of intense development at the time of publication of this book. It is likely that future modifications of Stan will impact some of the examples that we present here. The website for this book semiparametric-regression-with-r.net should be consulted for updates. Section 2.13 suggests further readings on Bayesian inference, MCMC, BUGS and Stan.

A Bayesian version of the penalized spline nonparametric regression model is:

$$\begin{gathered}
y_i|\boldsymbol{\beta}, \boldsymbol{u}, \sigma_\varepsilon^2 \stackrel{\text{ind.}}{\sim} N\Big(\beta_0 + \beta_1 x_i + \sum_{k=1}^K u_k z_k(x_i), \sigma_\varepsilon^2\Big), \ 1 \le i \le n, \\
\boldsymbol{u}|\sigma_u \sim N(0, \sigma_u^2 \boldsymbol{I}), \quad \beta_0, \beta_1 \stackrel{\text{ind.}}{\sim} N(0, \sigma_{\boldsymbol{\beta}}^2), \\
\sigma_u \sim \text{Half-Cauchy}(A_u), \quad \sigma_\varepsilon \sim \text{Half-Cauchy}(A_\varepsilon)
\end{gathered} \tag{2.20}$$

where $\boldsymbol{u}$ is the $K \times 1$ vector containing the u_k. The notation $x \sim \text{Half-Cauchy}(A)$ means that x has density function

$$p(x) = 2/[A\pi\{1 + (x/A)^2\}], \quad x > 0$$

where $A > 0$ is a scale parameter. As explained in Gelman (2006), Half-Cauchy priors possess attractive noninformativity properties.

It should be noted that (2.20) is the same as (2.13) except that now $\beta_0, \beta_1, \sigma_u$, and σ_ε are treated as random variables, and have prior distributions imposed on them.

The *hyperparameters* in (2.20) are $\sigma_{\boldsymbol{\beta}} > 0$, $A_u > 0$, and $A_\varepsilon > 0$. These need to be specified by the user. For some Bayesian analyses the analyst may be able to make an informed choice of the hyperparameters. However, this is usually not the case for model (2.20) and its various semiparametric regression extensions. Therefore noninformative priors are recommended in general. Throughout this book we will enforce noninformativity by setting hyperparameters such as $\sigma_{\boldsymbol{\beta}}$, A_u, and A_ε to very high positive numbers, but only after first standardizing the data to avoid scale issues. We then transform the results back to the original units for presentation of final results.

MCMC is a computer-intensive methodology and semiparametric regression analyses typically take minutes, and sometimes hours or days, to run—depending on the complexity of the model. Therefore, we do not recommend doing MCMC-based analyses in R from the command line. Instead we recommend working with scripts. The R script `WarsawAptsBayes.R` uses the `rstan` function `stan()` to fit (2.20) to the following variables in the data frame `WarsawApts`:

$$x_i = \text{standardized construction date of } i\text{th apartment},$$

$$y_i = \text{standardized area/price ratio of } i\text{th apartment}$$

for $1 \leq i \leq 409$ and with hyperparameters set to $\sigma_{\boldsymbol{\beta}} = A_u = A_\varepsilon = 10^5$. Thus, the data for input into `stan()` are created as follows:

```
> xOrig <- WarsawApts$construction.date
> yOrig <- WarsawApts$areaPerMzloty
> mean.x <- mean(xOrig)  ; sd.x <- sd(xOrig)
> mean.y <- mean(yOrig)  ; sd.y <- sd(yOrig)
> x <- (xOrig - mean.x)/sd.x ; y <- (yOrig - mean.y)/sd.y
> sigmaBeta <- 1e5 ; Au <- 1e5 ; Aeps <- 1e5
```

The design matrices (2.14) are then obtained and stored in `X` and `Z`. The number of columns in `Z` is coded as `ncZ`. Specification of (2.20) in Stan is done via the R object `npRegModel` as follows:

```
> npRegModel <-
+     'data
+     {
+        int<lower=1> n;              int<lower=1> ncZ;
+        vector[n] y;                 matrix[n,2] X;
+        matrix[n,ncZ] Z;             real<lower=0> sigmaBeta;
+        real<lower=0> Au;            real<lower=0> Aeps;
+     }
+     parameters
+     {
+        vector[2] beta;               vector[ncZ] u;
+        real<lower=0> sigmaeps;  real<lower=0> sigmau;
+     }
+     model
+     {
+        y ~ normal(X*beta + Z*u,sigmaeps);
+        u ~ normal(0,sigmau);   beta ~ normal(0,sigmaBeta);
+        sigmaeps ~ cauchy(0,Aeps);  sigmau ~ cauchy(0,Au);
+     }'
```

The code under the `data` heading declares each of the required dimension and data variables. These are to be inputted, as described below. The model parameters, for which inference is sought, are under the heading `parameters`. Finally, under the `model` heading is the Stan coding of the Bayesian nonparametric regression model (2.20). The code

```
y ~ normal(X*beta + Z*u,sigmaeps);
```

specifies the likelihood by giving the conditional distribution of the response vector y given the parameters and the predictor data. Observe that the second argument of the Stan function `normal()` specifies the *standard deviation*, not the variance. Also, even though y and `X*beta + Z*u` are vectors of length n, Stan sets the distributions in an element-wise fashion and independently. The code

```
u ~ normal(0,sigmau);
```

then specifies $\boldsymbol{u}|\sigma_u \sim N(\boldsymbol{0}, \sigma_u^2 \boldsymbol{I})$. Finally, the priors are specified via the code

```
beta ~ normal(0,sigmaBeta);
sigmaeps ~ cauchy(0,Aeps);   sigmau ~ cauchy(0,Au);
```

A Stan trick is being used here to specify the priors $\sigma_\varepsilon \sim \text{Half-Cauchy}(A_\varepsilon)$ and $\sigma_u \sim \text{Half-Cauchy}(A_u)$. The declarations `real<lower=0> sigmaeps` and `real<lower=0> sigmau` force these parameters to be confined to the positive half-line. Since Half-Cauchy density functions are proportional to Cauchy density functions on the positive half-line the `sigmaeps ~ cauchy(0,Aeps)` and `sigmau ~ cauchy(0,Au)` specifications achieve the desired effect.

The next step is to set up a list containing the data to be inputted into Stan:

```
allData <- list(n = length(x),ncZ = ncZ,y = y,X = X,Z = Z,
                sigmaBeta = sigmaBeta,Au = Au,Aeps = Aeps)
```

We then compile the code via the command:

```
stanCompilObj <- stan(model_code = npRegModel,data = allData,
                      iter = 1,chains = 1)
```

although this only needs to be done once within the same R session if `npRegModel` remains unchanged. Now that the model has been compiled, we run 3000 iterations using:

```
nWarm <- 1000  ;  nKept <- 2000  ;  nThin <- 2
stanObj <-  stan(model_code = npRegModel,data = allData,
                 warmup = nWarm,iter = (nWarm + nKept),
                 chains = 1,thin = nThin,refresh = 100,
                 fit = stanCompilObj)
```

The command `nWarm <- 1000` and the setting `warmup = nWarm` inside `stan()` means that the first 1000 iterations are used to "warm-up" the sampling scheme. The specifications:

`nKept <- 2000`, `nThin <- 2`, `iter = (nWarm+nKept)` and `thin = nThin` mean that the samples from 2000 iterations following the warm-up period are kept temporarily, but then thinned in such a way that only every second sample value is ultimately kept. Thinning a MCMC sample results in some loss of information. However, it aids visualization of MCMC output since, for unthinned output, the number of points and lines may be too high to assess chain behavior.

The resultant MCMC samples of size 1000 are extracted via:

```
MCMCsamples <- extract(stanObj,permuted = FALSE)
```

The object `MCMCsamples` is a three-dimensional array containing MCMC samples corresponding to each of the parameters specified in `npRegModel`, and a log-probability quantity. The first dimension corresponds to the sample indices, the second dimension corresponds to the chain indices, and the third dimension corresponds to the different parameters. For the current example `MCMCsamples` is a

$$1000 \times 1 \times 32 \quad \text{array}$$

since the final MCMC sample size is 1000, there is only one chain (`chains = 1`) and, with `ncZ=` 27, the number of parameters, including the log-probability quantity, is 32.

The MCMC sample for σ_ε is extracted using:

```
sigmaepsMCMC <- MCMCsamples[,1,
                  dimnames(MCMCsamples)$parameters == "sigmaeps"]
```

A more direct command that has the same effect is:

```
sigmaepsMCMC <- extract(stanObj,"sigmaeps",permuted = FALSE)
```

The command

```
sigmaepsOrigMCMC <- sd.y*sigmaepsMCMC
```

then puts the sample on the same scale as the original data. Figure 2.11 plots pertinent aspects of the MCMC sample stored in `sigmaepsOrigMCMC`:

1. a *time series* or *trace* plot,
2. a *lag-1* plot,
3. an *autocorrelation function (acf)* plot, and
4. a kernel density estimate of the posterior density function of σ_ε in the original units of the data.

The code and rationale behind Fig. 2.11 is now described. First the 4×1 layout and the trace plot in the upper panel are produced using:

```
> par(mfrow=c(4,1),mai=c(0.7,0.6,0.15,0))
> plot(sigmaepsOrigMCMC,type = "l",col = "tomato",bty = "l",
+        ylab=expression(sigma[epsilon]),cex.lab = 1.5,
+        cex.axis = 1.5)
```

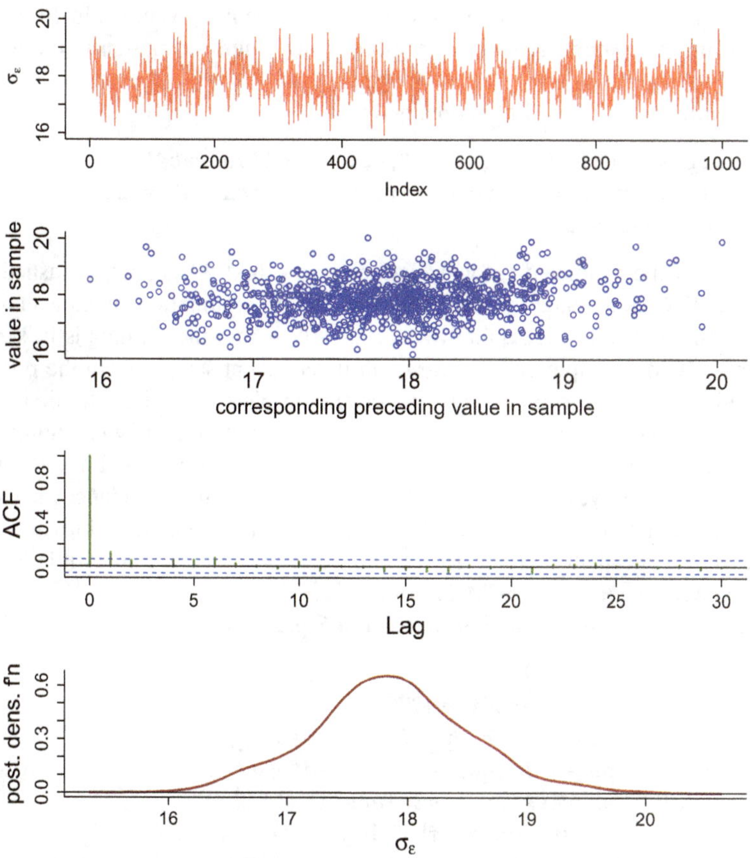

Fig. 2.11 Four plots based on the MCMC sample for the parameter σ_ε for `rstan` fitting of the Bayesian nonparametric regression model (2.20) to the Warsaw apartments running example regression dataset. The MCMC sample has been converted to the original units. First panel: trace plot. Second panel: lag 1 plot. Third panel: autocorrelation function plot. Fourth panel: approximate posterior density function based on kernel density estimation applied to the MCMC sample.

The horizontal axis variable is understood to be the indices 1 up to 1000 and the vertical axis variable are the entries of `sigmaepsOrigMCMC`. The specification `type="l"` means that the coordinate points are joined with line segments. This allows better appreciation of the time ordering in the MCMC sample.

In the second panel we plot the entries of `sigmaepsOrigMCMC`, with the exception of the first entry, against each entry's preceding value:

```
> plot(sigmaepsOrigMCMC[-1000],sigmaepsOrigMCMC[-1],
+       xlab = "corresponding preceding value in sample",
+       ylab = "value in sample",col = "mediumblue",
+       bty = "l",cex.lab = 1.8,cex.axis = 1.8)
```

This permits visual assessment of lag 1 autocorrelation in the sample. In this case no relationship is apparent, suggesting negligible lag 1 autocorrelation in the thinned chain.

The third panel of Fig. 2.11 is obtained via

```
> acf(sigmaepsOrigMCMC,bty="l",col = "olivedrab4",
+      ci.col = "blue",lwd = 2,main = "",cex.lab = 2,
+      cex.axis = 1.5)
```

and displays the estimated autocorrelations at the first 30 lags. The dashed blue lines are 95% confidence interval limits under the assumption of independence. The fact that the spikes are quite low and mostly within these limits is in keeping with this MCMC sample being close to an independent sample from the posterior distribution of σ_ε. In forthcoming autocorrelation plots, including the first one in Fig. 2.13, we will see examples of autocorrelation function plots with significantly high spikes and, therefore, MCMC samples with some dependence. For the purpose of approximate Bayes estimates and credible sets such dependence is usually innocuous because laws of large numbers, on which these approximations are based, also hold for dependent samples. However, as discussed below, autocorrelation may necessitate having a larger MCMC sample size.

The code that produces the fourth panel of Fig. 2.11 is

```
> library(KernSmooth)
> dest <- bkde(sigmaepsOrigMCMC,
+                 bandwidth=dpik(sigmaepsOrigMCMC))
> plot(dest,type = "l",col = "darkred",
+       xlab = expression(sigma[epsilon]),
+       ylab = "post. dens. f'n",bty = "l",
+       lwd = 2,cex.lab = 1.9,cex.axis = 1.5)
> abline(0,0)
```

The kernel density estimate object `dest` is obtained using the function `bkde()` from the package `KernSmooth` (Wand and Ripley 2015) with the bandwidth selected using the *direct plug-in* method (e.g. Wand and Jones 1995) via the function `dpik()`. This is an MCMC-based approximation of the posterior density function $p(\sigma_\varepsilon|\mathbf{y})$ on the original scale of the data.

When examining the trace plots, one should check that the initial behavior of the plot is similar to the plot's overall appearance—this indicates an adequate warm-up period. In contrast, an initial increase or decrease of the plot indicates that a longer warm-up is needed. The trace, lag 1, and autocorrelation function plots all serve to show how well the chain is converging. Ideally, the trace plot should show rapid changes, not slow undulations, the lag 1 plot should show little correlation between each iterate and the previous one, and the autocorrelation function plot should show a rapid decay to zero. If, instead, these plots show signs of slow convergence, then a larger Monte Carlo sample size might be warranted.

The MCMC-approximate *Bayes estimate*, with respect to squared-error loss, of σ_ε is

```
> print(mean(sigmaepsOrigMCMC))
```

```
[1] 17.86439
```

and a corresponding 95% credible interval is

```
> print(quantile(sigmaepsOrigMCMC,c(0.025,0.975)))
```

```
    2.5%    97.5%
16.58944 19.21266
```

This is the Bayesian analog of the frequentist 95% confidence interval. There is a 0.95 posterior probability that a parameter is in its 95% credible interval. From here onwards, all Bayes estimates are with respect to squared-error loss—corresponding to the mean of the posterior density function.

Of central interest, however, is Bayesian inference of the regression function f in (2.20) which is shown in Fig. 2.12. The curve in this figure is the Bayes estimate of f but after conversion to the original units. The shaded region signifies pointwise 95% credible intervals. The Bayes estimate of f shows a similar nonlinear form to the frequentist mixed model-based spline fit shown in Figs. 2.9 and 2.10. The dashed vertical lines in Fig. 2.12 pass through the quartiles of the construction date sample. The MCMC samples corresponding to these three vertical slices are assessed later in Fig. 2.13.

The coding for Fig. 2.12 is given in the R script `WarsawAptsBayes.R`. Figure 2.13 summarizes the MCMC output for the effective degrees of freedom, the error standard deviation, and the regression function at the first, second, and third quartiles of the predictor. This figure was created by the function `summMCMC()`, which is a generic function in the `HRW` package for producing chain diagnosis and summary plots such as those given in Fig. 2.11, but for several parameters. Note that the MCMC sample for the effective degrees of freedom shows some significant autocorrelation. As explained in the discussion concerning Fig. 2.11, this level of dependence is acceptable for approximate Bayesian inference.

The initial segments of the trace plots in Fig. 2.13 appear similar to the rest of the trace plots. This suggests that the Markov chain was already in its steady state by the end of the warm-up, so the warm-up period of 1000 iterations is adequate.

Numerical summaries can be formed for any chain of interest using the `monitor()` function. The following code provides illustration for the parameters summarized in Fig. 2.13:

```
> myMCMCarray <- array(0, dim = c(length(sigmaepsMCMC),1,5))
> myMCMCarray[,1,1] <- EDFMCMC
> myMCMCarray[,1,2] <- sigmaepsOrigMCMC
> myMCMCarray[,1,3] <- fhatOrigQ1MCMC
> myMCMCarray[,1,4] <- fhatOrigQ2MCMC
> myMCMCarray[,1,5] <- fhatOrigQ3MCMC
```

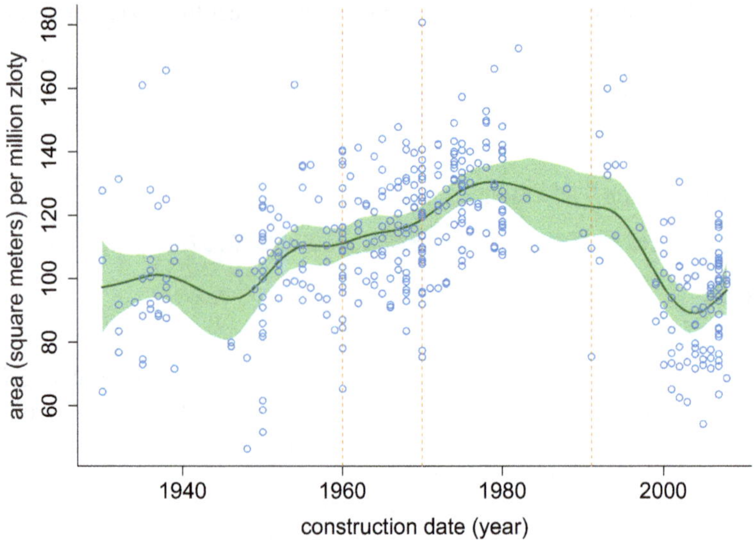

Fig. 2.12 The Bayes estimate of the mean area per million złoty conditional on construction date according to the Bayesian penalized spline model (2.20) fit to the Warsaw apartments running example regression dataset. The shaded region corresponds to pointwise 95% credible sets. The dashed vertical lines pass through the quartiles of the construction date sample.

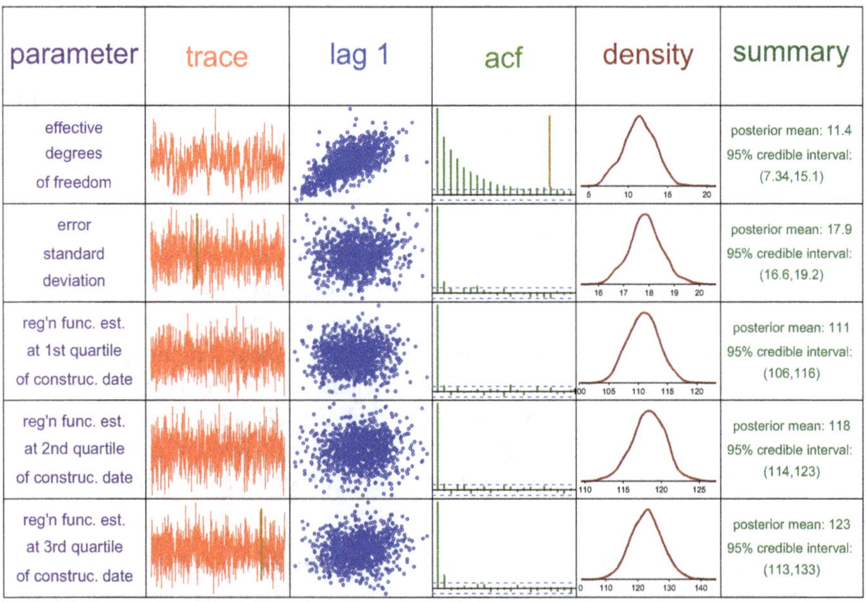

Fig. 2.13 Summary of the MCMC sample, thinned to every second iteration, from the posterior distribution of the parameters in the Bayesian nonparametric regression model fit to the Warsaw apartments running example regression dataset. The columns are: parameter, trace plot of MCMC sample, plot of sample against sample lagged by one iteration, sample autocorrelation function, kernel estimates of posterior density, and basic numerical summaries.

```
> monitorAnswer <- monitor(myMCMCarray,warmup=0,print=FALSE)
> dimnames(monitorAnswer)[[1]] <- c("EDF","err.st. dev.",
+        "f(Q1)","f(Q2)","f(Q3)")
> print(signif(monitorAnswer,4))
```

which produces:

```
                   mean se_mean     sd   2.5%    25%    50%
EDF               11.43 0.16220 2.0020   7.34  10.00  11.38
err. st. dev.     17.86 0.02588 0.6531  16.64  17.35  17.79
f(Q_1)           111.00 0.09281 2.5960 105.60 109.30 111.00
f(Q_2)           118.20 0.08245 2.2270 113.60 116.70 118.40
f(Q_3)           122.70 0.18850 5.1140 112.80 119.30 122.80

                    75%  97.5% n_eff   Rhat
EDF               12.84  15.09 152.3 0.9990
err. st. dev.     18.26  19.19 637.0 1.0000
f(Q_1)           112.70 116.10 782.2 1.0020
f(Q_2)           119.70 122.80 729.5 0.9992
f(Q_3)           125.90 133.00 735.9 1.0010
```

For each parameter, mean is the mean of the MCMC sample and, hence, the approximate Bayes estimate. Also, sd is the standard deviation of the MCMC sample and estimates the posterior standard deviation. The second column lists values of the se_mean is the *Monte Carlo standard error*, defined to be

$$\texttt{se\_mean} \equiv \texttt{sd}/\sqrt{\texttt{n\_eff}}$$

where sd is the sample standard deviation and n_eff is the *effective sample size* of the MCMC sample and takes into account loss of information due to autocorrelation. The Stan *User's Guide and Reference Manual* provides full details on the calculation of the effective sample size for a given MCMC sample. For the EDF parameter (the effective degrees of freedom of the penalized spline) the effective sample size is only about 152 even though the actual sample size is 1000. As a rule-of-thumb, we like to have effective Monte Carlo sample sizes of at least 100 for all parameters. For a correlated sample, the Monte Carlo sample size required might be much larger. That is why we used a Monte Carlo sample size of 1000. We see here that the rule-of-thumb criterion has been met for all parameters, so we can conclude that the Monte Carlo sample size of 1000 is sufficiently large. As the number of iterations increases, se_mean converges to zero and sd converges to the posterior standard deviation, so se_mean will eventually become negligible relative to sd.

The summary statistics for the posterior sample also include the 2.5 and 97.5% sample quantiles. The 2.5 and 97.5% sample quantiles are the endpoints of a 95% credible interval.

The `Rhat` column is superfluous in this single chain example, but is a meaningful statistic in the case of multiple chains, which is treated next in Sect. 2.10.1.

The full `WarsawAptsBayes.R` script can be run via the following commands:

```
> library(HRW) ; demo(WarsawAptsBayes,package = "HRW")
```

To access and, possibly, copy and edit `WarsawAptsBayes.R` note the following code provides its location on the computer on which `HRW` is installed:

```
> system.file("demo","WarsawAptsBayes.R",package = "HRW")
```

2.10.1 Multiple Chains Extension

A common recommendation in MCMC-based analyses is to run more than one Monte Carlo chain from different starting values and use the combined chains to assess convergence (e.g., Sect. 11.4 of Gelman et al. 2014). The `stan()` function in `rstan` has a default of four chains per parameter. These can be run in parallel on multiple cores. When multiple chains are available, formal convergence diagnostics such as those based on *Brooks–Gelman–Rubin statistics* (Gelman and Rubin 1992; Brooks and Gelman 1998), can be employed. One such convergence statistic is denoted by $\widehat{R}_{\text{interval}}$ (Brooks and Gelman 1998) and is a ratio of credible interval lengths based on the combined chains and the mean of such intervals based on individual chains.

Figure 2.14 provides illustration. The top panel shows the traces of three chains for the parameter σ_ε with distinctly different initial values. The middle panel shows the trace plots of the numerator and denominator of the $\widehat{R}_{\text{interval}}$ statistic and the bottom panel shows the statistic itself. As the three chains come together we see that the numerator and denominator traces get closer to each other and the ratio approaches unity. The upper dashed horizontal line in the bottom panel corresponds to $\widehat{R}_{\text{interval}} = 1.2$ and is a reasonable threshold for convergence being attained.

Figure 2.15 is an embellishment of Fig. 2.13 in that a column for Brooks–Gelman–Rubin $\widehat{R}_{\text{interval}}$ trace plots has been added; the column headed BGR. This plot is produced by the functions `summMCMC()` and `BGRinterval()` in the `HRW` package. The BGR column is included in the plot produced by `summMCMC()` whenever multiple chains are inputted. For this example we see that, for all chains, the $\widehat{R}_{\text{interval}}$ statistic stays close to 1 after the first few kept iterations—indicative of very good convergence.

MCMC analyses in the remainder of this book will involve single chains and the less formal convergence diagnosis basedon examining trace, lag 1, and

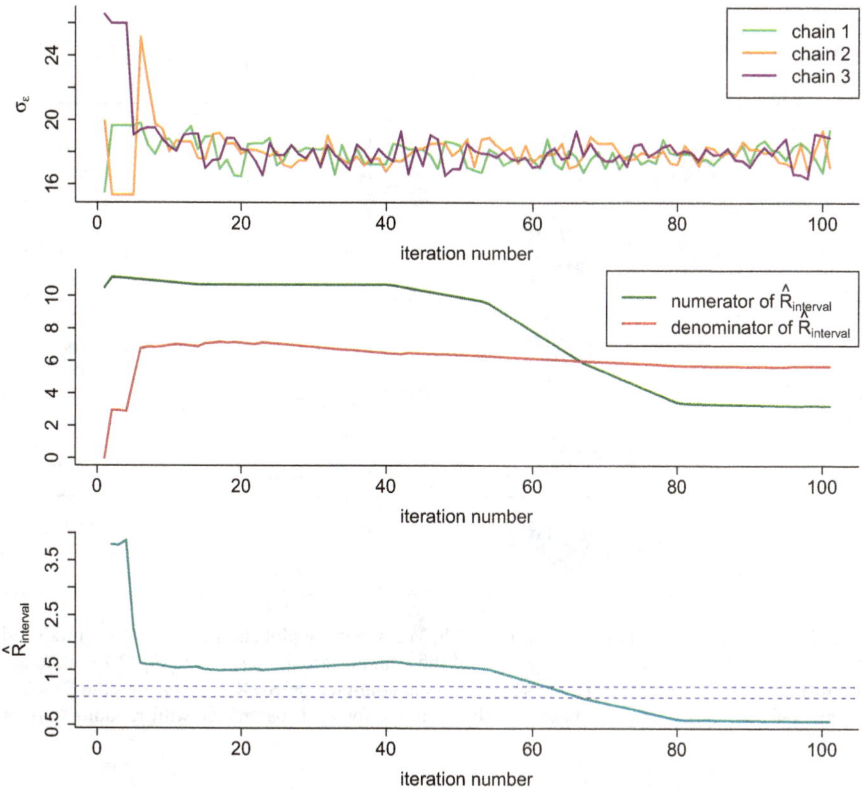

Fig. 2.14 Brooks–Gelman–Rubin convergence diagnosis for σ_ε in `rstan` fitting of the Bayesian nonparametric regression model (2.20) fit to the Warsaw apartments running example regression dataset. Top panel: trace plots of three MCMC samples (chains) with different starting values. Middle panel: trace plots of the numerator and denominator of the $\widehat{R}_{\text{interval}}$ statistic. Bottom panel: trace plot of the $\widehat{R}_{\text{interval}}$ statistic. The horizontal dashed lines pass through $\widehat{R}_{\text{interval}} = 1$ and $\widehat{R}_{\text{interval}} = 1.2$.

autocorrelation function plots. This means that the examples can be tried out without the additional effort required to run multiple chains. The extension of these examples to multiple chains and BGR plots is straightforward if a formal diagnosis is desired.

2.11 Choosing Between Different Penalized Spline Approaches

So far in this chapter we have presented three different approaches to penalized spline fitting:

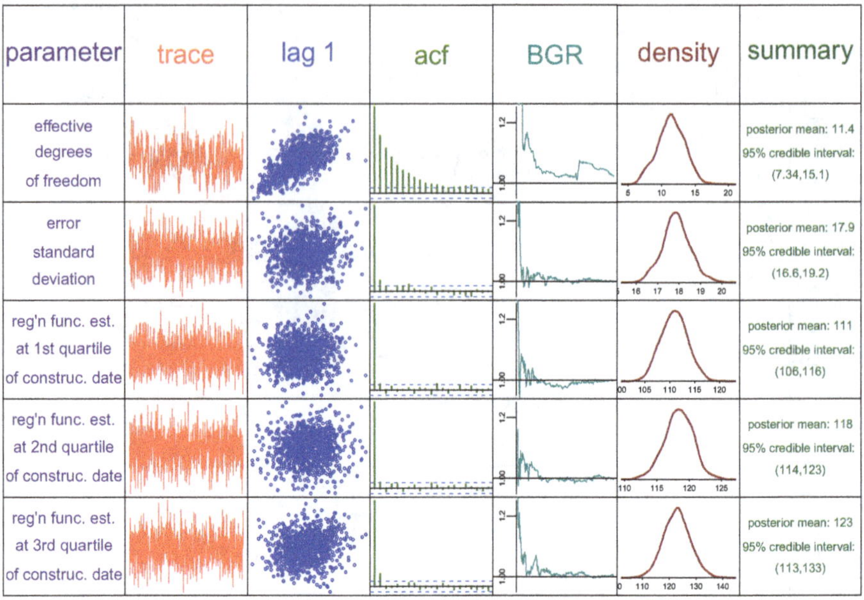

Fig. 2.15 A multiple chain extension of the MCMC summary plot shown in Fig. 2.13, produced using `summMCMC()` in the `HRW` package. Six of the columns are the same as in Fig. 2.13, although for different MCMC samples. There is an additional column labeled BGR that shows the Brooks–Gelman–Rubin $\widehat{R}_{\text{interval}}$ statistic based on three chains for each parameter with random starting values.

(A) penalized least-squares with smoothing parameter chosen by a model selection criterion such as GCV,
(B) couching the problem within the frequentist mixed models framework and estimating the coefficients via best linear unbiased prediction, with the smoothing parameter chosen via REML,
(C) couching the problem within the Bayesian mixed models framework and estimating all parameters via posterior means, made practical using MCMC.

How should one choose between these different approaches?

In the case of models with complete data satisfying the standard model assumptions (independence, normality, and homoscedasticity) reasonably, then (A) is the preferred approach since it is very quick and simple to perform in R. However, (B) and (C) come into their own when we move away from this utopia and complications such as grouping, non-normality, heteroscedasticity, missingness, and measurement error rear their ugly heads. R packages such as `mgcv` can handle some of these problems, but not all of them. In Chap. 4 we show how the mixed model approach is very useful for handling correlations arising from grouping, such as in longitudinal studies. In Chap. 6 we will see that the Bayesian mixed model approach with MCMC packages such as `rstan` allows arbitrarily complicated situations to be handled.

2.12 Penalized Splines with Factor Effects

Semiparametric regression combines the ideas of nonparametric and parametric regression. We now introduce what is perhaps the simplest semiparametric regression model where the nonparametrically modeled effect of a continuous predictor is combined with the parametrically modeled effect of a *factor*, which is a *categorical* predictor.

2.12.1 A Simple Semiparametric Additive Model

The `WarsawApts` data frame includes a categorical variable named `district` that indicates whether the apartment is in one of four districts of Warsaw: Mokotow, Srodmiescie, Wola, and Zoliborz. A question that can be raised is whether the relationship between the area/price ratio and construction date varies between districts. This can be explored by extending the nonparametric regression model (2.1) to allow each district to have its own intercept:

$$\begin{aligned}\texttt{(area/price)}_i &= \beta_1\, I(\texttt{Srodmiescie}_i) + \beta_2\, I(\texttt{Wola}_i)\\ &\quad +\beta_3\, I(\texttt{Zoliborz}_i) + f(\texttt{construction.date}_i) + \varepsilon_i.\end{aligned} \tag{2.21}$$

Here

$$I(\texttt{Srodmiescie}_i) = \begin{cases} 1 & \text{if the } i\text{th apartment is in Srodmiescie,}\\ 0 & \text{otherwise}\end{cases}$$

and similar definitions apply to $I(\texttt{Wola}_i)$ and $I(\texttt{Zoliborz}_i)$. Model (2.21) is a combination of a nonparametric effect for `construction.date` and parametric model for the factor `district`, and is called a *simple semiparametric regression model* by Ruppert et al. (2003). It is also *additive* since the effects of district and construction date are added together. The Mokotow district is used here as the reference level of the factor `district`, so that β_1 is the difference between the expected area/price ratio for apartments in the Srodmiescie and Mokotow districts constructed in the same year, and β_2 and β_3 have analogous interpretation. Model (2.21) can be fit in R using the following code:

```
> fitSimpSemi <- gam(areaPerMzloty ~ factor(district) +
+                    s(construction.date,bs = "cr",k = 27),
+                    data = WarsawApts)
> summary(fitSimpSemi)
```

which leads to the output:

```
Family: gaussian
Link function: identity

Formula:
areaPerMzloty ~ factor(district) + s(construction.date,
                bs = "cr",k = 27)

Parametric coefficients:
                             Estimate Std. Error t value Pr(>|t|)
(Intercept)                  113.7319   1.2804   88.826  < 2e-16
                                                              ***
factor(district)Srodmiescie -12.1574   2.1580   -5.634 3.38e-08
                                                              ***
factor(district)Wola           0.7244   2.5419    0.285    0.776
factor(district)Zoliborz      -1.2114   3.2364   -0.374    0.708
---
Signif. codes:  0 '***' 0.001 '**' 0.01 '*' 0.05 '.' 0.1 ' ' 1

Approximate significance of smooth terms:
                        edf Ref.df     F p-value
s(construction.date) 13.77  16.66 13.84  <2e-16 ***
---
Signif. codes:  0 '***' 0.001 '**' 0.01 '*' 0.05 '.' 0.1 ' ' 1

R-sq.(adj) =  0.416   Deviance explained =   44%
GCV =  298.7  Scale est. = 285.73    n = 409
```

From this output we see that only Srodmiescie appears to be significantly different from Mokotow. It is reasonable to conclude that mean area per million złoty is about 12 square meters lower in the Srodmiescie district compared to the Mokotow district. The area/price ratio in the Wola and Zoliborz are similar to that in the Mokotow district. The negative effect of Srodmiescie on the mean area/price ratio can be explained by noting that Srodmiescie is the central business district, or "downtown" area, of Warsaw.

The script `WarsawAptsSimpSemi.R` in the `HRW` package contains the code that produced Fig. 2.16 and it can be run by issuing:

```
> library(HRW) ; demo(WarsawAptsSimpSemi,package = "HRW")
```

The address of `WarsawAptsSimpSemi.R` on the computer on which `HRW` is installed is determined from:

```
> system.file("demo","WarsawAptsSimpSemi.R",package = "HRW")
```

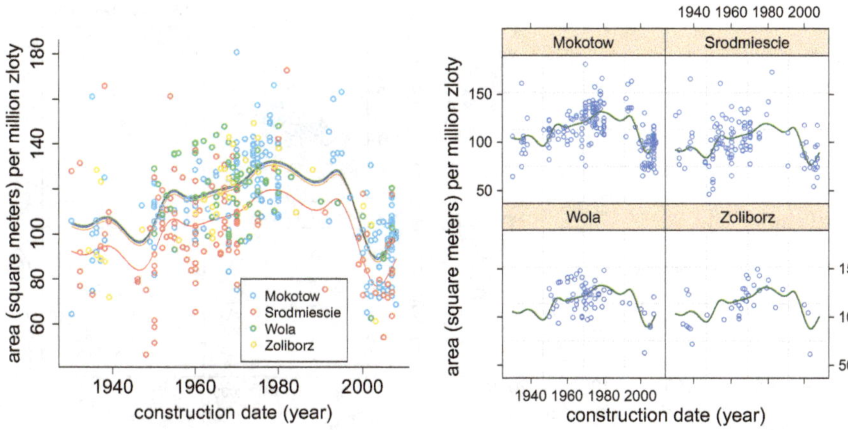

Fig. 2.16 Left panel: scatterplot of area/price ratio versus construction date with color-coding according to district for the Warsaw apartments running example. The curves are the estimated mean response for each district based on the `mgcv` package `gam()` fit of the simple semiparametric additive model (2.21). Right panel: R `lattice` graphics plot of the same data and curves as in the left panel. Each panel within the `lattice` plot shows the scatterplot and estimated mean response for each of the four districts of Warsaw. The left panel facilities comparisons between the districts whereas the right panel presents separate, unobstructed views of each district. The four curves are parallel because they come from a model without interactions. A model with interactions will be considered in the next section.

2.12.2 *A Simple Semiparametric Interaction Model*

Next we consider a more complex model that allows each district to have its own linear form and still keep the semiparametric relationship to be the same between area/price ratio and construction data, so that the variables `district` and `construction.date` interact in a parametric way. This model can be written as

$$\begin{aligned}
\texttt{(area/price)}_i = {} & f(\texttt{construction.date}_i) + \beta_1\, I(\texttt{Srodmiescie}_i) \\
& + \beta_2\, I(\texttt{Wola}_i) + \beta_3 I(\texttt{Zoliborz}_i) \\
& + \beta_4\, I(\texttt{Srodmiescie}_i) \times \texttt{construction.date}_i \qquad (2.22) \\
& + \beta_5\, I(\texttt{Wola}_i) \times \texttt{construction.date}_i \\
& + \beta_6\, I(\texttt{Zoliborz}_i) \times \texttt{construction.date}_i + \varepsilon_i .
\end{aligned}$$

(2.23)

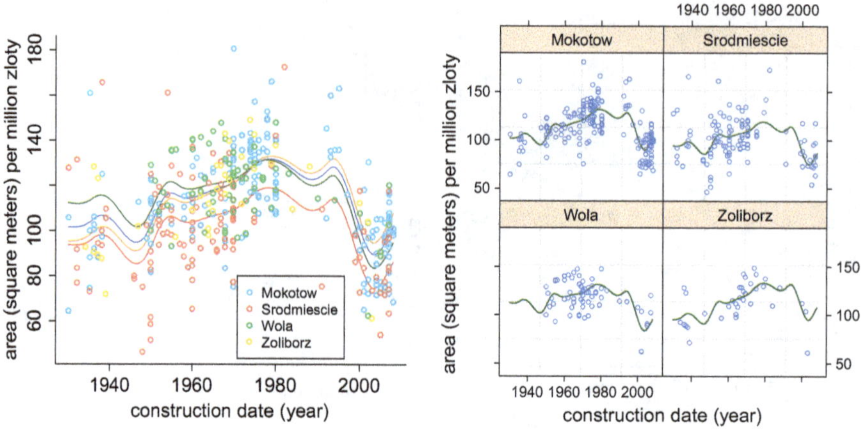

Fig. 2.17 Left panel: scatterplot of area/price ratio versus construction date with color-coding according to district for the Warsaw apartments running example. The curves are the estimated mean response for each district based on the `mgcv` package `gam()` fit of the interaction factor effect model (2.23). Right panel: R `lattice` graphics plot of the same data and curves as in the left panel. Each panel within the `lattice` plot shows the scatterplot and estimated mean response for each of the four districts of Warsaw. Notice that because the model includes interactions, the four curves are not parallel.

To fit model (2.23), we use the following code (contained in the script `WarsawAptsSimpSemiInt.R`):

```
> fitSimpSemiInt <- gam(areaPerMzloty ~
+                       factor(district)*construction.date +
+                       s(construction.date,bs = "cr",k = 27),
+                       data = WarsawApts)
```

The fits are shown in Fig. 2.17. We see that the linear differences between the districts are not enormous. For example, the area/price ratios in the Srodmiescie district are uniformly lower than in the other three districts. However, the area/price ratios in the Zoliborz and Mokotow districts are predicted to be smaller than in the Wola district for the buildings built between 1930 and 1980, but greater for the buildings erected after 1980.

Next, we compare the nonparametric regression model, the simple semiparametric additive model of Sect. 2.12.1, and the simple semiparametric interaction model. In the additive model, the effect of the factor `district` was added to the nonparametric curve for the effect of `construction.date`. In the interaction model, each district had its own linear component. We can compare all three models using an F-test as follows:

```
> fitNonParRegn <- gam(areaPerMzloty ~ s(construction.date,
+                        k = 20),method = "REML",
+                        data = WarsawApts)
> fitSimpSemiAdd <- gam(areaPerMzloty ~ s(construction.date,
+                        k = 20) + district,method = "REML",
+                        data = WarsawApts)
> fitSimpSemiInt <- gam(areaPerMzloty ~ s(construction.date,
+                        k = 20) + construction.date*district,
+                        method = "REML",data = WarsawApts)
> anova(fitNonParRegn,fitSimpSemiAdd,fitSimpSemiInt,test = "F")
```

```
Analysis of Deviance Table

Model 1: areaPerMzloty ~ s(construction.date, k = 20)
Model 2: areaPerMzloty ~ s(construction.date, k = 20)
             + district
Model 3: areaPerMzloty ~ s(construction.date, k = 20)
             + construction.date * district
  Resid. Df Resid. Dev     Df Deviance       F    Pr(>F)
1    394.20     126223
2    391.64     115501 2.5625  10722.2 14.3372 5.937e-08 ***
3    389.02     114686 2.6236    815.6  1.0651    0.3584
---
Signif. codes:  0 '***' 0.001 '**' 0.01 '*' 0.05 '.' 0.1 ' ' 1
```

The small p-value (5.9×10^{-8}) suggests that the additive model is an improvement over the nonparametric regression model. However, the linear interaction model does not provide improvement over the additive model (p-value$=0.3584$).

The code that produced Fig. 2.17 is in the script WarsawAptsSimpSemiInt.R in the HRW package. To run the code type:

```
> library(HRW) ; demo(WarsawAptsSimpSemiInt,package = "HRW")
```

To access WarsawAptsSimpSemiInt.R issue the following code to determine its location on the computer on which HRW is stored:

```
> system.file("demo","WarsawAptsSimpSemiInt.R",package = "HRW")
```

2.12.3 A Simple Factor-by-Curve Model

Another simple model that allows for an interaction between the district and construction date effects is

$$(\texttt{area/price})_i = f_{1-I(\texttt{Srodmiescie}_i)}(\texttt{construction.date}_i) + \varepsilon_i$$

where

$$f_0(x) \equiv \beta_0 + \beta_1 x + \sum_{k=1}^{K} u_{0k} z_k(x),$$

$$f_1(x) \equiv \beta_0 + \beta_0^{\text{contrast}} + (\beta_1 + \beta_1^{\text{contrast}})\, x + \sum_{k=1}^{K} u_{1k} z_k(x),$$

$$u_{0k} \overset{\text{ind.}}{\sim} N(0, \sigma_{u0}^2), \quad u_{1k} \overset{\text{ind.}}{\sim} N(0, \sigma_{u1}^2), \quad \text{and } \varepsilon_i \overset{\text{ind.}}{\sim} N(0, \sigma_\varepsilon^2), \quad 1 \le i \le n. \qquad (2.24)$$

which allows for two unconstrained penalized splines—one for Srodmiescie and one for the other three districts. The *contrast function*:

$$c(x) \equiv f_1(x) - f_0(x) = \beta_0^{\text{contrast}} + \beta_1^{\text{contrast}}\, x + \sum_{k=1}^{K} (u_{1k} - u_{0k})\, z_k(x)$$

represents the difference between mean area/price ratios outside of Srodmiescie and in Srodmiescie. Note that Srodmiescie is the central business district of Warsaw.

The R script `WarsawAptsSimpFacByCurv.R` fits a Bayesian version of (2.24) using `rstan`. The left-hand panel of Fig. 2.18 shows the Bayes estimates of f_0 and f_1, together with pointwise 95% credible sets. The right-handpanel shows the

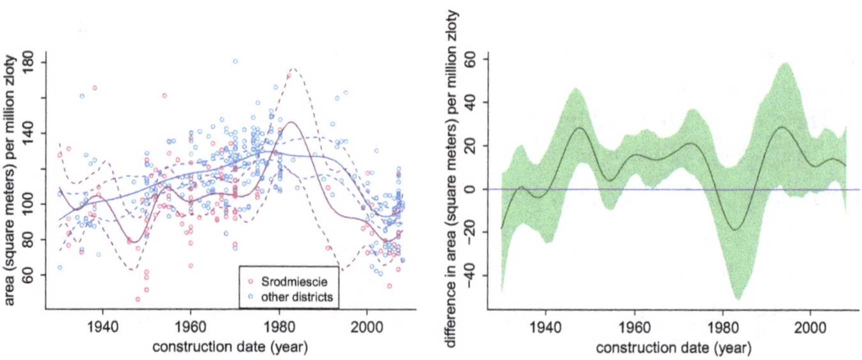

Fig. 2.18 Left panel: Scatterplot of area/price ratio versus construction date for the Warsaw apartments running example with color-coding according to whether or not the apartment is located in the Srodmiescie district. The curves are penalized spline fits according to the Bayesian simple factor-by-curve interaction model (2.24). The solid curves are Bayes estimates and the dashed curves are pointwise 95% credible sets. The fits are based on Markov chain Monte Carlo sampling using the R package `rstan`. Right panel: The estimated contrast function and corresponding pointwise 95% credible sets. The contrast function corresponds to non-Srodmiescie minus Srodmiescie.

same for the contrast function c. As one might expect, the mean area/price ratio is significantly higher outside of central Warsaw over much of the construction date axis and especially so for particular periods such as construction dates between 1960 and 1975.

The code that produced Fig. 2.18 is in the script `WarsawAptsSimpFacByCurv.R` in the `HRW` package and can be run by issuing:

```
> library(HRW) ; demo(WarsawAptsSimpFacByCurv,package = "HRW")
```

To view, copy, and edit `WarsawAptsSimpFacByCurv.R` note the location information provided by:

```
> system.file("demo","WarsawAptsSimpFacByCurv.R",
+             package = "HRW")
```

More general factor-by-curve interaction models are treated later in Sect. 3.6.

2.13 Further Reading

Semiparametric regression is a major area of research with hundreds of journal articles and several books on the topic published since the 1990s. Summaries and pointers to the wider literature are provided by Gurrin et al. (2005), Ruppert et al. (2003, 2009), Wood (2006a), and Hodges (2014). Key articles on low-rank penalized splines are Parker and Rice (1985), O'Sullivan (1986), Eilers and Marx (1996), Hastie (1996), and Wood (2003).

Pinheiro and Bates (2000) and McCulloch et al. (2008) are standard references on mixed models. Gałecki and Burzykowski (2013) provide detailed discussion of the `nlme` and `lme4` packages.

There are many excellent introductions to Bayesian analysis and MCMC including Carlin and Louis (2009), Gelman et al. (2014), Hoff (2010), and Lee (2012). Lunn et al. (2013) is strongly recommended for learning more about BUGS. Albert (2007) discusses Bayesian computations using R and mentions BUGS briefly. Stan is still in its early years and its main reference is Stan Development Team (2017). Semiparametric regression via BUGS is illustrated by Crainiceanu et al. (2005b) and Marley and Wand (2010). The R packages `BayesX` (Kneib et al. 2014) and `R2BayesX` (Umlauf et al. 2016) also provide support for Bayesian semiparametric regression. Illustrations are given in Umlauf et al. (2015).

As of this writing the main websites for BUGS and Stan are www.mrc-bsu.cam.ac.uk/bugs/ and `mc-stan.org`.

2.14 Exercises

1. Let T_1, T_2, and T_3 be three functions on [0, 1] given by

$$T_1(x) = 1, \qquad T_2(x) = x, \qquad T_3(x) = (x - \tfrac{1}{2})_+.$$

Recall that $x_+ \equiv \max(x, 0)$ for $x \in \mathbb{R}$. Also, let B_1, B_2, and B_3 be three functions on [0, 1] given by

$$B_1(x) = (1 - 2x)_+, \qquad B_2(x) = 1 - |2x - 1|, \qquad B_3(x) = (2x - 1)_+.$$

a. Obtain plots of the T_i and B_i, $i = 1, 2, 3$, by running the following R code:

```
> ng <- 101 ; xg <- seq(0,1,length = ng)
> T1g <- rep(1,ng)  ; T2g <- xg ;
> T3g <- (xg - 0.5)*(xg - 0.5>0)
> B1g <- (1 - 2*xg)*(1 - 2*xg>0)
> B2g <- 1 - abs(2*xg - 1)  ; B3g <- 2*T3g
> par(mfrow = c(2,1))
> plot(0,type = "n",xlim = c(0,1),ylim = c(0,1),
+        xlab = "x",ylab = "",bty = "l")
> lines(xg,T1g,col = 1) ; lines(xg,T2g,col = 2)
> lines(xg,T3g,col = 3)
> text(0.1,0.8,expression(T[1]),col = 1)
> text(0.4,0.5,expression(T[2]),col = 2)
> text(0.8,0.2,expression(T[3]),col = 3)
> plot(0,type = "n",xlim = c(0,1),ylim = c(0,1),
+        xlab = "x",ylab = "",bty = "l")
> lines(xg,B1g,col = 4) ; lines(xg,B2g,col = 5)
> lines(xg,B3g,col = 6)
> text(0.1,0.9,expression(B[1]),col = 4)
> text(0.4,0.9,expression(B[2]),col = 5)
> text(0.9,0.6,expression(B[3]),col = 6)
```

b. Find expressions for B_1, B_2, and B_3 in terms of T_1, T_2, and T_3.
Hints: The R plots, rather than algebraic expressions, may be useful. Also, what is $B_1 + B_2 + B_3$?

c. Obtain the 3×3 matrix $\boldsymbol{L}_{\text{TB}}$ such that

$$[B_1(x)\ B_2(x)\ B_3(x)] = [T_1(x)\ T_2(x)\ T_3(x)]\boldsymbol{L}_{\text{TB}}$$

for any $x \in [0, 1]$.

d. Find the determinant of $\boldsymbol{L}_{\text{TB}}$ and establish that $\boldsymbol{L}_{\text{TB}}$ is invertible. In linear algebra language, this implies that $\{B_1, B_2, B_3\}$ is an alternative *basis* for the *vector space* of functions spanned by $\{T_1, T_2, T_3\}$. It is known as the *linear B-spline basis*, and has better numerical properties that the truncated line basis $\{T_1, T_2, T_3\}$.

e. Based on a set of predictor values $x_1, \ldots, x_n$ let

$$\boldsymbol{X}_T \equiv \begin{bmatrix} T_1(x_1) & T_2(x_1) & T_3(x_1) \\ \vdots & \vdots & \vdots \\ T_1(x_n) & T_2(x_n) & T_3(x_n) \end{bmatrix} \quad \text{and} \quad \boldsymbol{X}_B \equiv \begin{bmatrix} B_1(x_1) & B_2(x_1) & B_3(x_1) \\ \vdots & \vdots & \vdots \\ B_1(x_n) & B_2(x_n) & B_3(x_n) \end{bmatrix}$$

be design matrices for the two bases in part d. Run the following code to confirm that ordinary least-squares regression with design matrix $\boldsymbol{X}_T$ leads to same fit as that with design matrix $\boldsymbol{X}_B$.

```
> par(mfrow=c(1,1))
> set.seed(1) ; n <- 100 ; x <- sort(runif(100))
> y <- cos(2*pi*x) + 0.2*rnorm(n)
> plot(x,y,col = "dodgerblue",bty = "l")
> XT <- cbind(rep(1,n),x,(2*x - 1)*(2*x - 1>0))
> XB <- cbind((1 - 2*x)*(1 - 2*x>0),1 - abs(2*x - 1),
+             (2*x - 1)*(2*x - 1>0))
> fitT <- lm(y~-1+XT) ; fitB <- lm(y~-1+XB)
> lines(x,fitted(fitT),col = "orange",lwd = 6)
> lines(x,fitted(fitB),col = "darkgreen",lwd = 2)
```

2. a. Issue the following commands in R to fit the following cubic regression model

$$y_i = \beta_0 + \beta_1 x_i + \beta_2 x_i^2 + \alpha_3 x_i^3 + \varepsilon_i,$$

to the running nonparametric regression example data, plot the fit, and examine the residuals:

```
> library(HRW); data(WarsawApts)
> x <- WarsawApts$construction.date
> y <- WarsawApts$areaPerMzloty
> fitCubic <- lm(y ~ poly(x,3,raw = TRUE))
> ng <- 101 ; xg <- seq(1.01*min(x) - 0.01*max(x),
>                       1.01*max(x) - 0.01*min(x),length = ng)
> fHatCubicg <- as.vector(cbind(rep(1,ng),xg,xg^2,xg^3)
+                               %*%fitCubic$coef)
> plot(x,y,col = "dodgerblue")
> lines(xg,fHatCubicg,col = "darkgreen",lwd = 2)
> plot(fitted(fitCubic),residuals(fitCubic),
+      col = "dodgerblue")
> abline(0,0,col = "slateblue",lwd = 2)
```

Comment on the adequacy of this model.

b. Issue the command:

```
trLin <- function(x,kappa) return((x-kappa)*(x>kappa))
```

to create the truncated line function with a knot at kappa (κ):

$$(x-\kappa)_+ = \begin{cases} 0, & x \le \kappa \\ x-\kappa, & x \ge \kappa. \end{cases}$$

c. Now consider the spline regression model

$$y_i = \beta_0 + \beta_1\, x_i + u_1\, (x_i - \kappa_1)_+ + u_2\, (x_i - \kappa_2)_+ + u_3\, (x_i - \kappa_3)_+ + \varepsilon_i$$

where κ_1, κ_2, and κ_3 are equally spaced knots over the range of the x_is. Issue the following R commands to fit the model to the data from part a. and plot the fit:

```
> knots <- seq(min(x),max(x),length = 5)[-c(1,5)]
> X <- cbind(1,x)
> for (k in 1:3) X <- cbind(X,trLin(x,knots[k]))
> fitTLQ <- lm(y ~ -1 + X)
> Xg <- cbind(1,xg)
> for (k in 1:3) Xg <- cbind(Xg,trLin(xg,knots[k]))
> fHatTLQg <- as.vector(Xg%*%fitTLQ$coef)
> plot(x,y,col = "dodgerblue")
> lines(xg,fHatTLQg,col = "darkgreen",lwd = 2)
> plot(fitted(fitTLQ),residuals(fitTLQ),col = "dodgerblue")
> abline(0,0,col = "slateblue",lwd = 2)
```

Comment on the adequacy of this model.

d. Write an R script that uses ordinary least-squares via `lm()` to fit the spline regression model

$$y_i = \beta_0 + \beta_1\, x_i + \sum_{K=1}^{20} u_k\, (x_i - \kappa_k)_+ + \varepsilon_i$$

where $\kappa_1, \ldots, \kappa_{20}$ are equally spaced knots over the range of the data. The script should then produce a scatterplot of the data, with the fitted curve added and then a plot of the residuals against the fitted values.

e. Re-do part d., but using penalized least-squares, given by (2.4), with $\lambda = 100$.

Hint: See the explicit form of penalized least-squares given in Sect. 2.6.

3. a. Issue the following commands in R to plot truncated line and B-spline basis functions with $K = 15$ knots:

```
> library(splines) ; par(mfrow = c(2,1))
> K <- 15 ; ng <- 1001
> knots <- seq(0,1,length = (K+2))[-c(1,(K+2))]
> xg <- seq(0,1,length = ng)
> ZTg <- outer(xg,knots,"-") ; ZTg <- ZTg*(ZTg>0)
> CTg <- cbind(1,xg,ZTg)
> plot(0,type = "n",xlim = range(xg),ylim = range(CTg),
+       xlab = "x",ylab = "basis function")
> for (j in 1:(2+K)) lines(xg,CTg[,j],col = j)
> points(knots,rep(0,K),col = "darkmagenta",pch = 18)
> allKnots <- c(rep(0,2),knots,rep(1,2))
> CBg <- spline.des(allKnots,xg,ord = 2,
+                   outer.ok = TRUE)$design
> plot(0,type = "n",xlim = range(xg),ylim = range(CBg),
+       xlab = "x",ylab = "basis function")
> for (j in 1:(2+K)) lines(xg,CBg[,j],col = j)
> points(knots,rep(0,K),col = "darkmagenta",pch = 18)
```

b. Re-issue the commands from part a. with higher values of K such as 25 and 35 to display larger bases of the same type.

c. For general K the matrix that transforms truncated line design matrices to B-spline basis design matrices is the $(K+2) \times (K+2)$ matrix:

$$\boldsymbol{L}_{\text{TB}} = (K+1)\begin{bmatrix} \frac{1}{K+1} & 0 & 0 & 0 & \cdots & 0 & 0 & 0 & 0 \\ -1 & 1 & 0 & 0 & \cdots & 0 & 0 & 0 & 0 \\ 1 & -2 & 1 & 0 & \cdots & 0 & 0 & 0 & 0 \\ 0 & 1 & -2 & 1 & \cdots & 0 & 0 & 0 & 0 \\ \vdots & \vdots & \vdots & \vdots & & \vdots & \ddots & \vdots & \vdots \\ 0 & 0 & 0 & 0 & \cdots & 1 & -2 & 1 & 0 \\ 0 & 0 & 0 & 0 & \cdots & 0 & 1 & -2 & 1 \end{bmatrix}.$$

Issue the following commands in R to verify that $\boldsymbol{L}_{\text{TB}}$ has the aforementioned transformation property:

```
> K <- 15 ; ng <- 1001
> knots <- seq(0,1,length = (K+2))[-c(1,(K+2))]
> xg <- seq(0,1,length = ng)
> ZTg <- outer(xg,knots,"-") ; ZTg <- ZTg*(ZTg>0)
> CTg <- cbind(1,xg,ZTg) ; LTB <- matrix(0,(K+2),(K+2))
> LTB[1,1] <- 1 ; LTB[2,1:2] <- (K+1)*c(-1,1)
> for (k in 3:(K+2)) LTB[k,((k-2):k)] <- (K+1)*c(1,-2,1)
> CBviaLTBg <- CTg%*%LTB
> plot(0,type = "n",xlim = range(xg),
+       ylim = range(CBviaLTBg),xlab = "x",
+       ylab = "basis function")
> for (j in 1:(2+K)) lines(xg,CBviaLTBg[,j],col = j)
> points(knots,rep(0,K),col = "darkmagenta",pch = 18)
```

d. Show that $|\boldsymbol{L}_{\text{TB}}| = (K+1)^{K+1}$. Since $|\boldsymbol{L}_{\text{TB}}| \neq 0$, the two bases span the same vector space for any K.

e. Let (x_i, y_i), $1 \leq i \leq n$, be a set of regression data and define the matrices

$$\boldsymbol{C}_T \equiv \begin{bmatrix} T_1(x_1) & \cdots & T_{K+2}(x_1) \\ \vdots & \ddots & \vdots \\ T_1(x_n) & \cdots & T_{K+2}(x_n) \end{bmatrix}, \quad \boldsymbol{D} \equiv \begin{bmatrix} \boldsymbol{0}_{2\times 2} & \boldsymbol{0}_{2\times K} \\ \boldsymbol{0}_{K\times 2} & \boldsymbol{I}_K \end{bmatrix} \quad \text{and } \boldsymbol{y} \equiv \begin{bmatrix} y_1 \\ \vdots \\ y_n \end{bmatrix}.$$

Show that the vector of fitted values for the penalized least-squares problem (2.4), with f given by (2.2), is

$$\begin{bmatrix} \widehat{f}(x_1; \lambda) \\ \vdots \\ \widehat{f}(x_n; \lambda) \end{bmatrix} = \boldsymbol{C}_T(\boldsymbol{C}_T^T \boldsymbol{C}_T + \lambda\, \boldsymbol{D})^{-1} \boldsymbol{C}_T \boldsymbol{y}.$$

f. If $\boldsymbol{C}_B$ is defined analogously to $\boldsymbol{C}_T$, but containing $B_j(x_i)$ values, then show that the vector of fitted values also equals

$$\boldsymbol{C}_B(\boldsymbol{C}_B^T \boldsymbol{C}_B + \lambda\, \boldsymbol{L}_{\text{TB}}^T \boldsymbol{D} \boldsymbol{L}_{\text{TB}})^{-1} \boldsymbol{C}_B \boldsymbol{y}.$$

g. Issue the commands:

```
> set.seed(1) ; n <- 500 ; x <- sort(runif(n))
> y <- (6*x + sin(4*pi*x^2))/(5*x+1) + 0.1*rnorm(n)
> plot(x,y,bty = "l",col = "dodgerblue")
```

to generate and plot some test nonparametric regression data. Add the $(K, \lambda) = (35, 0.01)$ fitted values based on both the truncated line basis expression from part e. and the B-spline basis expression from part f. and verify their equality for this example. The B-spline basis is preferable due to its better numerical stability.

4. Ensure that the package Ecdat (Croissant 2016) is installed in your R environment and issue the following R commands to obtain a scatterplot of data on inflation rate against logarithm of gross domestic product for Canada based on quarterly observations between 1950 and 1996:

```
> library(Ecdat) ; data(Tbrate) ;
> x <- as.data.frame(Tbrate)$y
> y <- as.data.frame(Tbrate)$pi
> plot(x,y,bty = "l",col =  "dodgerblue",
+        xlab = "inflation rate (percentage)",
+        ylab = "logarithm(gross domestic product)")
```

a. Use the function gam() in the package mgcv to fit and plot a penalized spline to these data with GCV choice of the smoothing parameter. Use gam.check() to ensure that the number of spline basis functions in your final answer is sufficient. An autocorrelation function plot of the residuals shows a small amount of serial correlation. For the purpose of this exercise we ignore this and assume independence of the errors.

b. Use the function `lme()` in the package `nlme` to fit and plot a mixed model-based penalized spline to these data with REML choice of smoothing parameter. Also add a variability band to the plot, corresponding to approximate pointwise 95% confidence interval coverage.

c. Use the function `stan()` in the package `rstan` to fit and plot a Bayesian mixed model-based penalized spline to these data. Also add a variability band to the plot, corresponding to approximate pointwise 95% credible interval coverage. Your answer should include some diagnostic plots of MCMC samples for key parameters. The script `WarsawAptsBayes.R` in the `HRW` package contains some relevant code.

5. Issue the R commands:

```
> library(HRW) ; library(mgcv) ; data(WarsawApts)
> x <- WarsawApts$construction.date
> y <- WarsawApts$areaPerMzloty
> plot(x,y,bty = "l",col = "dodgerblue")
> fitGAMcr <- gam(y ~ s(x,bs = "cr",k = 30))
> xg <- seq(min(x),max(x),length = 1001)
> fHatgGAMcr <- predict(fitGAMcr,newdata = data.frame(x = xg))
> lines(xg,fHatgGAMcr,col = "darkgreen")
```

to fit and plot a penalized spline fit to the running nonparametric regression example data with 30 cubic regression spline basis functions and GCV-based smoothing parameter selection.

a. Obtain fits with other types of penalized bases specified in the call to `s()` within the `gam()` function. Specifically, instead of `bs = "cr"` use:

   i. `bs = "gp"` (Gaussian process basis functions),
   ii. `bs = "ps"` (P-splines),
   iii. `bs = "tp"` (thin plate regression splines).

   Produce a plot that compares all four fits.

b. Obtain fits with other numbers of basis functions specified in the call to `s()` within the `gam()` function. Specifically, retain `bs = "cr"` but instead of `k = 30` use:

   i. `k = 40` (40 basis functions),
   ii. `k = 50` (50 basis functions),
   iii. `k = 60` (60 basis functions).

   Produce a plot that compares all four fits.

c. Obtain the fit with REML smoothing parameter selection, but with `bs = "cr"` and `k = 30` retained, via the call:

```
> fitGAMcrREML <- gam(y ~ s(x,bs = "cr",k = 30),
+                     method = "REML")
```

Produce a plot that compares the GCV-based and REML-based fits.

d. Based on the plots in parts a., b., and c., what can be said about the relative influence of the type of basis, number of basis functions, and method of smoothing parameter selection for this example?

6. Consider the nonparametric regression model

$$y_i = f(x_i) + \varepsilon_i, \quad \varepsilon_i \overset{\text{ind.}}{\sim} N(0, \sigma_\varepsilon^2), \quad 1 \le i \le n \tag{2.25}$$

and the hypothesis testing problem

$$H_0: \ f \text{ is linear} \quad \text{versus} \quad H_1: f \text{ is a smooth nonlinear function.} \tag{2.26}$$

a. Ensure that the package RLRsim (Scheipl and Bolker 2016) is installed in your R environment.

b. Start an R session and issue the commands:

```
> set.seed(1)
> x <- seq(0,1,length = 200)   ;   y <- x + rnorm(200)
```

to generate a dataset from (2.25) with $f(x) = x$, $\sigma_\varepsilon = 1$, and $n = 200$.

c. Issue the following commands to obtain p-values for three different tests of (2.26):

```
> library(mgcv) ; library(RLRsim)
> fitLine <- gam(y ~ x) ; fitDfltPenSpl <- gam(y ~ s(x))
> print(anova(fitLine,fitDfltPenSpl,
+             test = "F")$"Pr(>F)"[2])
> fitOLSspl <- gam(y ~ s(x,k = 5,sp = 0))
> print(anova(fitLine,fitOLSspl,test = "F")$"Pr(>F)"[2])
> fitGAMM <- gamm(y ~ s(x),method = "REML")
> print(exactRLRT(fitGAMM$lme)[2])
```

The three tests are (1) an F-test using the `gam()` default penalized spline fit, (2) an F-test using an ordinary least-squares fit with two linear and three spline basis functions, and (3) the exact restricted likelihood ratio test using the function `exactRLRT()` in RLRsim. Given that H_0 is true, comment on the results.

d. Replicate 1000 random datasets of according to the code in part b. For each replication perform the three tests from part c. and store the p-values. Plot histograms of the p-values for all three tests. Comment on the histograms.

e. Suppose that the significance level of the tests is set at 0.05. Using the p-values from d. tabulate the proportions of tests that reject H_0 for each of the three tests. Which tests have rejection probabilities that are close to the advertised significance level?

7. As in Exercise 6 we consider the nonparametric regression model (2.25) and hypothesis testing problem (2.26) but work data generated from with the family of regression functions

$$f(x;\theta) = x + \theta\phi(x; 1/2, 1/4), \quad 0 \le \theta \le 1$$

where $\phi(\cdot; \mu, \sigma)$ is the Normal density function with mean μ and standard deviation σ. Note that $f(x; 0)$ is linear but $f(x;\theta)$ is nonlinear for $0 < \theta \le 1$.

   a. Enter the following commands in R to obtain a plot of $f(\cdot;\theta)$ for $\theta \in \{0, 0.05, \ldots, 1\}$:
```
> theta <- seq(0,1,by = 0.05)
> xg <- seq(0,1,length = 1001)
> plot(xg,xg,type="l",ylim = c(0,2.15),bty = "l",
+   xlab = "x",ylab = expression(paste("f(x;",theta,")")))
> for (j in 2:length(theta))
+   lines(xg,xg + theta[j]*dnorm(xg,0.5,0.25),col = j)
```
   b. Consider the three tests from part c. of Exercise 6 and set the significance level of the tests to be 0.05. For each $\theta \in \{0, 0.05, \ldots, 1\}$ simulate 1000 random datasets according to the model

$$y_i = f(x_i;\theta) + \varepsilon_i, \quad \varepsilon_i \overset{\text{ind.}}{\sim} N(0, \sigma_\varepsilon^2), \quad 1 \le i \le n,$$

   with $n = 200$, $\sigma_\varepsilon = 1$ and the x_is equally spaced between 0 and 1 and apply all three tests to each dataset. Use the rejection proportions to approximate the power functions of each test. Plot the approximate power functions on the same set of axes.
   c. Based on the plot in part b., which test performs best in terms of achieving the advertised significance and being the most powerful?

8. As discussed in Sect. 2.7 GCV and REML are two methods for selecting the smoothing parameter of a penalized spline. Consider the nonparametric regression setting from Exercise 7:

$$y_i = f(x_i;\theta) + \varepsilon_i, \quad \varepsilon_i \overset{\text{ind.}}{\sim} N(0, \sigma_\varepsilon^2), \quad 1 \le i \le n \tag{2.27}$$

where the x_is are equally spaced between 0 and 1.

   a. Fix $(n, \theta, \sigma_\varepsilon) = (800, 0.5, 0.2)$ and generate a sample of (x_i, y_i) observations via the code:
```
> n <- 800 ; theta <- 0.5 ; sigmaEps <- 0.2
> set.seed(1) ; x <- seq(0,1,length = n);
> y <- x + theta*dnorm(x,0.5,0.25) + sigmaEps*rnorm(n)
```
   Next obtain GCV-based and REML-based penalized spline fits via the calls:
```
> library(mgcv) ; fitGCV <- gam(y ~ s(x,k = 27))
> fitREML <- gam(y ~ s(x,k = 27),method = "REML")
```

For a generic estimator $\widehat{f}$ of $f(\cdot;\theta)$, the *average squared error (ASE)* performance measure is

$$\text{ASE} \equiv \frac{1}{n}\sum_{i=1}^{n}\{\widehat{f}(x_i) - f(x_i;\theta)\}^2.$$

Implement the following code to compare the ASE values for the GCV-based and REML-based penalized spline estimates:

```
> fhatGCV <- fitted(fitGCV) ; fhatREML <- fitted(fitREML)
> fTrue <- x + theta*dnorm(x,0.5,0.25)
> ASEforGCV <- sum((fhatGCV - fTrue)^2)/n
> ASEforREML <- sum((fhatREML - fTrue)^2)/n
> print(c(ASEforGCV,ASEforREML))
```

If the value of `ASEforGCV` is lower than `ASEforREML`, then GCV is better than REML for this particular dataset and vice versa.

b. Fix $(\theta, \sigma_\varepsilon) = (0.5, 0.2)$ and allow the sample size to vary over the set $n \in \{25, 50, 100, 200, 400, 800, 1600, 3200\}$. For each value of n in this set generate 100 random datasets according to (2.27), obtain the GCV-based and REML-based penalized spline estimates of $f(\cdot;\theta)$ using the code from part a. and record the ASE performance measures for each fit.
c. Do the same as in part b. but fixing $(n, \sigma_\varepsilon) = (800, 0.2)$ and allowing θ to vary over the set $\theta \in \{0, 0.1, \ldots, 1\}$.
d. Do the same as in part b. but fixing $(n, \theta) = (800, 0.5)$ and allowing σ_ε to vary over the set $\sigma_\varepsilon \in \{0.1, 0.2, \ldots, 1\}$.
e. Obtain comparative graphical and statistical summaries of the ASE performance values and make some conclusions about the relative performance of GCV and REML.

9. The simple factor-by-curve interaction model for the Warsaw apartment data given by (2.24) has one curve for the Srodmiescie district and one curve for each of the remaining districts.

   a. Extend the model so that there is a separate curve for each of the four districts.
   b. Modify the code in the R script `WarsawAptsSimpFacByCurv.R` to fit the model formulated in part a. and obtain a color-coded plot showing fitted mean curves for each district and corresponding 95% pointwise credible sets.
   c. Obtain and plot contrast function estimates for all six pairs of districts. Include 95% pointwise credible sets with each contrast function estimate.

10. The dataset `TreasuryRate` in the `HRW` package contains two variables, `date` and `rate`. The latter is the daily one-month maturity U.S. Treasury rate from July 31, 2001 until July 10, 2013. There are occasional missing values due to holidays.

a. Issue the following commands in R to remove the missing values and obtain a time series plot of the rates:

```
> library(HRW) ; data(TreasuryRate)
> date <- (as.Date(TreasuryRate$date,"%m/%d/%Y")
+                   [!is.na(TreasuryRate$rate)])
> r <- TreasuryRate$rate[!is.na(TreasuryRate$rate)]
> plot(date,r,type = "l",bty = "l",col = "darkgreen",
+        xlab = "date",ylab="U.S. Treasury rate")
```

Let r_t be the tth value of the variable plotted on the vertical axis, $1 \leq t \leq 2987$.

b. Define $\Delta r_t \equiv r_t - r_{t-1}$ and consider the nonparametric *mean* and *variance* regression model

$$\Delta r_t = f(r_{t-1}) + \sqrt{g(r_{t-1})}\,\varepsilon_t,$$

where f and g are smooth functions and, as a working assumption, assume first that the ε_t are independent and identically distributed. Mixed-model based penalized spline models for the mean and variance functions are

$$f(r) = \beta_0 + \beta_1 r + \sum_{k=1}^{K_f} u_k\, z_k^f(r), \quad u_k \sim N(0, \sigma_u^2)$$

and

$$g(r) = \exp\left\{\gamma_0 + \gamma_1 r + \sum_{k=1}^{K_g} v_k\, z_k^g(r)\right\}, \quad v_k \sim N(0, \sigma_v^2)$$

where the z_k^f and z_k^g are O'Sullivan cubic spline bases of sizes K_f and K_g respectively. By modifying the code in `WarsawAptsBayes.R` fit a Bayesian penalized model with $K_f = 10$ and $K_g = 100$. Use hyperparameter values similar to those used in `WarsawAptsBayes.R`. Make the additional working assumption that the ε_t are normally distributed. Plot the estimates of the mean function f and the standard deviation function $\sqrt{g}$.

c. The tth standardized residual is

$$\widehat{\varepsilon}_t \equiv \{\Delta r_t - \widehat{f}(r_{t-1})\}/\sqrt{\widehat{g}(r_{t-1})}$$

and estimates ε_t. Plot $\widehat{\varepsilon}_t^2$ against t. If the assumption that the ε_t are independent and identically distributed holds, then the $\widehat{\varepsilon}_t^2$ should have a constant mean and be uncorrelated. Do you see evidence against this assumption?

d. The plot in part c. shows that there is a period of high volatility during the financial crisis of 2008–2009. The model for g does not account for extra variance due to this volatility. This extra variance could be modeled by assuming that ε_t follows a *generalized autoregressive conditional heteroscedasticity (GARCH)* process rather than being an independent process. See Ruppert and Matteson (2015) for details on GARCH processes and R code for fitting them.

Chapter 3
Generalized Additive Models

3.1 Introduction

The models fit in Chap. 2 have two limitations. First, the conditional distribution of the response, given the predictors, is assumed to be Gaussian. Second, only a single predictor is allowed to have a smooth nonlinear effect—the other predictors are modeled linearly. The first limitation is addressed by using *generalized linear models (GLMs)*, which remove the Gaussian assumption and allow the response variable to have other distributions such as those within the Binomial and Poisson families. The second limitation can be relaxed via the notion of *additive models*, which is explained in Sect. 3.3. The combination of an additive model and generalized regression is called a *generalized additive model (GAM)* and is the focus of this chapter.

GAMs were proposed in Hastie and Tibshirani (1986); Hastie and Tibshirani (1990) with accompanying software that is now packaged as `gam` (Hastie 2017a). The attractions of allowing flexible predictor effects, interpretability, and ready-to-use software have led to widespread use of GAMs in many areas of application. The `mgcv` package (Wood 2017) also supports GAM analyses and has the advantage of providing automatic smoothing parameter selection for multiple penalized spline components. As we will demonstrate in this chapter, GAMs are supported by other R packages including `gamlss` (Stasinopoulos and Rigby 2017), `polspline` (Kooperberg 2015) and `VGAM` (Yee 2017).

3.2 Generalized Linear Models

Generalized linear models (GLMs) are often used when the assumptions of a linear model are not met. Classical linear models make three assumptions about the conditional distribution of the response given the predictors:

J. Harezlak et al., *Semiparametric Regression with R*, Use R!,
https://doi.org/10.1007/978-1-4939-8853-2_3

1. the conditional mean of the response is a linear function of the predictors,
2. the conditional variance of the response is constant,
3. the conditional distribution of the response is Gaussian.

A GLM relaxes all three assumptions of a linear model. Assumption 1 is relaxed to the assumption that a monotone transformation of the mean response, known as the *link*, is a linear function of the predictors. A GLM also relaxes assumption 3, most commonly to the assumption that the conditional distribution of the response is in a one-parameter exponential family. Finally, assumption 2 is not made by a GLM. Instead, the conditional variance is imposed by the exponential family.

The general form of the one-parameter exponential family of densities is

$$p(y;\theta,\phi) = \exp\left\{\frac{y\theta - b(\theta)}{\phi} + c(y,\phi)\right\}, \tag{3.1}$$

where θ is the so-called natural parameter, ϕ is a dispersion parameter, and $b(\theta)$ and $c(y,\phi)$ are known functions. The relationship between the moments of y and $b(\theta)$ is: $E(y) = b'(\theta)$ and $\text{Var}(y) = b''(\theta)$. We assume that the conditional expectation of y, denoted by μ, depends on the predictors x_j, $1 \le j \le d$, via the relationship $g(\mu) = \eta$, where $g(\cdot)$ is the link, and η has a linear form so that

$$g(\mu) = \eta = \beta_0 + \beta_1 x_1 + \cdots + \beta_d x_d. \tag{3.2}$$

Since η can potentially be any real number, ideally g maps the domain of μ onto the entire real line. For example, for a binary response variable y, where y takes only the values 0 and 1, μ is in the interval (0, 1) so g should map (0, 1) to the real line. The *logit* function,

$$\text{logit}(x) \equiv \log\{x/(1-x)\}$$

and the *probit* function Φ^{-1}, where Φ is the Standard Normal cumulative distribution function, have this property. For a binary response, the expected response is also the probability that the response equals 1. The logit link transforms this probability to the log-odds.

For conditionally Poisson-distributed or Gamma-distributed responses, the link function should map the positive half-line to the entire real line. The log function is the usual choice for the link.

The *canonical link* has the special property that $\eta = \theta$ and so there exists a $(d+1)$-dimensional sufficient statistic regardless of the sample size. For the Binomial distribution, the canonical link is the logit link (see Table 3.1). It is not essential that one uses a canonical link. For example, probit regression uses a noncanonical link. For Gamma regression, the log link is often used although the reciprocal function is the canonical link. In Gamma regression, an advantage of the noncanonical log link is that it maps $(0,\infty)$, the set of possible values of the mean, onto the entire real line; the reciprocal link does not do this. One should use a link function that fits the

Table 3.1 Common exponential family distributions and corresponding link functions.

Family	Canonical link (mathematical form)	Other commonly used link (mathematical form)
Gaussian	identity (x)	
Binomial	logit $(\log\{x/(1-x)\})$	probit $(\Phi^{-1}(x))$
Poisson	log $(\log(x))$	
Gamma	reciprocal $(1/x)$	log $(\log(x))$

data well. In the case of binary regression, the logit and probit links are so similar that one often finds that they provide equally good fits. For Gamma regression, the log and reciprocal functions are sufficiently different that one might try both and compare the fits.

Table 3.1 summarizes the main GLM families and their corresponding link transformations.

In Sect. 2.9, likelihood ratio tests were discussed where random effects, if present, are treated as fixed effects and replaced by estimates, rather than being integrated out. Likelihood ratio tests of this type are also used for GLMs and GAMs and for their mixed-model counterparts, generalized linear mixed models, and generalized additive mixed models. Likelihood ratio tests for these models are usually implemented using the so-called *deviance* statistics.

We begin by defining a *scaled deviance*. The scaled deviance of a model is twice the logarithm of the likelihood ratio comparing that model to the saturated model. The saturated model is the model where each y_i has its own parameter, θ_i, and therefore its own mean, μ_i. The saturated model has the highest likelihood of all models. Usually, the saturated model provides a perfect fit, so that the fitted mean $\widehat{\mu}_i$ equals the response y_i for all cases. The effect of comparing the likelihood of a model to that of the saturated model is a simplification due to the cancellation of the terms $c(y, \phi)$ in their likelihoods, assuming that both models use the same value of the scale parameter ϕ.

If D_0 and D_1 are the scaled deviances of a reduced and full model, respectively, then twice the logarithm of the likelihood ratio for these models is $D_1 - D_0$. The log-likelihood of the saturated model appears in both D_0 and D_1 and is cancelled by the subtraction. Therefore, $D_1 - D_0$ is $2\log(\text{LR})$, the likelihood ratio test statistic discussed in Sect. 2.9.

The scaled deviance D of a model can be expressed as $\sum_{i=1}^{n} d_i$ where $d_i \geq 0$ is the contribution of the ith case to the scaled deviance. The deviance residual for the ith case is defined as $\text{sign}(y_i - \widehat{\mu}_i)\sqrt{d_i}$, so that D is the sum of the squared deviance residuals. Here, assuming model (3.2), the fitted mean $\widehat{\mu}_i$ is given by $g(\widehat{\mu}_i) = \widehat{\beta}_0 + \widehat{\beta}_1 x_{1i} + \cdots + \widehat{\beta}_d x_{di}$. To keep the notation simple, we do not show that μ_i depends on $\beta_0, \ldots, \beta_p$ (or $\widehat{\mu}_i$ on $\widehat{\beta}_0, \ldots, \widehat{\beta}_p$), but these facts should be kept in mind.

The deviance of a model is its scaled deviance times ϕ. For the logistic or Poisson regression models, $\phi = 1$ and the scaled deviance and deviance are equal. If, for example, y_i is Poisson distributed with mean μ_i, then the ith term of the log-likelihood is

$$-\mu_i + y_i \log(\mu_i) - \log(y_i!) = y_i \eta_i - \exp(\eta_i) - \log(y_i!) = \frac{y_i \eta_i - b(\eta_i)}{\phi} + c(y_i, \phi)$$

where $\eta_i = \log(\mu_i)$, $b(\eta_i) = \exp(\eta_i)$, $\phi = 1$, and $c(y) = c(y, \phi) = -\log(y!)$.

For the saturated model, the estimated mean is simply y_i. For another model with $\widehat{\mu}_i$ as the estimated mean, the deviance is

$$\begin{aligned} D &= \sum_{i=1}^{n} \{ -y_i + y_i \log(y_i) - \log(y_i!) \} - \{ -\widehat{\mu}_i + y_i \log(\widehat{\mu}_i) - \log(y_i!) \} \\ &= \sum_{i=1}^{n} (\widehat{\mu}_i - y_i) + y_i \log(y_i / \widehat{\mu}_i) = \sum_{i=1}^{n} y_i \log(y_i / \widehat{\mu}_i), \end{aligned}$$

since $\sum_{i=1}^{n} (\widehat{\mu}_i - y_i) = 0$.

As another example, if y_i is normally distributed with mean μ_i and variance σ^2, then the scaled deviance is $\sum_{i=1}^{n} (y_i - \widehat{\mu}_i)^2 / \sigma^2$. Also $\phi = \sigma^2$, so the deviance is the residual sum of squares, $\sum_{i=1}^{n} (y_i - \widehat{\mu}_i)^2$.

Maximum likelihood requires that one have a probability model for the data. If instead, only the mean and variance of y_i (conditional on the predictors) are specified, then one can estimate the regression parameters by quasi-likelihood. Quasi-likelihood assumes that the variance of y_i is equal to $\phi V(\mu_i)$ where V is a known function, μ_i is the mean of y_i, and ϕ is called the *scale parameter* The quasi-likelihood estimator of the regression parameters maximizes $Q \equiv \sum_{i=1}^{n} Q_i$ where

$$Q_i = \int_{y_i}^{\mu_i} \frac{y_i - u}{\phi V(u)} du.$$

The maximum does not depend on ϕ, but the standard errors do. To calculate the standard errors, Q is treated as a log-likelihood. The negative second derivative matrix of Q with respect to $(\beta_0, \ldots, \beta_d)$ is used as an information matrix and its inverse estimates the covariance matrix of $(\widehat{\beta}_0, \ldots, \widehat{\beta}_d)$. The latter is proportional to ϕ, so the standard errors, which are the square roots of the diagonal elements, are proportional to $\sqrt{\phi}$.

Once the regression parameters have been estimated, one can estimate ϕ by

$$\widehat{\phi} = \frac{1}{n - d - 1} \sum_{i=1}^{n} \frac{(y_i - \widehat{\mu}_i)^2}{V(\widehat{\mu}_i)}. \tag{3.3}$$

Often count data are overdispersed relative to the Poisson model. One remedy is to assume that y_i has a Negative Binomial distribution; see Sect. 3.2.2. Another approach is the quasi-Poisson model, which does not specify a probability distribution for y_i but only assumes that $\text{var}(y_i) = \phi\mu_i$ for some ϕ. The regression parameters are estimated by quasi-likelihood and ϕ by (3.3).

3.2.1 Example: Mortgage Applications in Boston

Binary regression is the special case of Binomial regression where the response can take only two possible values, so the conditional distribution of the response is the Bernoulli special case of the Binomial distribution. Our first illustration of GLM is a binary response example.

This example uses the `BostonMortgage` dataset introduced in Chap. 1. Recall that our response of interest is the variable `deny`, the status of the mortgage application—`deny` is coded as "yes" when the mortgage application was denied and "no" otherwise. For GLM fitting `deny` is then converted to a binary variable with "yes" equated with 1 and "no" equated with 0.

First, we look at the effect of race on the probability of a mortgage application being denied. We see that approximately 12% of mortgages are denied and the odds ratio of mortgage denial comparing Blacks to non-Blacks is 3.9:

```
> library(HRW) ; data(BostonMortgages)
> denyBin <- as.numeric(BostonMortgages$deny == "yes")
> blackIndic <- BostonMortgages$black == "yes"
> print(round(mean(denyBin),2))
```

```
[1] 0.12
```

```
> probabBlack <- mean(denyBin[blackIndic])
> probabNotBlack <- mean(denyBin[!blackIndic])
> oddsBlack <- probabBlack/(1 - probabBlack)
> oddsNotBlack <- probabNotBlack /(1 - probabNotBlack)
> print(round(oddsBlack/oddsNotBlack,1))
```

```
[1] 3.9
```

Next we fit a logistic regression model using `glm()` with `deny` as the response variable and `black` as the only predictor variable. The argument `family = binomial` is used for logistic regression; the default link is the logit function.

```
> fit1GLMBostMort <- glm(deny ~ black,family = binomial,
+                        data = BostonMortgages)
> summary(fit1GLMBostMort)
```

```
Call:
glm(formula = deny ~ black, family = binomial,
                        data = BostonMortgages)
```

```
Deviance Residuals:
    Min       1Q   Median       3Q      Max
-0.8160  -0.4409  -0.4409  -0.4409   2.1815

Coefficients:
            Estimate Std. Error z value Pr(>|z|)
(Intercept) -2.28227    0.07636 -29.888   <2e-16
blackyes     1.35356    0.14270   9.485   <2e-16

(Dispersion parameter for binomial family taken to be 1)

    Null deviance: 1744.2  on 2379  degrees of freedom
Residual deviance: 1663.4  on 2378  degrees of freedom
AIC: 1667.4

Number of Fisher Scoring iterations: 5
```

We see that the coefficient of `blackyes` is equal to the logarithm of the odds ratio and is highly significant. Next, we investigate whether or not the higher probability of mortgage denial for individuals with black ethnicity persists after adjusting for confounders. The following confounders, with abbreviations as used in the `BostonMortgages` data frame, are considered:

`dir`	ratio of the debt payments to total income,
`lvr`	ratio of the loan size to the assessed value of property,
`pbcr`	indicator public bad credit record (yes/no),
`self`	indicator of self-employment (yes/no),
`single`	indicator that the applicant is single (yes/no),
`ccs`	credit score, an ordinal variable ranging from 1 to 6, with lower being better.

```
> fit2GLMBostMort <- glm(deny ~ black + dir + lvr + pbcr
+                        + self + single + as.factor(ccs),
+                        family = binomial,data = BostonMortgages)
> summary(fit2GLMBostMort)

Call:
glm(formula = deny ~ black + dir + lvr + pbcr + self + single +
    as.factor(ccs), family = binomial, data = BostonMortgages)
```

```
Deviance Residuals:
    Min       1Q   Median       3Q      Max
-1.9299  -0.4721  -0.3319  -0.2347   3.0930

Coefficients:
                Estimate Std. Error z value Pr(>|z|)
(Intercept)      -6.7185     0.4811 -13.966  < 2e-16
blackyes          0.6894     0.1657   4.160 3.19e-05
dir               4.7210     0.7634   6.184 6.24e-10
lvr               2.5367     0.4664   5.439 5.37e-08
pbcryes           1.2664     0.1971   6.426 1.31e-10
selfyes           0.5741     0.2035   2.821 0.004783
singleyes         0.3917     0.1419   2.761 0.005763
as.factor(ccs)2   0.6600     0.1959   3.370 0.000753
as.factor(ccs)3   0.8174     0.2832   2.887 0.003894
as.factor(ccs)4   1.3792     0.3189   4.324 1.53e-05
as.factor(ccs)5   1.1290     0.2311   4.885 1.04e-06
as.factor(ccs)6   1.4319     0.2147   6.669 2.58e-11

(Dispersion parameter for binomial family taken to be 1)

    Null deviance: 1744.2  on 2379  degrees of freedom
Residual deviance: 1403.2  on 2368  degrees of freedom
AIC: 1427.2

Number of Fisher Scoring iterations: 6
```

All of the confounders are significant. Moreover, the odds ratio for `black` versus non-`black` is only $e^{0.6894} \approx 2.0$, compared to 3.9 when the confounders are not in the model. However, the effect of `black` is still significant, which indicates that black applicants are more likely to have a mortgage application denied even if one conditions on the confounders.

The ordinal variable `ccs` has been treated as a factor so that each level has its own mean. This requires five degrees of freedom. A linear effect for `ccs` involves only one degree of freedom but is not appropriate for an ordinal variable. A compromise between a linear fit and treating `ccs` as a factor is to fit a penalized spline and we investigate this in Sect. 3.3.3.

In this section we have used the base R function `glm()` for GLM fitting. An alternative is to use `gam()` from the package `mgcv` (Wood 2017). For example:

```
> library(mgcv)
> fit2GLMBostMortAlt <- gam(deny ~ black + dir + lvr + pbcr
+                      + self + single + as.factor(ccs),
+                      family = binomial,data = BostonMortgages)
```

Since `gam()` will be used later for the additive model extension, we will start using this function for GLM fitting as well. This will aid model comparison.

We will see later that the effects of `dir` and `lvr` are highly nonlinear. Also, `dir` and `lvr` have outlying cases. A GLM cannot accommodate the nonlinear effects of `dir` and `lvr` and are sensitive to the high leverage of their outliers. Therefore, the results in this section are presented merely to illustrate GLM fitting. To accommodate the nonlinear effects and outlying values of `dir` and `lvr` a GAM, rather than a GLM, is needed. GAM analyses of these data are presented in Sects. 3.3.3 and 3.6.1.

3.2.2 Example: Physician Offices Visits

The data frame `OFP` in the `Ecdat` (Croissant 2016) contains data on physician visits and demographic characteristics of 4406 individuals in the USA. The data were collected as part of the 1987 National Medical Expenditure Survey in the USA with details given in Deb and Trivedi (1997). We consider the response variable:

$$\texttt{ofp} \equiv \text{number of physician office visits}$$

which is a count variable. We start with a Poisson GLM, but we will see that the data are *overdispersed* compared to a Poisson distribution, that is, the conditional variance of the response given the predictors is larger than the conditional mean. To deal with overdispersion, we will use two strategies. First, we will fit a modified Poisson model with an overdispersion parameter. Second, we will use a Negative Binomial model for the conditional distribution of the response. Poisson regression models can be fit using the function `gam()` in the `mgcv` package. To fit the Negative Binomial distribution we will use the function `gamlss()` within the package `gamlss` (Stasinopoulos and Rigby 2017). The Negative Binomial distribution can also be fit with `gam()` in the `mgcv` package.

After preliminary exploratory data analysis, we selected the subset of the patients whose age was 95 years or less. Also, for illustrative purposes, the predictors were selected casually. With the abbreviation used in the `OFP` data frame, they are:

`age`	age in decades which we convert to years,
`school`	number of years of education,
`adldiff`	indicator that person has a condition that limits activities of daily living (0/1),
`black`	indicator that the person is African-American (yes/no),
`sex`	indicator of gender (male/female),

`maried` (*sic*)	indicator that the person is married (yes/no),
`privins`	indicator of having private insurance coverage (yes/no),
`medicaid`	indicator of having Medicaid coverage (yes/no),
`region`	region of the USA (midwest/noreast/west/other),
`hlth`	self-perceived health status (excellent/poor/other).

The following code creates the data frame that we will use:

```
> library(Ecdat) ; data(OFP) ; OFPforAna <- OFP
> OFPforAna$age <- 10*OFPforAna$age
> OFPforAna <- OFPforAna[OFPforAna$age<=95,]
```

After subsetting on age, the number of observations is $n = 4394$. A Poisson GLM is fit as follows:

```
> library(mgcv)
> fit1GLMOFP <- gam(ofp ~ age + school + black + sex + maried +
+     adldiff + privins + medicaid + region + hlth,
+     family = poisson,data = OFPforAna)
> print(summary(fit1GLMOFP))

Family: poisson
Link function: log

Formula:
ofp ~ age + school + black + sex + maried + adldiff + privins +
    medicaid + region + hlth

Parametric coefficients:
               Estimate Std. Error z value Pr(>|z|)
(Intercept)    1.444856   0.086863  16.634  < 2e-16
age           -0.004246   0.001085  -3.914 9.07e-05
school         0.025973   0.001913  13.577  < 2e-16
blackyes      -0.098457   0.022470  -4.382 1.18e-05
sexmale       -0.054165   0.014165  -3.824 0.000131
mariedyes     -0.048654   0.014472  -3.362 0.000774
adldiff        0.172812   0.016649  10.380  < 2e-16
privinsyes     0.331427   0.019682  16.839  < 2e-16
medicaidyes    0.309938   0.025143  12.327  < 2e-16
regionnoreast  0.095057   0.017677   5.377 7.56e-08
regionmidwest -0.022209   0.016758  -1.325 0.185065
regionwest     0.107607   0.018016   5.973 2.33e-09
hlthexcellent -0.494783   0.030231 -16.367  < 2e-16
hlthpoor       0.476775   0.017555  27.160  < 2e-16
```

```
R-sq.(adj) =  0.0624   Deviance explained = 8.26%
UBRE = 4.6145  Scale est. = 1          n = 4394
```

Even though the majority of the predictors is highly significant, we can only explain 8.3% of the deviance and the adjusted R^2 is only 6.24%. This type of behavior, where highly significant predictors are not very predictive of the response, is seen commonly when, as here, the sample size is so large that small effects can be detected with high probability. Moreover, the p-values in the summary above are too small because we have not yet accounted for overdispersion.

A well-known property of the Poisson distribution is that the variance equals the mean. In practice, count data often exhibit overdispersion meaning that the variance exceeds the mean. Underdispersion is possible in principle but not common in practice. To accommodate either overdispersion or underdispersion, most software for count data introduces a scale parameter such that the variance is equal to the squared scale parameter times the mean. Poisson dispersion corresponds to the scale parameter being equal to 1. When the scale parameter does not equal 1, the model is often called *quasi-Poisson*.

Note that by fitting the Poisson regression model with the `scale` argument set at its default value, we fixed the scale parameter to be 1. To check for overdispersion, we will fit the GLM with an estimated scale parameter. If the `scale` argument is set to a negative number, then `gam()` takes the scale parameter to be unknown and estimates it. When `family = poisson` is specified, if the `scale` argument of `gam()` is positive, then the scale parameter is set to that value and the `quasipoisson` family is used. One can adjust for overdispersion by specifying the family `quasipoisson` directly instead of using the family `poisson` with `scale = -1`.

```
> library(mgcv)
> fit2GLMOFP <- gam(ofp ~ age + school + black + sex + maried +
+                   adldiff + privins + medicaid + region + hlth,
+                   family = poisson,scale = -1,data = OFPforAna)

> print(summary(fit2GLMOFP))

Family: poisson
Link function: log

Formula:
ofp ~ age + school + black + sex + maried + adldiff + privins +
    medicaid + region + hlth

Parametric coefficients:
              Estimate Std. Error t value Pr(>|t|)
(Intercept)   1.444856   0.232112   6.225 5.27e-10
age          -0.004246   0.002898  -1.465 0.143042
school        0.025973   0.005112   5.081 3.91e-07
```

```
blackyes        -0.098457  0.060043  -1.640 0.101125
sexmale         -0.054165  0.037851  -1.431 0.152494
mariedyes       -0.048654  0.038670  -1.258 0.208399
adldiff          0.172812  0.044488   3.884 0.000104
privinsyes       0.331427  0.052594   6.302 3.23e-10
medicaidyes      0.309938  0.067186   4.613 4.08e-06
regionnoreast    0.095057  0.047236   2.012 0.044240
regionmidwest   -0.022209  0.044779  -0.496 0.619937
regionwest       0.107607  0.048141   2.235 0.025452
hlthexcellent   -0.494783  0.080782  -6.125 9.87e-10
hlthpoor         0.476775  0.046909  10.164  < 2e-16


R-sq.(adj) =  0.0624   Deviance explained = 8.26%
GCV = 5.6441  Scale est. = 7.1404     n = 4394
```

The estimated scale is equal to 7.14 which, as expected, is much greater than 1. Comparing the summary of `fit1GLMOFP` with that of `fit2GLMOFP`, we see that after accounting for overdispersion, the standard errors are much larger and the p-values are also larger. Some of the changes in p-value are dramatic. For example, `age` has a p-value of 0.0001 when scale is fixed at 1 but the p-value increases over 1000-fold to 0.14 when scale is estimated.

3.3 Generalized Additive Models

In many cases, the relationships between η and the predictors are not well represented by the simplistic linear relationship in (3.2). To accommodate possibly nonlinear relationships, a smooth function of x_j, $j = 1, \ldots, d$, is used in place of a linear function. Thus, the GAM extension to modeling η is

$$\eta = \beta_0 + f_1(x_1) + \cdots + f_d(x_d), \tag{3.4}$$

where the f_j, $1 \leq j \leq d$, are smooth functions. Thus, η is no longer linear in the predictor variables as in a GLM, but instead η is an *additive* function of the predictors. Most commonly the f_j are fitted using penalized splines as described in Chap. 2. In the case of $d = 2$ and O'Sullivan splines (3.4) becomes

$$\eta = \beta_0 + \beta_1 x_1 + \sum_{k=1}^{K_1} u_{1k} z_{1k}(x_1) + \beta_2 x_2 + \sum_{k=1}^{K_2} u_{2k} z_{2k}(x_2). \tag{3.5}$$

where $z_{1k}(\cdot)$, $1 \leq k \leq K_1$, is an O'Sullivan spline basis over the range of the x_1 data and the $z_{2k}(\cdot)$ are defined analogously. Differing amounts of penalization are applied to the u_{1k} and the u_{2k}. Note that (3.5) identifies f_1 according to the definition

$$f_1(x_1) \equiv \beta_1 x_1 + \sum_{k=1}^{K_1} u_{1k} z_{1k}(x_1)$$

which means that it can only be interpreted as an effect on η due to changes in x_1. A common convention used by R GAM software, such as `gam()` in the package `gam` and `gam()` in the package `mgcv`, is to plot estimates of the f_j with vertical centering around zero with respect to the predictor data. For an estimate of $\widehat{f_1}$ of f_1 this entails plotting

$$\widehat{f_1}(x_1) - \frac{1}{n}\sum_{i=1}^{n} \widehat{f_1}(x_{1i}) \text{ against } x_1.$$

An alternative is to plot the slice of the estimated surface in the x_1 direction with all other predictors set to their averages. In the $d = 2$ case this involves plotting

$$\widehat{\beta_0} + \widehat{f_1}(x_1) + \widehat{f_2}(\overline{x}_2) \text{ against } x_1$$

where $\overline{x}_2 \equiv \frac{1}{n}\sum_{i=1} x_{2i}$. In the Gaussian response case, this has the advantage of the function estimate being vertically aligned with the response data. The same applies to other GAMs but after application of the inverse link transformation to the function estimate. This approach involves use of the `predict()` function applied to the `gam()` object, and is demonstrated in the scripts `OFPGAMfit.R` and `BostMortfit.R` within the `HRW` package.

3.3.1 *Example: Test Scores of Children in California School Districts*

We now illustrate GAM fitting using the `Caschool` dataset in the `Ecdat` (Croissant 2016) package. This dataset contains 420 cross-sectional observations collected during the 1998–1999 school year in California school districts. We will model the average mathematics scores as an additive function of district characteristics. We will not be trying to build the best possible predictive model for `mathscr` (average math scores), but only want to illustrate GAM fitting using a few chosen variables for which the association with `mathscr` is not sufficiently explained by a linear relationship. Selection of better models is discussed in Sect. 3.4 We consider the following four predictors:

`calwpct`	percentage of children qualifying for CalWORKs,
`log.avginc`	natural logarithm of the average district income,
`compstu`	number of computers per student,
`expnstu`	expenditure per student.

According to a California Department of Social Services website CalWORKs is a "welfare program that gives cash aid and services to eligible needy California families."

The following code uses `gam()` in the `mgcv` package (Wood 2017) to produce a GAM fit:

```
> library(mgcv) ; library(Ecdat) ; data(Caschool)
> Caschool$log.avginc <- log(Caschool$avginc)
> fitGAMCaschool <- gam(mathscr ~ s(calwpct) + s(log.avginc) +
+                          s(compstu) + s(expnstu), data = Caschool)
```

In `gam()`, a predictor that is the argument of the `s()` function is modeled as having a smooth function effect using penalized splines. A summary of the fit can be obtained via:

```
> summary(fitGAMCaschool)

Family: gaussian
Link function: identity

Formula:
mathscr ~ s(calwpct) + s(log.avginc) + s(compstu) + s(expnstu)

Parametric coefficients:
            Estimate Std. Error t value Pr(>|t|)
(Intercept) 653.3426     0.5495    1189   <2e-16

Approximate significance of smooth terms:
                edf Ref.df      F p-value
s(calwpct)    3.826  4.751 19.976 < 2e-16
s(log.avginc) 3.942  4.938 19.829 < 2e-16
s(compstu)    3.480  4.396  3.290 0.00968
s(expnstu)    4.311  5.336  0.847 0.48521

R-sq.(adj) =  0.639   Deviance explained = 65.3%
GCV = 132.02  Scale est. = 126.82    n = 420
```

We see in the output that the `family` of conditional distributions of the responses is Gaussian and the `link` is the identity function. Such a Gaussian response model is often labeled an *additive model* rather than a *generalized additive model*. However, we will not make this distinction and use the term GAM for all additive models.

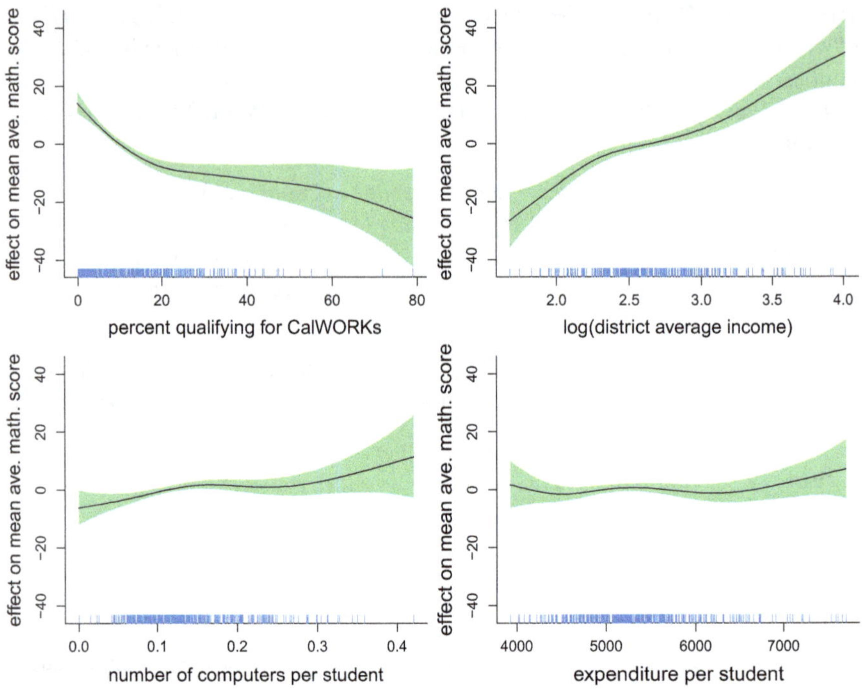

Fig. 3.1 Estimated smooth function components for the GAM fit to the California schools data obtained via the R function `gam()` in the package `mgcv` and stored in object **`fitGAMCaschool`** according to code given in the text. The response variable is average mathematics score. Each component function is vertically centered about zero. The shaded regions indicate approximately 95% pointwise confidence intervals. The tick marks indicate the values of the predictors.

A plot of the fitted penalized splines for each predictor is shown in Fig. 3.1. When `gam()` output is plotted, the same vertical scale is used for each effect, which allows a quick comparison among the sizes of the effects. This feature can be overridden by using the `select` argument of `plot.gam()` to plot the effects one-at-a-time. The script `CaSchoolGAMfit.R` in the `HRW` package contains the code used to produce Fig. 3.1. It can be run from the command line using:

```
> library(HRW) ; demo(CaSchoolGAMfit,package = "HRW")
```

and its location for possible copying and modifying is determined by:

```
> system.file("demo","CaSchoolGAMfit.R",package = "HRW")
```

We see that `log.avginc` has the largest effect with average `mathscr` increasing steadily with this predictor. Furthermore, `calwpct` has the next largest effect and its effect is nonlinear with an initial steep decrease before leveling off. Interestingly, `compstu` and `expnstu` appear to have little or no effect.

A question that arises frequently in practical data analysis is the adequacy of a linear model. To address this issue, we start by fitting a GLM using the `gam()`

function. This is done by having each predictor appear by itself rather than as the argument of the function s().

```
> library(mgcv) ; library(Ecdat) ; data(Caschool)
> Caschool$log.avginc <- log(Caschool$avginc)
> fitGLMCaschool <- gam(mathscr ~ calwpct + log.avginc
+                   + compstu + expnstu,data = Caschool)
```

The GLM just fit can be compared with the GAM fit by an F-test:

```
> anova(fitGLMCaschool,fitGAMCaschool,test = "F")

Analysis of Deviance Table

Model 1: mathscr ~ calwpct + log.avginc + compstu + expnstu
Model 2: mathscr ~ s(calwpct) + s(log.avginc) + s(compstu)
                   + s(expnstu)
  Resid. Df Resid. Dev    Df Deviance      F    Pr(>F)
1    415.00      58685
2    399.58      51163 15.42   7522.1 3.8466 1.499e-06
```

The small p-value is consistent with the nonlinear relationships seen in Fig. 3.1, and both the p-value and this figure indicate that the GAM is preferred over the GLM. One might consider dropping the variable expnstu, since in Fig. 3.1 it appears to have little or no effect on mathscr. In Sect. 3.4 we choose the predictors in a more principled fashion and, sure enough, there is good evidence that expnstu has at most only a small effect.

3.3.2 Example: Physician Office Visits

Next, we return to the physician office visit data and fit a model where the continuous predictors age and school are nonlinearly associated with ofp. We estimate scale as before by specifying scale = -1.

```
> fitGAMOFP <- gam(ofp ~ s(age) + s(school) + adldiff +  black
+    + sex + maried + privins + medicaid + region + hlth,
+          family = poisson,scale = -1,data = OFPforAna)

> summary(fitGAMOFP)

Family: poisson
Link function: log

Formula:
ofp ~ s(age) + s(school) + adldiff + black + sex + maried +
      privins +  medicaid + region + hlth
```

```
Parametric coefficients:
              Estimate Std. Error t value Pr(>|t|)
(Intercept)    1.39387    0.06037  23.090  < 2e-16
adldiff        0.18308    0.04440   4.124 3.80e-05
blackyes      -0.10365    0.06008  -1.725   0.0845
sexmale       -0.05997    0.03796  -1.580   0.1142
mariedyes     -0.05000    0.03856  -1.297   0.1947
privinsyes     0.33150    0.05275   6.285 3.60e-10
medicaidyes    0.30226    0.06730   4.491 7.26e-06
regionnoreast  0.10505    0.04733   2.219   0.0265
regionmidwest -0.01053    0.04493  -0.234   0.8147
regionwest     0.11267    0.04814   2.340   0.0193
hlthexcellent -0.49962    0.08081  -6.182 6.89e-10
hlthpoor       0.47569    0.04685  10.152  < 2e-16

Approximate significance of smooth terms:
            edf Ref.df     F  p-value
s(age)    2.453  3.086 3.558    0.013
s(school) 2.405  3.029 9.876 1.66e-06

R-sq.(adj) =  0.065   Deviance explained = 8.66%
GCV = 5.6266  Scale est. = 7.1209    n = 4394
```

Next, we compare the GLM and GAM fits:

```
> anova(fit2GLMOFP,fitGAMOFP,test = "F")

Analysis of Deviance Table

Model 1: ofp ~ age + school + black + sex + maried + adldiff
               + privins + medicaid + region + hlth
Model 2: ofp ~ s(age) + s(school) + adldiff + black + sex
               + maried + privins + medicaid + region + hlth

  Resid. Df Resid. Dev     Df Deviance      F  Pr(>F)
1    4380.0      24642
2    4375.9      24534 4.1149   108.26 3.6948 0.00479
```

The p-value is small which suggests that the effects of age and `school` should be modeled as smooth functions using a GAM.

The nonlinear effects of age and `school` on the number of physician visits can be viewed in Fig. 3.2. Since we are using Poisson regression, we are modeling the log-mean. The response scale plots in the bottom panels and correspond to slices, as described in Sect. 3.3, with each of the categorical variables set to its mode and the other continuous variable set to its mean. To run `OFPGAMfit.R` from the R command line issue:

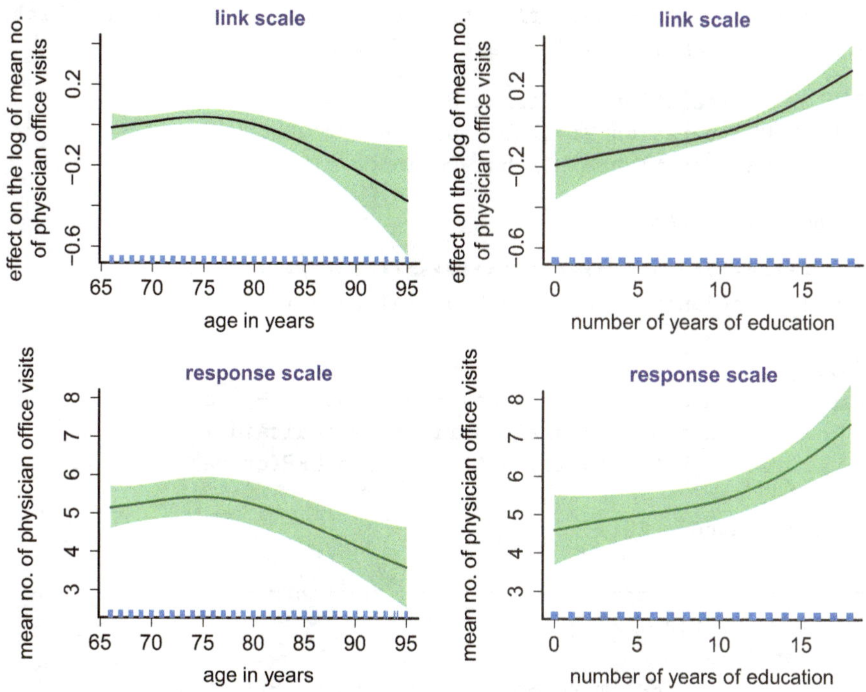

Fig. 3.2 Estimated smooth function components for the Poisson GAM fit to the physician office visits data obtained via the R function `gam()` in the package `mgcv` and stored in object `fitGAMOFP` *according to code given in the text.* The response variable is the number of physician office visits. Each component function in the upper panels is vertically centered about zero. The top panels show the components on the link (logarithm) scale and were obtained using the `plot()` function applied to the `gam()` fit object. The bottom panels show the components on the response scale and were obtained using the `predict()` function, with details in the script `OFPGAMfit.R`. The shaded regions indicate approximately 95% pointwise confidence intervals. The tick marks indicate the values of the predictors, which are slightly jittered.

```
> library(HRW) ; demo(OFPGAMfit,package = "HRW")
```

To obtain its location for copying and modifying type:

```
> system.file("demo","OFPGAMfit.R",package = "HRW")
```

We see that thes expected number of physician office visits increases very slowly from age 65 to 75 and then decreases at an increasingly rapid rate. The expected number of visits increases with the number of years of education, with the rate of increase being higher past 10 years of education.

Another way to model count data that properly accounts for overdispersion is to assume that the response has a conditional Negative Binomial distribution. To fit this more flexible model, we use the function `gamlss()` in the `gamlss` package (Stasinopoulos and Rigby 2017).

```
> library(gamlss)
> fitGAMlssOFP <- gamlss(ofp ~ pb(age) + pb(school) + black +
```

```
+   sex + maried + adldiff + privins + medicaid + region + hlth,
+          family = NBI,data = OFPforAna)

GAMLSS-RS iteration 1: Global Deviance = 24568.53
GAMLSS-RS iteration 2: Global Deviance = 24568.53
GAMLSS-RS iteration 3: Global Deviance = 24568.53

> summary(fitGAMlssOFP)

******************************************************
Family:  c("NBI", "Negative Binomial type I")

Call:
gamlss(formula = ofp ~ pb(age) + pb(school) + black +
    sex + maried + adldiff + privins + medicaid +
    region + hlth, family = NBI, data = OFPforAna)

Fitting method: RS()

-------------------------------------------------------
Mu link function:  log
Mu Coefficients:
               Estimate Std. Error t value Pr(>|t|)
(Intercept)    1.236738   0.218573   5.658 1.63e-08 ***
pb(age)       -0.001992   0.002725  -0.731  0.46465
pb(school)     0.027473   0.004702   5.843 5.50e-09 ***
blackyes      -0.112693   0.054104  -2.083  0.03732 *
sexmale       -0.060312   0.035902  -1.680  0.09305 .
mariedyes     -0.031337   0.037032  -0.846  0.39748
adldiff        0.188545   0.042834   4.402 1.10e-05 ***
privinsyes     0.339849   0.046940   7.240 5.27e-13 ***
medicaidyes    0.319242   0.065748   4.856 1.24e-06 ***
regionnoreast  0.102450   0.044810   2.286  0.02228 *
regionmidwest -0.008772   0.041237  -0.213  0.83155
regionwest     0.119663   0.046177   2.591  0.00959 **
hlthexcellent -0.514541   0.062338  -8.254  < 2e-16 ***
hlthpoor       0.486394   0.049533   9.820  < 2e-16 ***
---
Signif. codes:
0 '***' 0.001 '**' 0.01 '*' 0.05 '.' 0.1 ' ' 1

-------------------------------------------------------
Sigma link function:  log
Sigma Coefficients:
            Estimate Std. Error t value Pr(>|t|)
(Intercept)  -0.1064     0.0271  -3.925  8.8e-05 ***
```

```
---
Signif. codes:
0 '***' 0.001 '**' 0.01 '*' 0.05 '.' 0.1 ' ' 1

-------------------------------------------------------
NOTE: Additive smoothing terms exist in the formulas:
 i) Std. Error for smoothers are for the linear effect only.
ii) Std. Error for the linear terms maybe are not accurate.
-------------------------------------------------------
No. of observations in the fit:  4394
Degrees of Freedom for the fit:  17.68339
      Residual Deg. of Freedom:  4376.317
                      at cycle:  3

Global Deviance:     24568.53
            AIC:     24603.89
            SBC:     24716.86
*******************************************************
```

Note that the function pb() is used to specify penalized spline components in calls to gamlss(). Also family = NBI specifies a Negative Binomial response distribution with the so-called Type I parametrization in which the variance exceeds the mean by a multiple of the squared mean. The plot function for an R object of class gamlss gives four plots for checking quantile residuals. The plots are in Fig. 3.3. The residuals for a gamlss object are randomized quantile residuals (Dunn and Smyth 1996) and will have an exact Standard Normal distribution if the model is correct and is evaluated at the true parameters. For large samples, estimated parameters will be close to the true values and the residuals will be approximately Standard Normal if the model is correct. Randomization is used when the response is discrete, as it is here. Otherwise, the residuals would also have a discrete distribution and so cannot be normally distributed. We see no indication of modeling problems in the residuals. They appear to be normally distributed and homoscedastic. Plotting a gamlss object also causes the following summary statistics to be printed:

```
*******************************************************
         Summary of the Randomised Quantile Residuals
                           mean   =  -0.004607413
                       variance   =  1.00183
               coef. of skewness  =  0.1373671
               coef. of kurtosis  =  3.681591
Filliben correlation coefficient  =  0.997155
*******************************************************
```

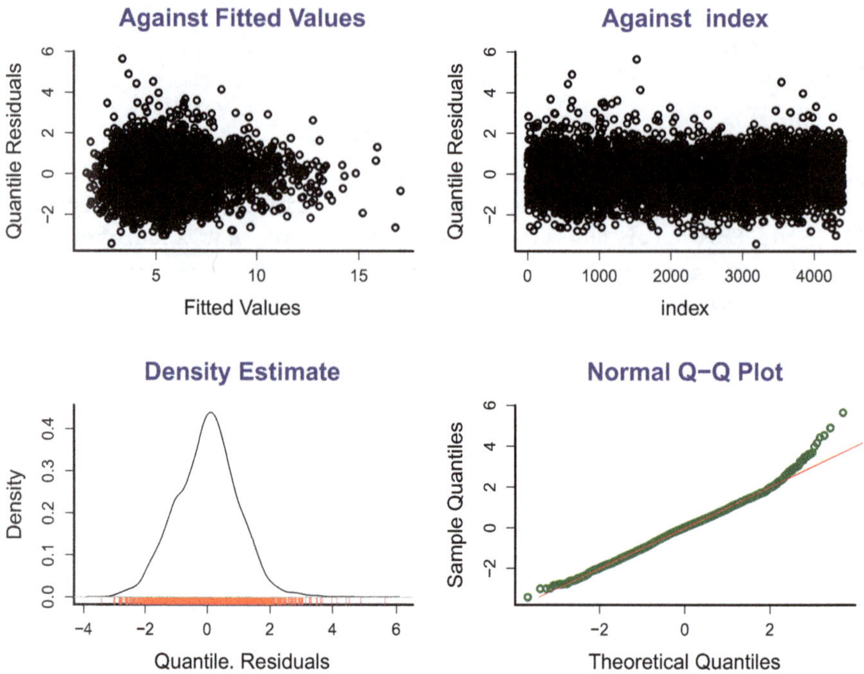

Fig. 3.3 Quantile residual plots for the Negative Binomial `gamlss()` fit with response `ofp` stored in the object `fitGAMlssOFP`. Top left panel: the quantile residuals versus fitted values. Top right panel: quantile residuals versus index. Bottom left panel: kernel density estimate of the quantile residuals. Bottom right panel: Normal quantile-quantile plot of the quantile residuals. The plots suggest a slight right skewness of the quantile residuals.

3.3.3 *Example: Mortgage Applications in Boston*

We now return to the Boston mortgage applications example. Recall that we are developing a model to predict whether a mortgage application will be denied and to test whether race is a factor in application denial after adjusting for confounders.

The model of Sect. 3.2.1 will be modified so that the effects of debt payments to income ratio (`dir`) and loan size to assessed value of the property (`lvr`) are modeled as penalized spline rather than linear functions:

```
> fit1GAMBostMort <- gam(deny ~ black + s(dir) + s(lvr)
+     + pbcr + self + single + as.factor(ccs),family = binomial,
+      data = BostonMortgages)
> summary(fit1GAMBostMort)

Family: binomial
Link function: logit
```

```
Formula:
deny ~ black + s(dir) + s(lvr) + pbcr + self + single
       + as.factor(ccs)

Parametric coefficients:
                Estimate Std. Error z value Pr(>|z|)
(Intercept)      -3.2833     0.1483 -22.140  < 2e-16
blackyes          0.6136     0.1718   3.571 0.000356
pbcryes           1.2500     0.2033   6.148 7.83e-10
selfyes           0.4549     0.2110   2.156 0.031073
singleyes         0.3819     0.1453   2.628 0.008597
as.factor(ccs)2   0.6643     0.2008   3.309 0.000936
as.factor(ccs)3   0.8193     0.2919   2.807 0.004998
as.factor(ccs)4   1.3336     0.3268   4.080 4.50e-05
as.factor(ccs)5   1.1561     0.2389   4.840 1.30e-06
as.factor(ccs)6   1.4982     0.2205   6.794 1.09e-11

Approximate significance of smooth terms:
         edf Ref.df Chi.sq  p-value
s(dir) 5.893  7.093  83.81 3.03e-15
s(lvr) 4.188  5.238  49.02 4.39e-09

R-sq.(adj) =  0.222   Deviance explained = 23.1%
UBRE = -0.41961  Scale est. = 1         n = 2380
```

The penalized spline fits to `dir` and `lvr` are shown in Fig. 3.4. As discussed in Sect. 3.3.2, once we fit a GLM or a GAM, we have a choice of plotting the relationship between the response and a predictor on the scale of the fitted model, e.g., log-odds for the logistic regression, or on the scale of the response, e.g., the probability scale for logistic regression. It is useful to look at both plots.

In logistic regression, the transformation from the logit scale to the probability scale is the cumulative distribution function of the Logistic distribution and is the inverse of the logit function in Table 3.1. Therefore, to plot on the probability scale, we use the function `plogis()`, which computes the Logistic cumulative distribution function

$$F_{\text{Logistic}}(x) \equiv \frac{e^x}{1+e^x}.$$

In Fig. 3.4 the effects of `dir` and `lvr` on the logit scale are in the top panels, with vertical centering about zero according to the default settings of `plot.gam()`. The effects on the probability scale are in the bottom panels and correspond to slices, as explained in Sect. 3.3, with each of the categorical variables set to its mode and the other continuous variable set to its mean. We can see from these plots that the best chance of not being denied a mortgage is when one's debt payments to income ratio is between 0.2 and 0.4 and the loan size to assessed value is low. Both low

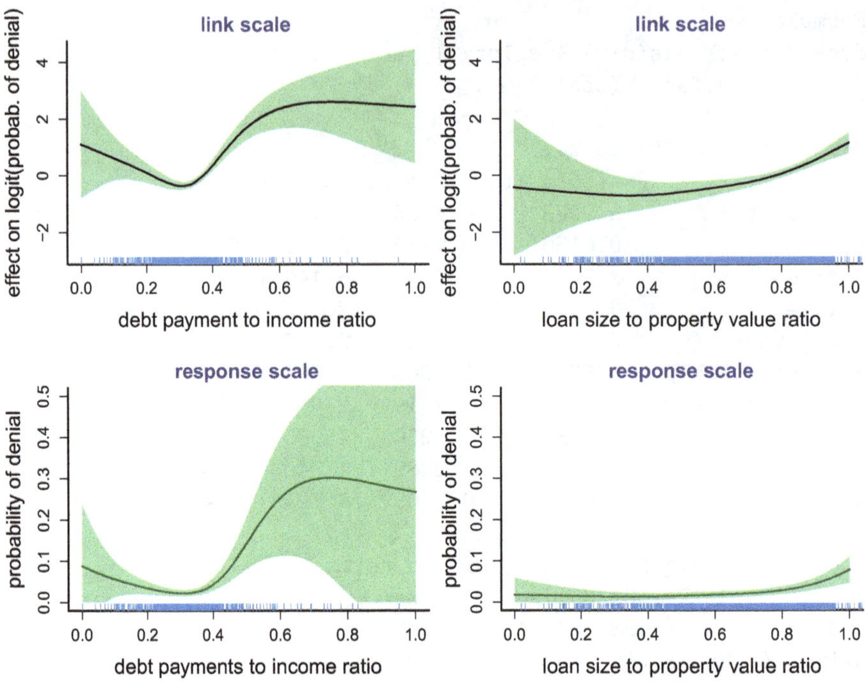

Fig. 3.4 Estimated smooth function components for the logistic GAM fit to the Boston mortgage application data obtained via the R function `gam()` in the package `mgcv` and stored in object `fit1GAMBostMort` according to code given in the text. The response variable is the indicator of mortgage application denial. Each component function in the upper panels is vertically centered about zero. The top panels show the components on the link (logit) scale and were obtained using the `plot()` function applied to the `gam()` fit object. The bottom panels show the components on the response scale and were obtained using the `predict()` function, with details in the script `BostonMortGAMfit.R`. The shaded regions indicate approximately 95% pointwise confidence intervals. The tick marks indicate the values of the predictors. As mentioned in Sect. 1.3.2, there are outlying values of debt payments to income ratio and of loan size to property value ratio. The outlying cases were used for estimation but, to improve visualization, are not included in these plots.

debt payments to income ratios and especially high values of this ratio increase the probability of application denial. The script `BostMortGAMfit.R` provides the code needed to produce Fig. 3.4. It can be run using:

```
> library(HRW) ; demo(BostMortGAMfit,package = "HRW")
```

and then located on the computer on which `HRW` is installed via:

```
> system.file("demo","BostMortGAMfit.R",package = "HRW")
```

A test comparing the linear and smooth fit gives us strong evidence that the association studied is nonlinear.

```
> anova(fit2GLMBostMort,fit1GAMBostMort,test = "Chisq")
```

```
Analysis of Deviance Table

Model 1: deny ~ black + dir + lvr + pbcr + self + single
                + as.factor(ccs)
Model 2: deny ~ black + s(dir) + s(lvr) + pbcr + self + single
                + as.factor(ccs)
  Resid. Df Resid. Dev     Df Deviance  Pr(>Chi)
1    2368.0     1403.2
2    2359.9     1341.2 8.0816   61.991 2.072e-10
```

So far the credit rating ccs has been modeled as a factor. One can see in the output that the coefficients increase nearly monotonically with ccs; this makes sense since higher values of ccs indicate worse credit. We will now consider modeling the effect of ccs as a smooth term. We specified a spline basis of dimension 4 (k = 4) because ccs takes only six distinct values.

```
> fit2GAMBostMort <- gam(deny ~ black + s(dir) + s(lvr)
+     + pbcr + self + single + s(ccs,k = 4),family = binomial,
+     data = BostonMortgages)

> summary(fit2GAMBostMort)

Family: binomial
Link function: logit

Formula:
deny ~ black + s(dir) + s(lvr) + pbcr + self + single + s(ccs,
    k = 4)

Parametric coefficients:
            Estimate Std. Error z value Pr(>|z|)
(Intercept)  -2.8503     0.1197 -23.820  < 2e-16
blackyes      0.6213     0.1713   3.627 0.000287
pbcryes       1.2407     0.2004   6.190 6.03e-10
selfyes       0.4557     0.2107   2.162 0.030598
singleyes     0.3918     0.1448   2.706 0.006806

Approximate significance of smooth terms:
         edf Ref.df Chi.sq  p-value
s(dir) 5.925  7.125  83.65 3.59e-15
s(lvr) 4.189  5.239  48.70 5.10e-09
s(ccs) 1.838  2.219  57.63 1.32e-12

R-sq.(adj) =  0.221   Deviance explained =   23%
UBRE = -0.42133  Scale est. = 1          n = 2380
```

Note that the effective degrees of freedom (see Sect. 2.6) of ccs was reduced from 5 to 1.838 by modeling that variable with a smooth function rather than as a factor.

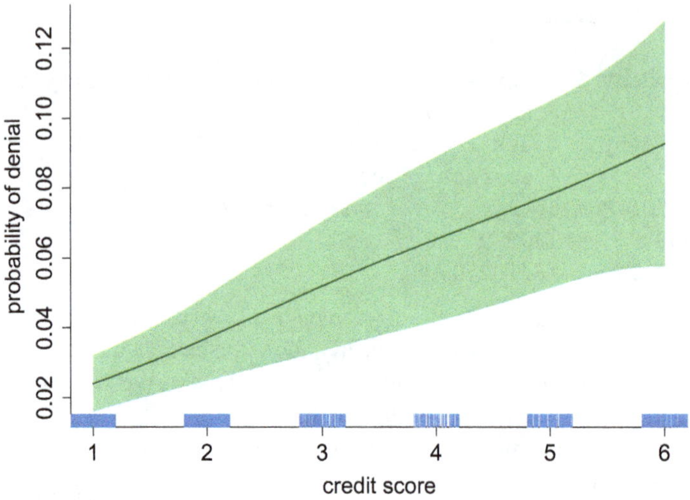

Fig. 3.5 Estimated smooth function of credit score for the logistic GAM fit to the Boston mortgage data obtained via the R function `gam()` in the package `mgcv` and stored in object `fit2GAMBostMort` according to code given in the text. The function estimate is on the probability scale. The shaded region indicates approximately 95% pointwise confidence intervals. The tick marks indicate the values of the predictor, with jittering, and show that credit score takes only integer values from 1 to 6. A lower credit score indicates a better credit rating. The curves were obtained using the `predict()` function applied to the `gam()` object.

To compare modeling `ccs` as a smooth function versus as a factor, a chi-squared test can be used.

```
> anova(fit2GAMBostMort,fit1GAMBostMort,test = "Chisq")
Analysis of Deviance Table

Model 1: deny ~ black + s(dir) + s(lvr) + pbcr + self + single
                + s(ccs,k = 4)
Model 2: deny ~ black + s(dir) + s(lvr) + pbcr + self + single
                + as.factor(ccs)
  Resid. Df Resid. Dev     Df Deviance Pr(>Chi)
1    2360.4     1343.3
2    2357.7     1341.2 2.7478   2.1546   0.4926
```

The large p-value suggests that `ccs` can be modeled as a smooth function rather than as a factor. The smooth function effect of `ccs` is plotted in Fig. 3.5 as a slice of overall surface with the other predictors set to mean and modal values. It shows that the probability that a mortgage is denied increases more rapidly with `ccs` between 1 and 3 compared to when `ccs` is greater than 3.

For this model the above output shows that `blackyes` is still highly significant, indicating possible racial discrimination. We return to this issue in Sect. 3.6.1 where our final analysis of the Boston mortgage applications data is presented.

3.4 Model Selection

A practical problem in GAM analysis is choosing among the multitude of possible models for given set of predictors. If a candidate predictor is continuous, then a choice needs to be made concerning whether it enters the model linearly, nonlinearly, or not at all. Therefore there are 3^d possible ways in which d continuous predictors may enter a GAM. For example, the California schools data frame has as many as $d = 7$ potential continuous predictors: `calwpct`, `mealpct`, `compstu`, `expnstu`, `str`, `log.avginc`, and `elpct`. Four of these predictors were used and defined in Sect. 3.3.1. The three additional predictors are:

`mealpct`	percentage of students qualifying for a reduced-price lunch,
`str`	student to teacher ratio,
`elpct`	percentage of English learners.

So there are $3^7 = 2187$ ways in which the 7 continuous predictors can enter a GAM. Which among these should be chosen? The R computing environment facilitates sifting through possibly very large candidate predictor sets. The approaches used can be divided into two types:

- stepwise strategies, which involves stepping through a subset of the 3^d possible models based on rules for adding and dropping predictors,
- penalty-based strategies, which involves including all of the candidate predictors in the fitting of the model but with penalties that annihilate the contributions from less important predictors.

The remainder of this section contains demonstrations of each of these approaches via use of relevant R functions. It should be pointed out that, around the time of this writing, tools are being developed for post-selection inference in particular regression contexts. See, for example, Taylor and Tibshirani (2015) and Tibshirani et al. (2017). This section does not delve into this new stream of research.

3.4.1 *Stepwise Model Selection*

Examples of R functions that use stepwise strategies for model selection are `bruto()` in the package `mda` (Hastie 2017b), `step.Gam()` in the package `gam()` (Hastie 2017a) and `polymars` in the package `polspline()` (Kooperberg 2015).

When faced with a large number of candidate models, a reasonable strategy is to use a stepwise procedure to select the form of the model and then call upon `gam()` in the `mgcv` package to fit the final model with GCV or REML choice of smoothing parameters. We now illustrate this approach for the California schools example. First we load the data and the `gam()` package using:

```
> library(gam) ; library(Ecdat) ; data(Caschool)
> Caschool$log.avginc <- log(Caschool$avginc)
```

Then obtain an initial fit using:

```
> fitInitial <- gam::gam(mathscr ~ calwpct + mealpct + compstu
+           + expnstu + str + log.avginc + elpct,data = Caschool)
```

Note that here we are using the version of the function gam() in the gam package (Hastie 2017a) and not the function with the same name in the mgcv package (Wood 2017). This fact is highlighted by the gam::gam() coding. Use of the function step.Gam() is exemplified by:

```
> stepFit <- step.Gam(fitInitial, scope =
+     list("calwpct" =     ~ 1 + calwpct + s(calwpct,2),
+          "mealpct" =     ~ 1 + mealpct + s(mealpct,2),
+          "compstu" =     ~ 1 + compstu + s(compstu,2),
+          "expnstu" =     ~ 1 + expnstu + s(expnstu,2),
+          "str" =         ~ 1 + str + s(str,2),
+          "log.avginc" =  ~ 1 + log.avginc + s(log.avginc,2),
+          "elpct" =       ~ 1 + elpct + s(elpct,2)))
```

Note that the scope argument in step.Gam() allows one to specify how each predictor may enter the model. For example the code:

```
"calwpct" = ~ 1 + calwpct + s(calwpct,2)
```

signifies that calwpct may either be not present (1), enter the model linearly (calwpct), or enter the model as a penalized spline with two effective degrees of freedom (s(calwpct,2)). A stepwise search based on the notion of *regimens* and a version of Akaike's Information Criterion (Hastie 1992) is used to choose among the models visited in the steps. The selected model can be obtained via:

```
> print(names(stepFit$"model")[-1])
```

which, in the current example, leads to

```
[1] "s(calwpct,2)" "mealpct" "s(compstu,2)" "s(log.avginc,2)"
[5] "elpct"
```

Since it is plausible that other effective degrees of freedom values are appropriate for the nonlinear components, a reasonable next step is to refit with GCV smoothing parameter choice. This involves the *other* version of the gam() function, from the mgcv package. Here we use mgcv:::gam to make it clear that a different gam() function is being called.

```
> detach("package:gam") ; library(mgcv)
> fitStepCaschool <- mgcv:::gam(mathscr ~   mealpct + elpct
+                       + s(calwpct,k = 27) + s(compstu,k = 27)
+                       + s(log.avginc,k = 27),data = Caschool)
```

The estimated functions corresponding to fitStepCaschool are shown in Fig. 3.6. The script CaSchoolStepFit.R in the HRW package provides the code used to produce Fig. 3.6. To run it type:

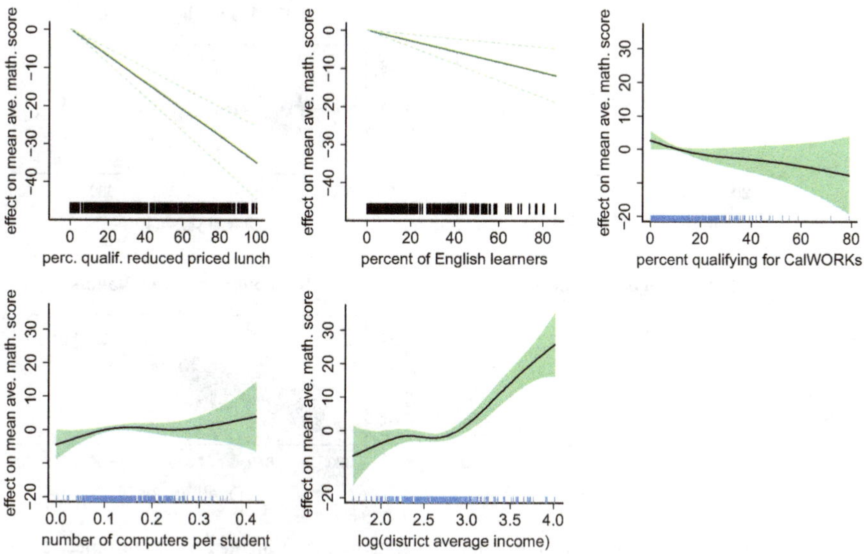

Fig. 3.6 Components of the GAM fit to the California schools data with GCV smoothing parameter selection using the function `gam()` in the `mgcv` package after the form of the model was selected via the function `step.Gam()` in the `gam` package. The bands around the fits correspond to approximately 95% pointwise confidence intervals. Each component function is vertically centered about zero. The tick marks at the base of each panel show the predictor data.

```
> library(HRW) ; demo(CaSchoolStepFit,package = "HRW")
```

To locate it on the computer on which `HRW` is installed and to possibly copy and modify it, issue the command:

```
> system.file("demo","CaSchoolStepFit.R",package = "HRW")
```

This 5-term GAM differs from the 4-term GAM for the same data presented in Sect. 3.3.1 in that `expnstu` is no longer present in the model but `mealpct` and `elpct` enter the model linearly. The deviance explained by this new `step.Gam()`-selected model is 73.9%, which is substantially higher than the 65.4% deviance explained by the Sect. 3.3.1 model.

Next, we conduct a residual check of this final model for average mathematics test score in California school districts. The command:

```
> gam.check(fitStepCaschool,cex.lab = 2,cex.main = 2)
```

```
Method: GCV   Optimizer: magic
Smoothing parameter selection converged after 10 iterations.
The RMS GCV score gradient at convergence was 0.0004017524.
The Hessian was positive definite.
Model rank =  81 / 81
```

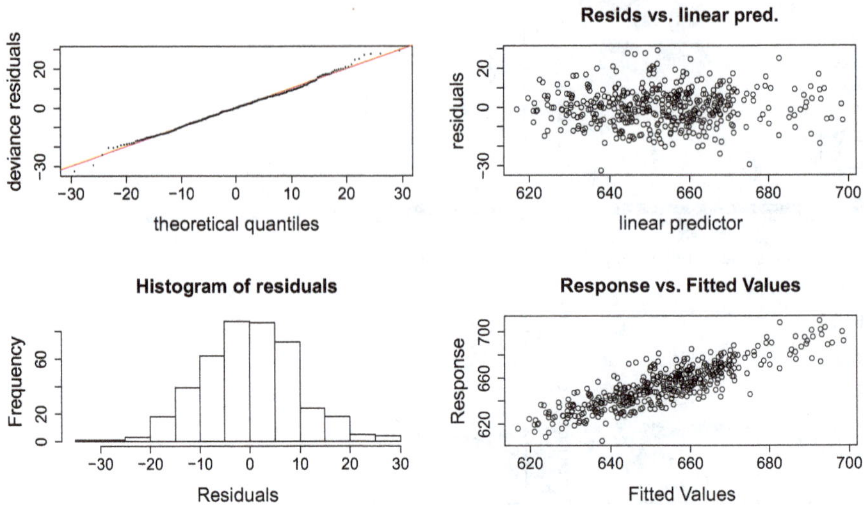

Fig. 3.7 Residual plots produced by the function `gam.check()` applied to the fit object `fitStepCaschool`.

```
Basis dimension (k) checking results. Low p-value (k-index<1)
may indicate that k is too low, especially if edf is close to k'.

                    k'   edf k-index p-value
s(calwpct)      26.00  2.05    0.98    0.41
s(compstu)      26.00  2.81    0.99    0.39
s(log.avginc) 26.00  4.10    0.99    0.43
```

leads to the plots shown in Fig. 3.7 and indicates good compliance with the model assumptions. In addition, the numbers of basis functions used for each penalized spline seem to be more than adequate.

Exercise 9 illustrates use of the function `polymars()` in the `polspline` package for additive model selection. The `polymars()` function differs from the `step.Gam()` and `gam()` functions in that the (truncated line) spline basis functions are not penalized. Instead subsets of the basis functions are selected. This is commonly labeled as a *regression spline* approach.

3.4.2 *Penalty-Based Model Selection*

Penalty-based model selection methods are a newer alternative to stepwise methods. They are particularly useful for datasets with many candidate predictors. The main penalty-based model selection strategy uses the *least absolute shrinkage and selection operator (LASSO)* (Tibshirani 1996).

For additive models, there are penalty-based model selection algorithms similar to the LASSO that are available in several R packages, `cosso` (Zhang and Lin 2013) which uses the *component selection and smoothing operator (COSSO)* (Lin and Zhang 2006), `gamsel` (Chouldechova et al. 2018), `mgcv`, and `SAM` (Zhao et al. 2014). We will discuss the first three of these starting with `cosso`.

At the time of this writing, the `cosso` package was not installable on all operating systems supported by R due to a problem with another package, `Rglpk`, on which it depends. This may require the GNU Linear Programming Kit to be installed. The `News, Software Updates and Errata` page on the book website (`semiparametric-regression-with-r.net`) contains instructions for installing the GNU Linear Programming Kit. This serves as a reminder that this website should be consulted regularly for news and updates on the examples and software in this book.

To understand the essential idea behind model selection by penalization, we begin with a quick introduction to the LASSO approach. Consider the linear model

$$y_i = \beta_0 + \sum_{j=1}^{d} \beta_j x_{ji} + \varepsilon_i, \quad 1 \le i \le n,$$

where x_{ji} denotes the ith observation for predictor x_j. Suppose that estimates $\widehat{\beta}_0, \ldots, \widehat{\beta}_d$ are obtained by minimizing

$$\sum_{i=1}^{n} \left(y_i - \beta_0 - \sum_{j=1}^{d} \beta_j x_{ji} \right)^2 + \lambda \sum_{j=1}^{d} |\beta_j|^q,$$

for some $\lambda > 0$ and $q > 0$. Penalty-based model selection is based on the rule:

$$x_j \text{ is selected if } \widehat{\beta}_j \neq 0.$$

If $q = 2$, then we are using an ℓ_2 penalty and the $\widehat{\beta}_j$, $1 \le j \le d$, correspond to ridge regression which shrinks each of $\widehat{\beta}_1, \ldots, \widehat{\beta}_d$ towards zero. However, the ℓ_2 penalty does not shrink any of the coefficients exactly to zero and therefore selects *all* of the predictors. If $q = 1$, so that an ℓ_1 penalty is used, then some of $\widehat{\beta}_1, \ldots, \widehat{\beta}_d$ will be exactly zero for large enough λ. The larger the value of λ, the fewer the number of selected variables. In summary, with an ℓ_2 penalty there is shrinkage but no useful model selection, whereas with an ℓ_1 penalty we have both shrinkage and a plausible model selection strategy.

The penalized least-squares estimators used so far in this book for semiparametric regression have used ℓ_2 penalties. In contrast, COSSO uses an ℓ_1-type penalty to select and shrink component functions. Consider the additive model

$$y_i = \beta_0 + f_1(x_{1i}) + \cdots f_d(x_{di}) + \varepsilon_i.$$

For this model, COSSO minimizes the objective function

$$n^{-1}\sum_{i=1}^{n}\left(y_i-\beta_0-\sum_{j=1}^{d}f_j(x_{ji})\right)^2+\tau^2\sum_{j=1}^{d}\|f_j\|_{\text{cosso}}, \tag{3.6}$$

where $\tau \geq 0$ is a smoothing parameter and $\|\cdot\|_{\text{cosso}}$ is the following ℓ_1-type norm:

$$\|f_j\|_{\text{cosso}} \equiv \left[\left(\int_0^1 f_j(t)\,dt\right)^2+\left(\int_0^1 f_j'(t)\,dt\right)^2+\int_0^1\{f_j''(t)\}^2\,dt\right]^{1/2},$$

where we assume that x_j has been scaled so that f_j is defined on the interval [0, 1]. It can be shown that $\|f_j\|_{\text{cosso}} = 0$ if and only if f_j is identically zero.

Note that in (3.6) the penalty uses the sum of the norms, not squared norms, of the component functions. To minimize (3.6), Lin and Zhang (2006) minimize instead

$$n^{-1}\sum_{i=1}^{n}\left(y_i-\beta_0-\sum_{j=1}^{d}f_j(x_{ji})\right)^2+\lambda_0\sum_{j=1}^{d}\theta_j^{-1}w_j^2\|f_j\|_{\text{cosso}}^2,\ \text{such that}\ \sum_{j=1}^{d}\theta_j\leq M. \tag{3.7}$$

Here $\lambda_0 > 0, \theta_1, \ldots, \theta_d > 0$, and $M > 0$ are smoothing parameters and $w_1, \ldots, w_d$ are weights with default values of 1. It is not necessary to have both smoothing parameters λ_0 and M, but they are introduced for computational reasons.

Minimizing (3.7) is equivalent to minimizing

$$n^{-1}\sum_{i=1}^{n}\left(y_i-\beta_0-\sum_{j=1}^{d}f_j(x_{ji})\right)^2+\lambda_0\sum_{j=1}^{d}\theta_j^{-1}w_j^2\|f_j\|_{\text{cosso}}^2+\lambda\sum_{j=1}^{d}\theta_j, \tag{3.8}$$

where λ depends on M. Assume now that $w_j = 1$ for all j. Notice that (3.8) is minimized over $\theta_1, \ldots, \theta_d$ by $\theta_j = \|f_j\|_{\text{cosso}}\lambda^{-1/2}$, so replacing the θ_j by their minimizing values yields objective function (3.6) for some τ.

The COSSO algorithm to minimize (3.7) starts with $\theta_j = 1$ for all j, so that the constraint $\sum_{j=1}^{d}\theta_j \leq M$ is superfluous and not imposed, and selects λ_0 by five-fold cross-validation. Then λ_0 is fixed at this value and M can be selected by GCV using the function `tune.cosso()`.

For fixed λ_0 and M, objective function in (3.7) is minimized over $f_1, \ldots, f_d$ with θ_j replaced by $\|f_j\|_{\text{cosso}}\lambda^{-1/2}$. The `cosso` package uses an efficient algorithm for implementing this minimization. For details, see the references in Sect. 3.7.

An alternative to COSSO is available in the `mgcv` package where the function `gam()` will select variables if the argument `select` is set to `TRUE`. Another alternative is provided by the `gamsel` package. The functions for fitting GAMs in this package use an ingenious penalty that allows shrinkage either to zero or to a linear function. See Chouldechova and Hastie (2015) for a description of this penalty and its implementation in an extremely fast algorithm.

Neither COSSO nor `gam()` are fast enough to work well with high-dimensional datasets, say, datasets with more than 10–20 predictors. We recommend the `gamsel` package when the number of predictors is large.

3.4.2.1 Example: Tests Scores of California School Children

We will use the dataset of test scores of school children in California to illustrate variable selection by the `cosso` package, the function `gam()` with model selection, and the `gamsel` package.

Model Selection with `cosso()`
First we define a matrix `X` of predictor variables.

```
library(Ecdat) ; data(Caschool)
Caschool$log.avginc <- log(Caschool$avginc)
mathScore <- Caschool$mathscr
X <- data.frame(Caschool$calwpct,Caschool$log.avginc,
                Caschool$mealpct,Caschool$elpct,
                Caschool$expnstu,Caschool$str)
names(X) <- c("percent qualifying for CalWORKs",
              "log(average district income)",
              "perc. qualif. for reduced price lunch",
              "percent of English learners",
              "expenditure per student",
              "student:teacher ratio")
```

Next COSSO is applied using the function `cosso()` in the `cosso` package and selected output is printed. Because random five-fold cross-validation is used to select λ_0, the seed is fixed so that the results are reproducible. The selected value of λ_0 is returned as `fitCOSSO$tune$OptLam`. With this value of λ_0, the components functions are estimated at each value of M on the grid `fitCOSSO$tune$Mgrid`. The norms of the estimated component functions at each grid-value of M are in `fitCOSSO$tune$L2norm`. A norm equal to zero indicates a variable that was not selected. The values of M and the corresponding norms were put into a matrix `COSSOoutMat` which is printed below.

```
> library(cosso) ; set.seed(3)
> fitCOSSO <- cosso(X,mathscr,scale = TRUE)
> COSSOoutMat <- data.frame(fitCOSSO$tune$Mgrid,
+                           fitCOSSO$tune$L2norm)
> fitCOSSO$tune$OptLam
[1] 7.258344e-05
> names(COSSOoutMat) <- c("M","calwpct","log.aveinc","mealpct",
+                         "elpct","expnstu","str")
> print(round(COSSOoutMat,1))
```

```
      M   calwpct  log.aveinc  mealpct   elpct expnstu  str
1  0.0     0.0        0.0        0.0      0.0     0      0
2  0.2     0.0        0.0       15.5      0.0     0      0
3  0.4     0.0        0.0       15.6      0.0     0      0
4  0.7     0.0        0.0       15.7      0.0     0      0
5  1.0     0.0        0.0       15.7      0.0     0      0
6  1.2     0.0        0.0       15.7      0.0     0      0
7  1.4     0.0        0.0       15.7      0.0     0      0
8  1.7     0.0        0.0       14.9      2.4     0      0
9  2.0     0.0        4.4       11.6      3.6     0      0
10 2.2     2.9        5.0       10.1      4.5     0      0
11 2.5     3.2        5.4        9.8      4.6     0      0
12 3.0     3.5        5.8        9.5      4.8     0      0
13 3.5     3.6        6.1        9.4      4.8     0      0
14 4.0     3.7        6.3        9.3      4.9     0      0
```

We see that the predictors *expenditures per student* (`expnstu`) and *student to teacher ratio* (`str`) are never selected, since their norms are always zero. The predictor *percentage qualifying for a reduced-price lunch* (`mealpct`) enters first. If M is 2.20 or greater, then all of the first four predictors are in the model.

We now use `tune.cosso()` to select M. The output of this function includes `OptM` which is the location of a *local* minimum of GCV.

```
> tuneCOSSO <- tune.cosso(fitCOSSO) ; print(tuneCOSSO)

$OptM
[1] 1.45

$M
 [1] 0.200 0.450 0.700 0.775 0.850 0.950 1.200 1.450 1.700
[10] 3.950

$cvm
 [1] 111.0006 110.0570 109.6784 109.6116 109.5581 109.5028
 [7] 109.4186 108.0508 108.2666  99.4271

$cvsd
 [1] 8.464088 8.448788 8.447729 8.448858 8.450390 8.452898
 [7] 8.460558 8.364579 8.178194 7.463247
```

The component `$cvm` is the minimum value of the CV estimate of mean-squared error for a given value of M, that is, the minimum of (3.7) over $\theta_1, \ldots, \theta_d$ with M fixed. (Recall that λ_0 has already been fixed as well.) Also, `$cvsd` is the standard deviation of the estimate in `$cvm`.

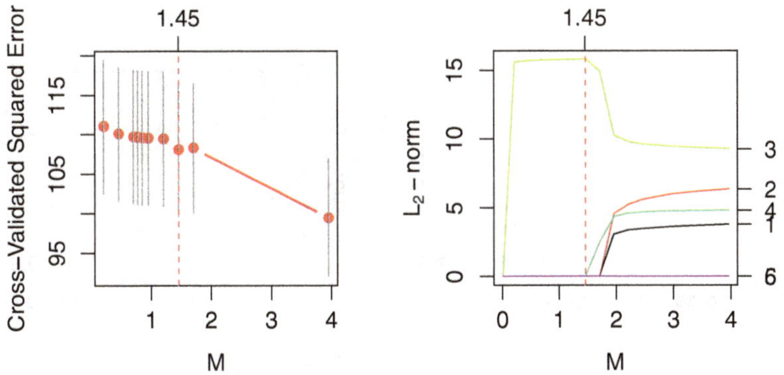

Fig. 3.8 Output from `tune.cosso()` for mathematics test scores of California school children. Left panel: cross-validation squared error for each value of M on a grid. Right panel: trace plot of the norms of the component functions.

In the output above, the function `tune.cosso()` selects M by k-fold cross-validation. The default value 5 of k is used. Figure 3.8 was produced by `tune.cosso()`. In the left panel of this figure we see that the cross-validated squared norm has its global minimum at the largest value of M on the grid, 3.95. There is a local minimum at 1.45.

The value of `OptM` returned by `tune.cosso()` is a *local minimum* of the cross-validated squared error. In our experience, it is better to use the *global minimum* of GCV rather than `OptM`. One should check the plot of the cross-validated squared error, e.g., the left panel of Fig. 3.8, to see if the global minimum occurs far from `OptM`. If it does, one can select the global minimum, e.g., by using the following code, which finds the global minimum, called `globalMinM`, of the cross-validated squared error:

```
> globalMinLocation <- which(tuneCOSSO$cvm
+                                  == min(tuneCOSSO$cvm))
> globalMinM <- tuneCOSSO$M[globalMinLocation]
```

Model selection should not be completely automated. It is recommended that one investigate the fits from both `OptM` and the global minimum of the cross-validated squared error whenever they are different. For this dataset and the particular seed that we have used, `OptM` is not the location of the global minimum, as can be seen from `$cvm` in the output above as well as Fig. 3.8. With some experimentation with the same data but the seed varying, we found that `OptM` is sometimes the location of the global minimum, e.g., if the seed is 1 instead of 3. The location of the global minimum itself is stable and remains at 3.95 as the seed varies, but `OptM` varies considerably.

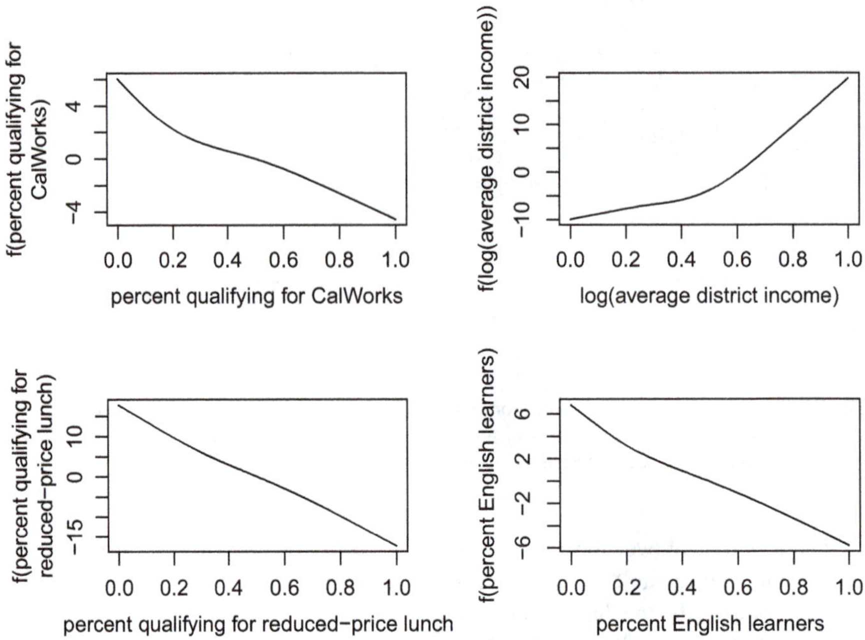

Fig. 3.9 Estimated component functions selected by COSSO for mathematics test scores of California school children. Expenditures per student and student teacher ratio were not selected.

The component functions estimated with the value of M achieving the global minimum are plotted using:

```
> plot.cosso(fitCOSSO,M = globalMinM,plottype = "Func")
```

and the result is displayed in Fig. 3.9. We see that four predictors were selected. If we used M equal to `OptM`, that is 1.45, instead of the global minimizer, 3.95, then only one predictor, *percent qualifying for reduced-price lunch*, would have been selected. Selecting only one predictor seems drastic and does not agree with the results below using `gam()` or `cv.gamsel()` to select the predictors; both of these functions select five of the six predictors, although they disagree on which predictor is not selected.

Model Selection with `gam()`

The `mgcv` function `gam()` will select variables if its argument `select` is set to TRUE as illustrated by the following code:

```
> library(mgcv) ; library(HRW)
> fitGAM <- gam(mathscr ~ s(calwpct) + s(log.avginc)
+                  + s(mealpct) + s(elpct) + s(expnstu)
+                  + s(str),data = Caschool,select = TRUE)
```

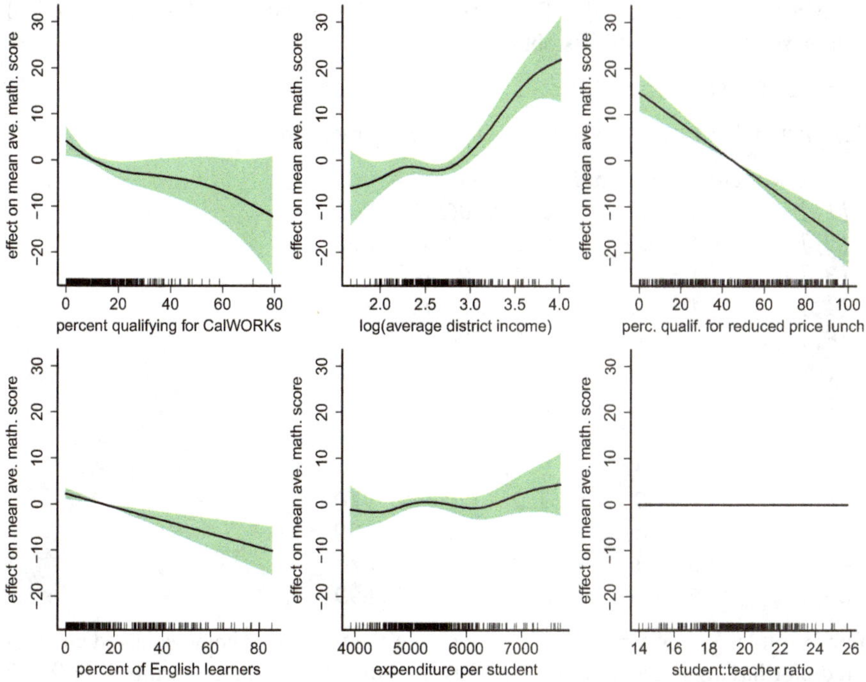

Fig. 3.10 Components of the GAM fit to the California schools data with GCV smoothing parameter selection using the function `gam()` in the `mgcv` package and with variables selected using the argument `select = TRUE`. The sixth component function was shrunk to zero. The bands around the fits correspond to approximately 95% pointwise confidence intervals. Each component function is vertically centered about zero. The tick marks at the base of each panel show the predictor data.

A type of penalty-based model selection, described in Marra and Wood (2011), is used. The fitted functions are in Fig. 3.10. We see that, like with COSSO, *student:teacher ratio* is not selected. Unlike COSSO, *expenditure per student is selected*, although its effect is small.

Model Selection with `cv.gamsel()`

Finally, we explain model selection using the function `cv.gamsel()` in the `gamsel` package. First issue the commands:

```
> library(gamsel) ; set.seed(919)
> fitCVgamsel  <- cv.gamsel(X,y = mathScore)
```

Because cross-validation is a random algorithm, a seed has been set so that the results are reproducible, although the fitted functions do not vary much if the seed is changed. The `cv.gamsel()` fit object, `fitCVgamsel`, contains a path of solutions parameterized by a regularization parameter λ which, in this case, takes values in a geometric sequence of length 50 between $\lambda = 7.899$ and $\lambda = 789.9$. A numerical summary of the path of fits is obtained using:

```
> print(fitCVgamsel$gamsel.fit)
```

The first and last five rows of the resulting data frame are:

```
        NonZero Lin NonLin   %Dev  Lambda
 [1,]         0   0      0 0.0000 789.900
 [2,]         1   1      0 0.1161 719.000
 [3,]         1   1      0 0.2122 654.500
 [4,]         1   1      0 0.2919 595.800
 [5,]         1   1      0 0.3580 542.400
   .          .   .      . .            .
   .          .   .      . .            .
   .          .   .      . .            .
[46,]         6   4      2 0.7312  11.500
[47,]         6   3      3 0.7320  10.470
[48,]         6   3      3 0.7328   9.532
[49,]         6   3      3 0.7337   8.677
[50,]         6   1      5 0.7346   7.899
```

For example, $\lambda = 595.8$ results in only one of the predictors being selected and that predictor having a linear effect. Towards the other end of the spectrum, $\lambda = 9.532$ leads to all six predictors being selected of which three are linear and three are nonlinear.

Figure 3.11 shows the estimated functions for four different λ values with the color-coding:

blue indicating a zero effect,
green indicating a linear effect, and
red indicating a nonlinear effect.

We see from Fig. 3.11 that, depending on the value of λ, some predictors can be in or out of the model and, if in the model, may have a linear or nonlinear effect.

With a view to selecting a single model objectively from the full λ-parameterized path we issue the command:

```
plot(fitCVgamsel)
```

This leads to the plot shown in Fig. 3.12, in which the cross-validation estimate of mean-squared error (CV-MSE) is plotted against $\log(\lambda)$ over the 50 λ values. The numbers above the plotting frame in Fig. 3.12 indicate the numbers of predictors in the model for varying values of λ. The solid vertical lines with stems indicate plus and minus one standard error and conveys the variability inherent in CV-MSE. The two dashed vertical lines indicate the values of λ for which:

- the overall minimum of CV-MSE occurs, and
- the largest value of λ for which CV-MSE is within one standard deviation of its minimum.

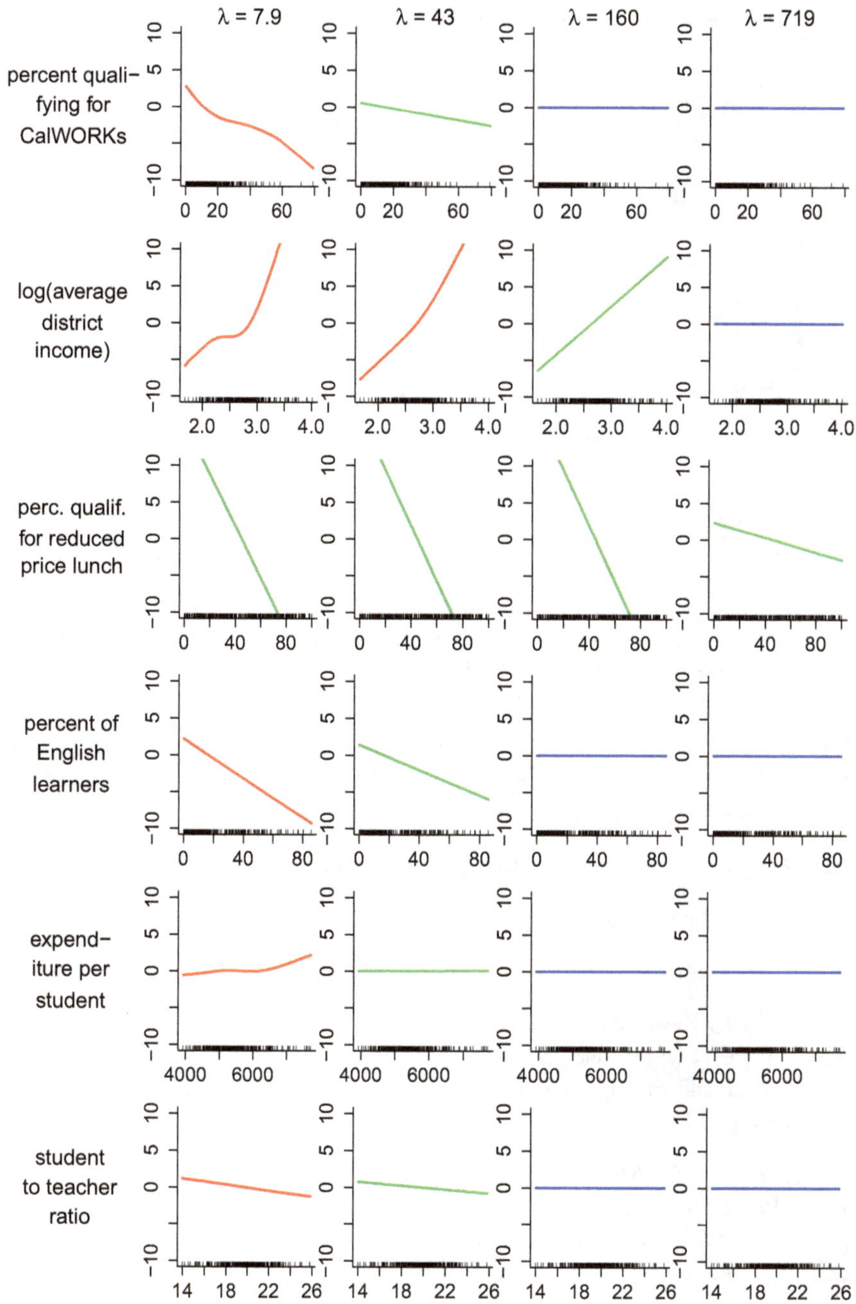

Fig. 3.11 Plots of the estimated functions from the `cv.gamsel()` fit object, with each column corresponding to a different value along the λ regularization path. The curves are color-coded according to: blue signifies a zero effect, green signifies a linear effect, and red signifies a nonlinear effect.

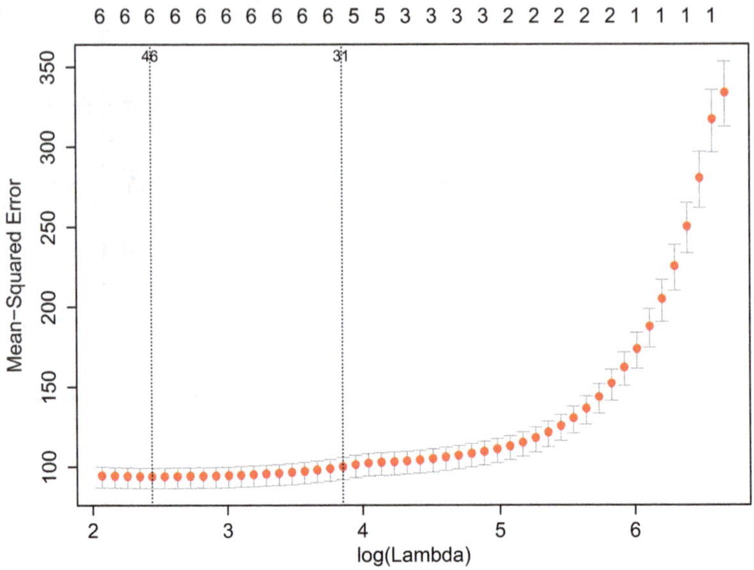

Fig. 3.12 Cross-validation estimate of mean-squared error (CV-MSE) from `cv.gamsel()`. CV-MSE is minimized at index 46 from largest to smallest. The largest value of λ where CV-MSE is within one standard error of its minimum has index 31.

The numbers at the top of the vertical dashed lines correspond to the row numbers of the data frame `fitCVgamsel$gamsel.fit`. These numbers can be extracted from the `cv.gamsel` fit object via the code:

```
print(fitCVgamsel$index.min) ; print(fitCVgamsel$index.1se)
```

The corresponding λ values are given by:

```
print(fitCVgamsel$lambda.min) ; print(fitCVgamsel$lambda.1se)
```

Figure 3.13 shows the fitted functions when λ is set to its largest value such that CV-MSE is within one standard deviation of its minimum. This corresponds to $\lambda = 47.1$ and the right-hand vertical dashed line of Fig. 3.12. The code that produced Fig. 3.13 is:

```
par(mfrow = c(2,3))
for (iPred in 1:6)
   plot(fitCVgamsel$gamsel.fit,newx = X,which = iPred,
        index = fitCVgamsel$index.1se,col.main = "navy",
        main = names(X)[iPred],rugplot = TRUE,bty = "l",
        ylim = c(-8,16))
```

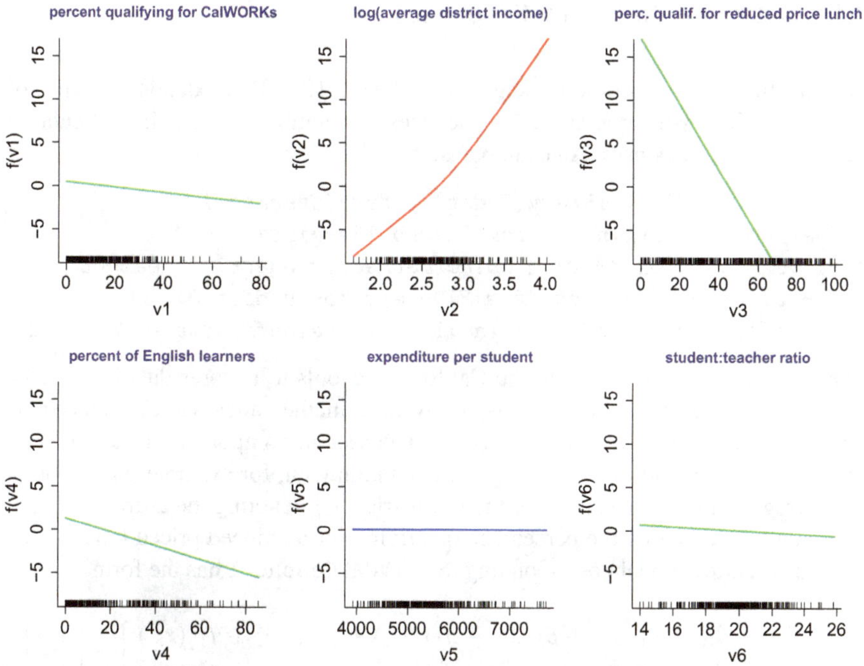

Fig. 3.13 Components of the fit from the `cv.gamsel()` function when λ equals `fitCVgamsel$lambda.1se` (which is 47.1 in this example). The green curves have been shrunk to a linear fit and the blue curve has been shrunk to zero. The red curve is the only nonlinear effect for this selected value of λ.

Five of the six predictors are selected. The predictor not selected is *expenditure per student*. The *student:teacher ratio* is selected but its effect is small. Of the five selected predictors, four of the fitted functions are shrunk to a linear fit. Only *log(average district income)* has a nonlinear effect. We see that the effect is convex with its derivative increasing especially in the middle range of that variable. This is in agreement with the fits by `cosso()` and `gam()`. Therefore, the effect on the mathematics score of increasing *log(average district income)* is strongest at higher values of *log(average district income)*. One needs to be cautious about interpreting this effect. If the predictor is *average district income* rather than *log(average district income)*, then the fitted function is concave rather than convex.

In summary, the results from `cosso()`, `gam()`, and `gamsel()` agree that the effects of *expenditure per student* and *student:teacher ratio* are all small, although they disagree about which of these effects are exactly zero. Of course, this is not unexpected. It is difficult to distinguish a small effect from one that is exactly zero.

3.5 Extension to Vector Responses

The function vgam() in the package VGAM (Yee 2010, 2017) extends the types of GAMs handled by the gam and mgcv packages in a number of ways. In particular, it supports *vector* responses. An example is:

```
> library(VGAM) ; library(Ecdat) ; data(Caschool)
> Caschool$log.avginc <- log(Caschool$avginc)
> fitVGAMCaschool <- vgam(cbind(mathscr,readscr) ~ mealpct
+    + elpct + s(calwpct,df = 3) + s(compstu,df = 3)
+    + s(log.avginc,df = 4),family = gaussianff,data = Caschool)
```

This fit extends the model fit to the California schools test scores data in Sect. 3.4 but with the scalar response mathscr (average mathematics score) replaced by the vector response (mathscr, readscr). The second component of the response vector is average reading score. In generic notation with, for example, y_{1i} denoting the average mathematics score for the ith district, y_{2i} denoting the average reading score, and x_{1i} denoting the percentage qualifying for a reduced priced lunch in the same district, the model corresponding to fitVGAMCaschool has the form

$$\begin{bmatrix} y_{1i} \\ y_{2i} \end{bmatrix} = \begin{bmatrix} \beta_{10} + \beta_{11}\,x_{1i} + \beta_{12}\,x_{2i} + f_{13}(x_{3i}) + f_{14}(x_{4i}) + f_{15}(x_{5i}) \\ \beta_{20} + \beta_{21}\,x_{2i} + \beta_{22}\,x_{2i} + f_{23}(x_{3i}) + f_{24}(x_{4i}) + f_{25}(x_{5i}) \end{bmatrix} + \begin{bmatrix} \varepsilon_{1i} \\ \varepsilon_{2i} \end{bmatrix}$$

where

$$\begin{bmatrix} \varepsilon_{1i} \\ \varepsilon_{2i} \end{bmatrix} \overset{\text{ind.}}{\sim} N(\mathbf{0}, \boldsymbol{\Sigma}_\varepsilon), \quad 1 \le i \le 420.$$

Here f_{1j} and f_{2j}, $j = 3, 4, 5$, are smooth functions and $\boldsymbol{\Sigma}_\varepsilon$ is an unstructured 2×2 covariance matrix. The extension to vector responses with more than two components has an analogous form.

Unlike gam() in the package mgcv, vgam() does not support automatic smoothing parameter selection via GCV or REML. Here we have used effective degrees of freedom values similar to those chosen for the scalar response GAMs. The command:

```
> summary(fitVGAMCaschool)
```

leads to the output

```
Call:
vgam(formula = cbind(mathscr, readscr) ~ mealpct + elpct
     + s(calwpct, df = 3) + s(compstu, df = 3)
     + s(log.avginc, df = 4), family = gaussianff,
       data = Caschool)
Number of linear predictors:    2
```

```
Names of linear predictors: mathscr, readscr

(Estimated) Dispersion Parameter for gaussianff family: 79.41081

Residual deviance:  65752.15 on 813.999 degrees of freedom

Log-likelihood: -28264306 on 813.999 degrees of freedom

Number of iterations:  2

DF for Terms and Approximate Chi-squares for Nonparametric
Effects

                         Df Npar Df Npar Chisq     P(Chi)
(Intercept):1             1
(Intercept):2             1
mealpct:1                 1
mealpct:2                 1
elpct:1                   1
elpct:2                   1
s(calwpct, df = 3):1      1       2      381.69 0.0000e+00
s(calwpct, df = 3):2      1       2      166.08 8.6474e-37
s(compstu, df = 3):1      1       2      568.89 0.0000e+00
s(compstu, df = 3):2      1       2      229.67 0.0000e+00
s(log.avginc, df = 4):1  1       3     3099.17 0.0000e+00
s(log.avginc, df = 4):2  1       3     1685.84 0.0000e+00
```

which shows significance of all terms via approximate chi-squared tests.

The `plot.vgam()` function for `vgam()` facilitates comparisons by putting the additive fits corresponding to different responses on the same sets of axes. This is illustrated in Fig. 3.14. One can see that the effects of the predictors are similar for the two responses. See `CaSchoolVGAMfit.R` in the `HRW` package for the code that produced Fig. 3.14. It can be run and located using:

```
> library(HRW) ; demo(CaSchoolVGAMfit,package = "HRW")
> system.file("demo","CaSchoolVGAMfit.R",package = "HRW")
```

The commands:

```
> plot(residuals(fitVGAMCaschool),col = "dodgerblue",bty = "l",
+      xlab = "mathematics score residuals",
+      ylab = "reading score residuals",
+      cex.lab = 1.5,cex.axis = 1.5)
> abline(h=0,col = "slateblue") ; abline(v=0,col = "slateblue")
```

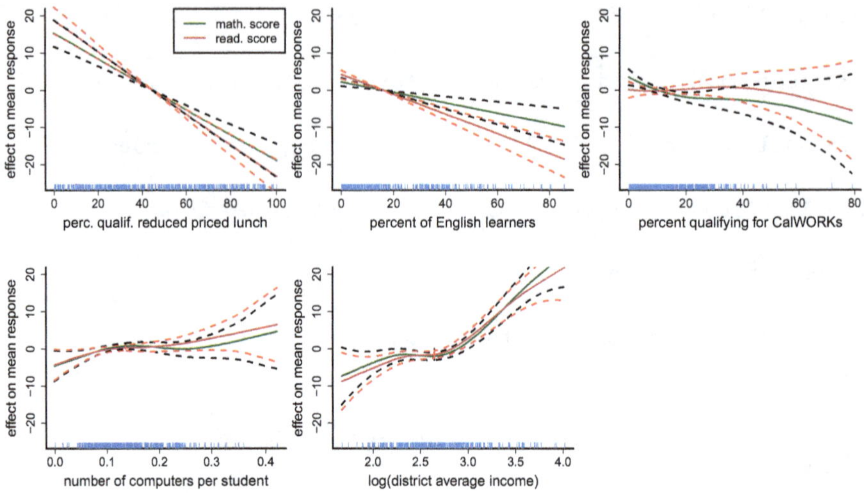

Fig. 3.14 Estimated smooth function components for the vector GAM fit to the California schools data obtained via the R function `vgam()` in the package `VGAM` and stored in object `fitVGAMCaschool` according to code given in the text. The response variables are average mathematics score and average reading. Each component function is vertically centered about zero. The dashed curves indicate approximately 95% pointwise confidence intervals. The tick marks indicate the values of the predictors.

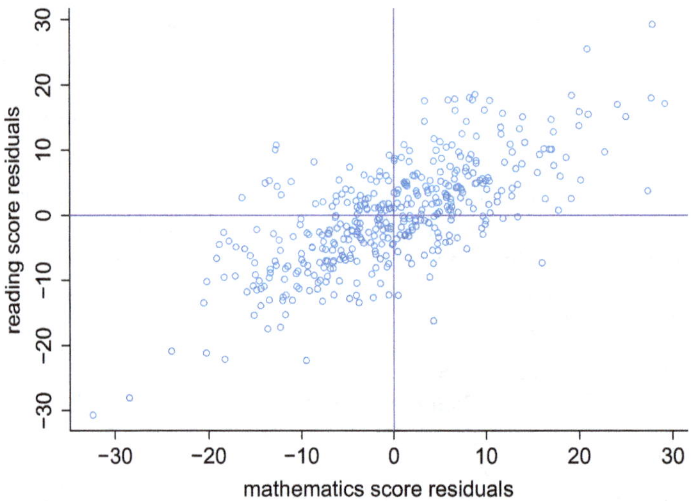

Fig. 3.15 Scatterplot of the bivariate residuals from the vector GAM fit stored in `fitVGAM-Caschool`.

lead to scatterplots of the bivariate residuals shown in Fig. 3.15. This plot is indicative of strong correlation between the two response variables and a vector GAM approach being appropriate.

A quick estimate of $\boldsymbol{\Sigma}_\varepsilon$ is:

```
> var(residuals(fitVGAMCaschool))
```

```
         [,1]     [,2]
[1,] 91.31114 54.96917
[2,] 54.96917 65.61523
```

This estimate does not adjust for the degrees of freedom lost due to mean function estimation but since the sample size is 420, such an adjustment has a very minor effect. The corresponding correlation coefficient is:

```
> cor(residuals(fitVGAMCaschool))[1,2]
```

```
[1] 0.7101579
```

which confirms the high positive correlation between average mathematics and reading scores even after removing the effects of predictors.

3.6 Extension to Factor-by-Curve Interactions

We now return to the Warsaw apartment data described in Chaps. 1 and 2. In Sect. 2.12.3 we modeled the dependence of the area/price ratio on year of construction using a simple factor-by-curve interaction with one curve for the Srodmiescie (central business) district and another curve for the other three districts. Stated differently, we used a factor-by-curve interaction model where the factor had two levels: Srodmiescie and the other three districts aggregated.

We now generalize this model by letting each of the four districts have its own curve, so that for the ith apartment

$$\begin{aligned}(\texttt{area/price})_i &= \sum_{\ell=1}^{4} f_\ell(\texttt{construction.date}_i)\, I(\texttt{district}_i = \ell) + \varepsilon_i,\\ \varepsilon_i &\overset{\text{ind.}}{\sim} N(0,\sigma_\varepsilon^2), \quad 1 \le i \le 409, \end{aligned} \tag{3.9}$$

where f_j is a smooth function describing the relationship between the construction date and the ratio of floor area to price in district ℓ, $\ell = 1, 2, 3, 4$. Model (3.9) can be fit easily using the function gam() via:

```
> library(HRW) ; library(mgcv) ; data(WarsawApts)
> fit1GAMWarsaw <- gam(areaPerMzloty ~ as.factor(district) +
+          s(construction.date,by = as.factor(district), k = 25),
+          data = WarsawApts,method = "REML")
```

The argument "by = as.factor(district)" inside "s()" specifies a separate smooth fit for each level of the factor district so that model (3.9) is used. We use REML smoothing parameter selection since, in this case, GCV tends to overfit

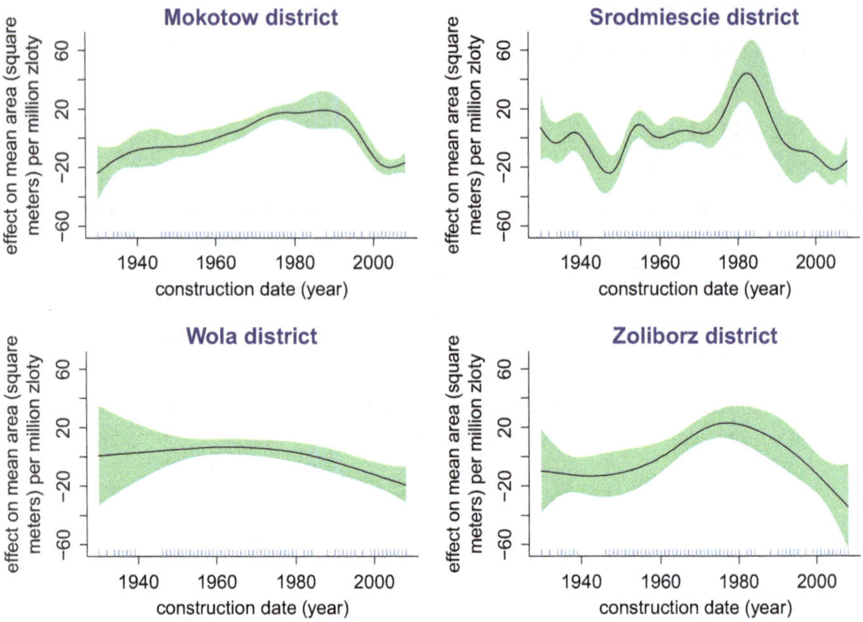

Fig. 3.16 Estimated smooth function components for the simple factor-by-curve interaction model fit to the Warsaw apartments data obtained via the R function `gam()` in the package `mgcv` and stored in object `fit1GAMWarsaw` according to code given in the text. The response variable is area (square meters) per million złoty. The panels correspond to functions of construction date for each district. Each component function is vertically centered about zero. The shaded regions indicate approximately 95% pointwise confidence intervals. The tick marks indicate the values of the predictor.

the data. We now print the summary and plot the fitted curves. The fitted curves are shown in Fig. 3.16.

```
Family: gaussian
Link function: identity

Formula:
areaPerMzloty ~ as.factor(district) + s(construction.date,
                  by = as.factor(district), k = 25)

Parametric coefficients:
                               Estimate Std. Error t value
(Intercept)                     113.491      1.275  89.037
as.factor(district)Srodmiescie  -11.968      2.250  -5.320
as.factor(district)Wola           2.527      2.811   0.899
as.factor(district)Zoliborz      -3.869      3.710  -1.043
                               Pr(>|t|)
(Intercept)                     < 2e-16
```

```
as.factor(district)Srodmiescie 1.78e-07
as.factor(district)Wola           0.369
as.factor(district)Zoliborz       0.298

Approximate significance of smooth terms:
                                                  edf
s(construction.date):as.factor(district)Mokotow     7.676
s(construction.date):as.factor(district)Srodmiescie 11.358
s(construction.date):as.factor(district)Wola        2.072
s(construction.date):as.factor(district)Zoliborz    3.616
                                                 Ref.df
s(construction.date):as.factor(district)Mokotow     9.366
s(construction.date):as.factor(district)Srodmiescie 13.593
s(construction.date):as.factor(district)Wola        2.542
s(construction.date):as.factor(district)Zoliborz    4.442
                                                      F
s(construction.date):as.factor(district)Mokotow    17.473
s(construction.date):as.factor(district)Srodmiescie  5.233
s(construction.date):as.factor(district)Wola        4.612
s(construction.date):as.factor(district)Zoliborz    6.362
                                                  p-value
s(construction.date):as.factor(district)Mokotow     < 2e-16
s(construction.date):as.factor(district)Srodmiescie 1.48e-08
s(construction.date):as.factor(district)Wola        0.00492
s(construction.date):as.factor(district)Zoliborz   2.99e-05

R-sq.(adj) =  0.458   Deviance explained = 49.5%
-REML = 1737.5  Scale est. = 265.47    n = 409
```

Next, we expand the model to include a second continuous predictor, surface area, and another factor: number of rooms. The expanded model is additive, with a penalized spline for surface area and district-specific penalized splines for construction date. We do not include an interaction between surface area and district or interactions between the number of rooms and either of the continuous predictors, although these effects could be added easily. The fit is displayed in Fig. 3.17 and the output is as follows:

```
> fit2GAMWarsaw <- gam(areaPerMzloty ~ as.factor(district) +
+                    s(construction.date,k = 25,by = district)
+                    + as.factor(n.rooms) + s(surface,k = 25),
+                    data = WarsawApts,method = "REML")

Family: gaussian
Link function: identity

Formula:
areaPerMzloty ~ as.factor(district) + s(construction.date, k = 25,
    by = district) + as.factor(n.rooms) + s(surface, k = 25)
```

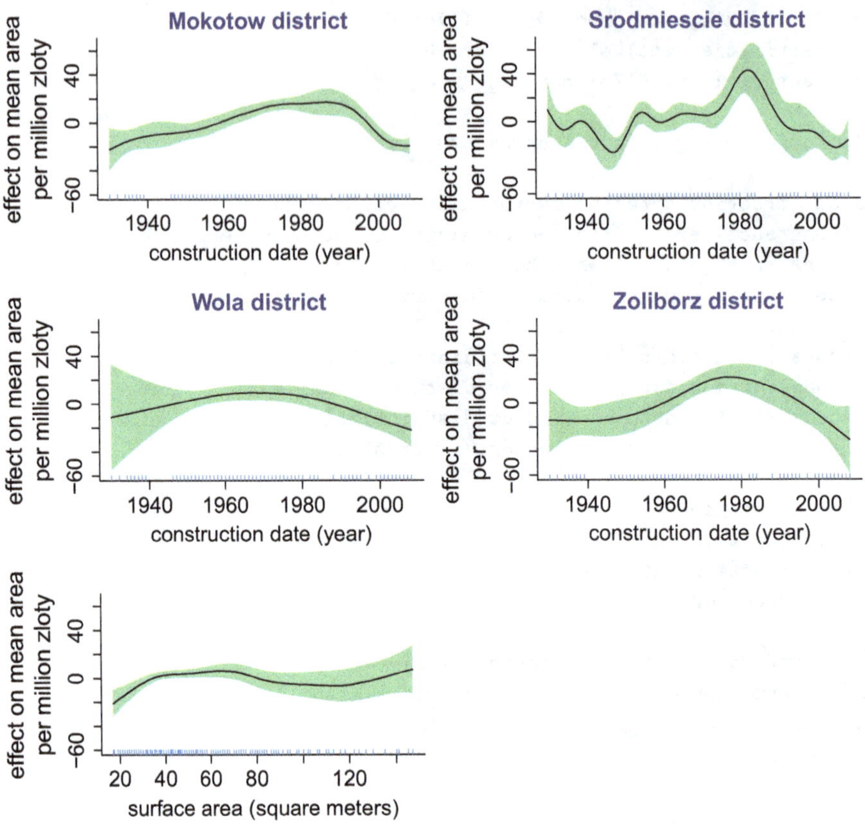

Fig. 3.17 Estimated smooth function components for the expanded factor-by-curve interaction model fit to the Warsaw apartments data obtained via the R function `gam()` in the package `mgcv` and stored in object **`fit2GAMWarsaw`** according to code given in the text. The response variable is area (square meters) per million złoty. The upper four panels correspond to functions of construction date for each district and the bottom panel is the estimated function of surface area. Each component function is vertically centered about zero. The shaded regions indicate approximately 95% pointwise confidence intervals. The tick marks indicate the values of the predictor.

```
Parametric coefficients:
                                Estimate Std. Error t value  Pr(>|t|)
(Intercept)                      116.417      2.965  39.269   < 2e-16
as.factor(district)Srodmiescie  -10.997      2.234  -4.922  1.29e-06
as.factor(district)Wola           3.388      2.956   1.146    0.2525
as.factor(district)Zoliborz      -2.202      3.647  -0.604    0.5464
as.factor(n.rooms)2              -5.635      3.400  -1.658    0.0982
as.factor(n.rooms)3              -3.738      4.713  -0.793    0.4283
as.factor(n.rooms)4              -1.976      7.016  -0.282    0.7784
```

```
Approximate significance of smooth terms:
                                           edf Ref.df      F  p-value
s(construction.date):districtMokotow     6.770  8.308 14.363  < 2e-16
s(construction.date):districtSrodmiescie 11.581 13.831  4.847 2.04e-08
s(construction.date):districtWola        2.483  3.068  5.449  0.00108
s(construction.date):districtZoliborz    3.511  4.309  6.829 1.57e-05
s(surface)                               5.743  7.201  3.728  0.00052

R-sq.(adj) =  0.492   Deviance explained = 53.7%
-REML = 1720.4  Scale est. = 248.7      n = 409
```

The results suggest that the area/price ratio does not depend on the number of rooms. A small *p*-value for `surface` suggests that the area/price ratio depends on the surface area. As seen in Fig. 3.17, area/price increases with surface area, at least for values of surface up to about 70 square meters. The data are sparse for values of `surface` above 70 square meters and the variability band widens, so the effect of `surface` above 70 square meters is uncertain.

We used REML to select tuning parameters for both examples in this section. If the default method, GCV, is used, then there is less smoothing and noticeably wigglier fitted curves. Figure 3.17 can be reproduced by running the code in the script `WarsawAptsGAMfit.R`, which is part of the `HRW` package. These commands run the script:

```
> library(HRW) ; demo(WarsawAptsGAMfit,package = "HRW")
```

and these commands locate the script for possible copy and editing:

```
> system.file("demo","WarsawAptsGAMfit.R",package = "HRW")
```

3.6.1 Example: Mortgage Applications in Boston

We now illustrate factor-by-curve interactions for the Boston mortgage applications data and allow the smooth function effects of debt payments to income ratio (`dir`) and loan size to property value ratio (`lvr`) to depend on the level of the binary variable indicator of self-employed (`self`). The required R code is:

```
> fitFacByCurvBostMort <- gam(deny ~ black + factor(self)
+                        + s(dir,by = factor(self))
+                        + s(lvr,by = factor(self)) +
+                        + pbcr + self + single + s(ccs,k = 4),
+                          family = binomial,data = BostonMortgages)

> summary(fitFacByCurvBostMort)

Family: binomial
Link function: logit
```

```
Formula:
deny ~ black + factor(self) + s(dir, by = factor(self)) + s(lvr,
    by = factor(self)) + +pbcr + self + single + s(ccs, k = 4)

Parametric coefficients:
                 Estimate Std. Error z value Pr(>|z|)
(Intercept)       -2.8284     0.1219 -23.203  < 2e-16
blackyes           0.6273     0.1721   3.646 0.000266
factor(self)yes    0.5548     0.2443   2.271 0.023143
pbcryes            1.2457     0.2014   6.184 6.26e-10
selfyes            0.0000     0.0000      NA       NA
singleyes          0.3844     0.1452   2.648 0.008090

Approximate significance of smooth terms:
                          edf Ref.df Chi.sq  p-value
s(dir):factor(self)no   4.767  5.740 75.321 5.19e-14
s(dir):factor(self)yes  1.000  1.000  5.252   0.0219
s(lvr):factor(self)no   3.936  4.898 38.622 2.35e-07
s(lvr):factor(self)yes  3.650  4.528 12.434   0.0188
s(ccs)                  1.635  1.990 55.924 5.53e-13

Rank: 44/45
R-sq.(adj) =  0.222   Deviance explained = 23.2%
UBRE = -0.42042  Scale est. = 1         n = 2380
```

Plots of smooth effects of dir and lvr for the two levels of self are shown in Fig. 3.18, and are produced by the script BostMortFacByCurv.R in the HRW package. The two sets of curves look somewhat similar. To see if there is a statistically significant difference, we compare the new model and the old with a chi-squared test.

```
> anova(fit2GAMBostMort,fitFacByCurvBostMort,test = "Chisq")

Analysis of Deviance Table

Model 1: deny ~ black + s(dir) + s(lvr) + pbcr + self + single
                + s(ccs, k = 4)
Model 2: deny ~ black + factor(self) + s(dir, by = factor(self))
                + s(lvr, by = factor(self)) + pbcr + self
                + single + s(ccs, k = 4)
  Resid. Df Resid. Dev     Df Deviance Pr(>Chi)
1    2360.4     1343.3
2    2356.8     1339.4 3.5731   3.9068   0.3553
```

The results of the test suggest that the incorporation of this interaction did not improve the fit much if at all.

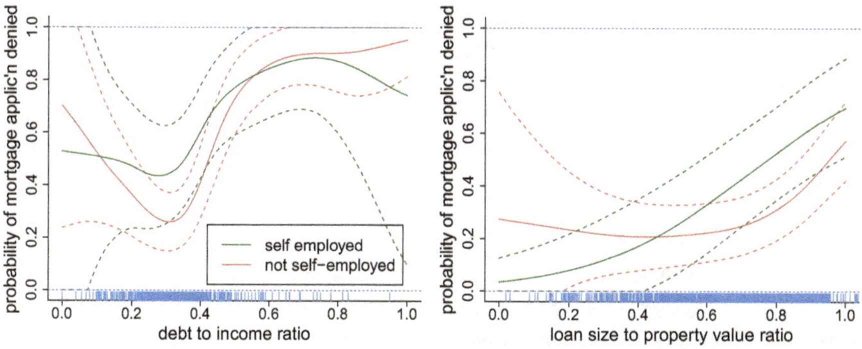

Fig. 3.18 Estimated smooth function components for the simple factor-by-curve interaction model fit to the Boston mortgages data obtained via the R function `gam()` in the package `mgcv` and stored in object `fitFacByCurvBostMort` according to code given in the text. The response variable is indicator of denial of mortgage application. The factor-by-curve plot for debt payments to income ratio in the left panel is such that loan size to property value ratio and credit score are set to their average and each of the other predictors is set to "yes." The right panel plot is analogous except that debt payments to income ratio and loan size to property value ratio are interchanged. The shaded regions indicate approximately 95% pointwise confidence intervals. The tick marks indicate the values of the predictor.

We see that in every model with confounders that has been considered, the variable `black` is still significant. However, its effect is considerably smaller than when it is the only predictor, in agreement with the results of Munnell et al. (1996). After adjusting for confounders, the odds ratio for denial comparing a Black or Hispanic applicant to an applicant who is not Black or Hispanic is between 1.8 and 1.9, depending upon the model. We saw in Sect. 3.2.1 that if there is no adjustment for confounders, then the odds ratio is about 3.9. Thus, the data provide substantial evidence of racial discrimination during the period 1997–1998 when they were collected.

The script `BostMortFacByCurv.R` can be run from the command line as follows:

```
> library(HRW) ; demo(BostMortFacByCurv,package = "HRW")
```

It can be copied and edited from the location given by:

```
> system.file("demo","BostMortFacByCurv.R",package = "HRW")
```

3.7 Further Reading

There are many books on GLMs. Some of the main ones are McCullagh and Nelder (1989), Faraway (2006), Fahrmeir and Kneib (2011), Fahrmeir and Tutz (1994), and McCulloch et al. (2008). Books that provide good coverage of GAMs include

Faraway (2006), Hastie and Tibshirani (1990), Ruppert et al. (2003), and Wood (2006a). The Faraway (2006) book is also geared towards implementation in R.

The gamlss package has far more functionality than we have room to describe here. A detailed account of this package is given by Stasinopoulos and Rigby (2008).

The component selection and smoothing operator were proposed by Lin and Zhang (2006) and sparse additive models by Ravikumar et al. (2009). Hastie et al. (2015) is a comprehensive introduction to sparse models.

Often, subject-matter considerations suggest that one or more of the component functions of a GAM satisfy a shape constraint such as being monotonically increasing. Using the scam (Pya 2017), a user can require a component function to be monotonically decreasing, monotonically increasing, convex, concave, or a combination of these constraints such as monotonically increasing and concave.

3.8 Exercises

1. Ensure that the package AER (Kleiber and Kleiber 2017) is installed in your R environment.

   a. Issue the following commands to fit a Gaussian GAM with `price` as the response variable and each of the other variables as predictors:
   ```
   > library(AER) ; data(HousePrices) ; library(mgcv)
   > fitGaussAM <-  gam(price ~ s(lotsize,k = 27) + bedrooms
   +               + factor(bathrooms) + factor(stories)
   +               + factor(driveway)
   +               + factor(recreation)
   +               + factor(fullbase) + factor(gasheat)
   +               + factor(aircon) + garage + factor(prefer),
   +                 data = HousePrices,family = gaussian)
   ```
   b. Issue the command `gam.check(fitGaussAM)` to check whether or not the residuals are consistent with the model assumptions.
   c. Obtain the *Gamma* GAM fit with the same response and predictors as used to obtain `fitGaussAM` in part a. This involves specifying `family = Gamma` in the call to `gam()`. Let `fitGammaAM` be the corresponding fit object.
   d. Issue the command `gam.check(fitGammaAM)` to check whether or not the residuals are consistent with the model assumptions.
   e. Choose between `fitGaussAM` and `fitGammaAM` according to which model has the better residual plots. Then obtain numerical and graphical summaries based on the chosen model.
   f. Consider a house with a lot size of 5000 square feet, three bedrooms, two bathrooms, two stories, a driveway, no recreation room, a finished basement, hot water heating, no air conditioning, two garage places, and located outside of the preferred neighborhood of Windsor, Canada. Use the model from part e. to predict the price of the house if sold during the same period that the

data were collected (July–September, 1987). Assume that the price data are in Canadian dollars.

g. Obtain an approximately 95% confidence interval for the mean price of houses with the same features as the house described in part f.

2. Ensure that the package gam (Hastie 2017a) is installed in your R environment.

a. Issue the following R commands to generate data from a Gaussian GAM:

```
> set.seed(1) ; n <- 500 ; error <- rnorm(n,0,0.5)
> x1 <- runif(n) ; x2 <- runif(n) ; x3 <- runif(n)
> x4 <- runif(n) ; x5 <- runif(n) ; x6 <- runif(n)
> x7 <- runif(n) ; x8 <- runif(n) ; x9 <- runif(n)
> f4 <- function(x) return(x + dnorm(x,0.5,0.25))
> f5 <- function(x) return(x + 0.5*dnorm(x,0.5,0.25))
> f6 <- function(x) return(x + 0.1*dnorm(x,0.5,0.25))
> y <- x1 + x2 + x3 + f4(x4) + f5(x5) + f6(x6) + error
```

The code `dnorm(x,0.5,0.25)` corresponds to evaluation of the function $\phi(x; 0.5, 0.25)$ where $\phi(\cdot; \mu, \sigma)$ is the Normal density function with mean μ and standard deviation σ. Note that the model contains x1, x2 and x3 as linear effects, x4, x5, and x6 as smooth function effects. The model does not contain x7, x8, or x9.

b. Issue the following commands to perform `step.Gam()` model selection:

```
> library(gam)
> gamObj <- gam(y ~ x1 + x2 + x3 + x4 + x5 + x6 + x7
+                  + x8 + x9)
> stepFit <- step.Gam(gamObj,scope =
+            list("x1" = ~ 1 + x1 + s(x1,df = 2),
+                 "x2" = ~ 1 + x2 + s(x2,df = 2),
+                 "x3" = ~ 1 + x3 + s(x3,df = 2),
+                 "x4" = ~ 1 + x4 + s(x4,df = 2),
+                 "x5" = ~ 1 + x5 + s(x5,df = 2),
+                 "x6" = ~ 1 + x6 + s(x6,df = 2),
+                 "x7" = ~ 1 + x7 + s(x7,df = 2),
+                 "x8" = ~ 1 + x8 + s(x8,df = 2),
+                 "x9" = ~ 1 + x9 + s(x9,df = 2)))
```

c. Issue the following command to view the selected model:

```
> print(names(stepFit$"model")[-1])
```

For example, if the printed object is

```
"x1" "x2" "x3" "s(x4,df = 2)" "s(x5,df = 2)" "x6" "x9"
```

then `step.Gam()` has correctly chosen x1, x2 and x3 as entering the model linearly and has also correctly chosen x4 and x5 as entering the model nonlinearly. However, `step.Gam()` has incorrectly chosen x6 as entering the model linearly and has incorrectly chosen x9 as being in the model.

d. Replicate 100 random datasets according to the code in part a. and, for each one, apply the code in parts b. and c. to obtain the model chosen by step.Gam(). Record the proportions of correct choices for each of the nine candidate predictors.

e. Repeat part d. but with `error <- rnorm(n,0,1)`, corresponding to a doubling of the error standard deviation.
f. Make numerical and graphical summaries of the results from the simulation studies conducted in parts d. and e.

3. The dataset `ozone` in the gss (Gu 2017) package contains data on ozone concentration and eight meteorological quantities in the Los Angeles area for 330 days of 1976.

   a. Ensure that gss is installed in your R environment and enter the commands:
      `> library(gss) ; data(ozone) ; pairs() ; help(ozone)`
      to make the `ozone` data frame available to the current R session, visualize the data, and to show the variable names and their definitions. The variable of interest is `upo3`, corresponding to daily ozone concentrations in Upland, California, USA
   b. Ensure that the package gam (Hastie 2017a) is installed in your R environment. Use the function `step.Gam()` to select predictors and their forms (i.e., linear versus nonlinear) among Gaussian GAMs with `upo3` as the response variable.
   c. Based on the form of the model chosen in part b., use the function `gam()` in the package mgcv (Wood 2017) to fit the model with GCV-based smoothing parameter selection.
   d. Apply the function `gam.check()` to the fit object from part c. to ensure the number of basis functions is sufficient. Increase the numbers of basis functions until the `k-index` test is passed. Also check the residuals and comment on their behavior.
   e. Obtain numerical and graphical summaries of the final fit from part d.

4. Ensure that the packages aplore3 (Braglia 2016) and gam (Hastie 2017a) are installed in your R environment.

   a. Issue the following commands to load and obtain a listing of the variables in the data frame `icu`:
      `> library(aplore3) ; data(icu) ; help(icu)`
      The data frame consists of 21 variables for 200 patients admitted to an intensive care unit (ICU).
   b. Use the function `step.Gam()` in the gam package to select a logistic GAM with the response variable being the indicator of the patient dying.
   c. Use the function `gam()` in the mgcv package to re-fit the model selected in part b. with GCV used for selection of the smoothing parameters of the penalized spline components of the model. Obtain numerical and graphical summaries of the selected model.
   d. Consider a 79-year-old white woman who enters the ICU as an emergency patient in a coma. The woman has no history of previous ICU admission, cancer, chronic renal failure, and is free of infection at the time of

admission. Cardiopulmonary resuscitation is not needed and she has no fractures. Finally, at the time of admission she has the following medical measurements:

Systolic blood pressure	228 millimeters of mercury
Heart rate	94 beats per minute
PO_2 from initial blood gases	49.8 millimeters of mercury
pH from initial blood gases	7.27 units
PCO_2 from initial blood gases	55.3 millimeters of mercury
Bicarbonate from initial blood gases	16.1 millimoles per liter
Creatinine from initial blood gases	2.2 milligrams per deciliter.

Estimate the woman's survival probability.

5. Ensure that the package `rstan` (Guo et al. 2017) and its dependencies are installed in your `R` environment.

   a. Using the package `rstan`, fit a Bayesian version of the GAM fit in part d. of Exercise 3. The script `WarsawAptsBayes.R` from Chap. 2 may be of use.
   b. Obtain MCMC convergence diagnostic plots for key model components and adjust the MCMC sample sizes if warranted.
   c. Obtain Bayes estimates and 95% credible sets for the coefficients of the predictors that enter the model linearly.
   d. For each predictor plot the Bayes estimates of the additive model component with each other predictors set at its average. Include corresponding 95% pointwise credible sets.

6. The `gam()` function in the `mgcv` package can also be used to perform nonparametric density function estimation (e.g. Eilers and Marx 1996). This exercise provides illustration.

   a. Load a univariate sample of flow cytometry data via the commands:

```
> library(HRW) ; data(plankton)
> x <- plankton$redFluorBlueLight
```

   b. Issue the following commands to bin the data into 500 equally spaced bins and then plot the bin counts against the bin centers:

```
> numBins <- 500
> xLow <- 1.05*min(x) - 0.05*max(x)
> xUpp <- 1.05*max(x) - 0.05*min(x)
> binLims <- seq(xLow,xUpp,length = numBins + 1)
> binCenters <- (binLims[-(numBins + 1)]
+                   + 0.5*diff(binLims)[1])
> binCounts <- as.vector(table(cut(x,binLims)))
> binData <- data.frame(binCenters,binCounts)
> plot(binCenters,binCounts,col = "dodgerblue")
```

c. Fit a Poisson penalized spline model to the data plotted in part b. Obtain the estimated mean function over a fine plotting grid. Then obtain a nonparametric density function estimate via trapezoidal rule normalization. The justification for use of the Poisson distribution is explained in Sect. 8 of Eilers and Marx (1996). Use the following R code to achieve this:

```
> library(mgcv)
> fitGAM <- gam(binCounts ~ s(binCenters,k = 47),
+                          family = poisson,data = binData)
> ng <- 1001
> xg <- seq(xLow,xUpp,length = ng)
> phatUnng <- exp(predict(fitGAM,
+                     newdata = data.frame(binCenters = xg)))
> phatg <- 2*phatUnng/(sum(phatUnng[-ng]
+                           + phatUnng[-1])*diff(xg)[1])
> plot(xg,phatg,type = "n",xlim = range(xg),
+       ylim = range(phatg),bty = "l",
+       xlab = "red fluorescence under blue light",
+       ylab = "estimated density function")
> lines(xg,phatg,col = "darkgreen",lwd=2)
> abline(h=0,col = "slateblue")
> rug(x,col="dodgerblue")
```

d. Use the same approach as in parts b. and c. to obtain and plot density estimates for the samples in each of the objects

   i. `plankton$greenFluorBlueLight` and
   ii. `plankton$redFluorRedLight`.

7. Ensure that the packages `kernlab` (Karatzoglou et al. 2016) and `mgcv` (Wood 2017) are installed in your R environment.

   a. Issue the following commands to load the `spam` data frame and to print out its variable names:

   ```
   > library(kernlab) ; data(spam) ; help(spam)
   > print(names(spam))
   ```

   The dataset consists of 58 variables on 4601 e-mail messages. The variable `type` codes whether or not the e-mail message is spam. Classification rules for the type of message based on the other variables are of interest.

   b. Issue the following commands to divide the `spam` data frame into test and training samples:

   ```
   > set.seed(1) ; nTest <- 1000
   > indsTest <- sample(1:nrow(spam),nTest,replace = FALSE)
   > indsTrain <- setdiff(1:nrow(spam),indsTest)
   > spamTest <- spam[indsTest,]
   > spamTrain <- spam[indsTrain,]
   ```

   c. Fit the following logistic GLM, which includes all possible predictors of `type`, to the training data and summarize the fit:

```
> fitTrainFullGLM <- glm(type ~ .,family = binomial,
+                                  data = spamTrain)
> print(summary(fitTrainFullGLM))
```

d. The summary table produced by the code in part c. shows that not all candidate predictors are significant. Use a model selection strategy such as that provided by the function `step.Gam()` in the `gam` package to select a subset of predictors. Fit the logistic GLM based on this subset.
e. Consider the rule for which a new e-mail message is classified as being spam if and only if the message's estimated probability of `type` = `spam`, according to the logistic GLM from part d., exceeds 1/2. Obtain the confusion matrix based on the test data in `spamTest` and estimate the misclassification rate. (The *confusion matrix* is a two-way tabulation of actual and predicted class and is also called the *error matrix*.)
f. Using the function `gam()` in the `mgcv` package, fit a logistic GAM to the training data with the same variables as used in part d., but with each variable entering the model as a penalized spline. Define the classification rule based on the GAM analogously to that defined in part e. for the GLM fit. Obtain the confusion matrix and estimate the misclassification rate.
g. Is one rule significantly better than the other?

8. As in Exercise 7, this exercise is concerned with classification of e-mail messages based on the data frame `spam`. Ensure that the package `gamsel` (Chouldechova et al. 2018) is installed in your R environment.

a. Many of the variables are highly right-skewed and so benefit from the concave transformation such as the log function. The variables also have zero values so a positive number should be added before taking logs. Issue the following commands to transform the predictor variables and to divide the `spam` data frame into test and training samples:

```
> transfSpam <- spam
> transfSpam[,-58] <- log(transfSpam[,-58] + 1)
> set.seed(1) ; nTest <- 1000
> indsTest <- sample(1:nrow(transfSpam),nTest,
                     replace = FALSE)
> indsTrain <- setdiff(1:nrow(transfSpam),indsTest)
> transfSpamTest <- transfSpam[indsTest,]
> transfSpamTrain <- transfSpam[indsTrain,]
```

b. Use a model selection strategy based on the function `cv.gamsel()` in the `gamsel` package to select a subset of predictors.
c. Use the function `glm()` to fit a logistic GLM regression model with response variable `type` and the predictors selected in part b.
d. Consider the rule for which a new e-mail message is classified as being spam if and only if the message's estimated probability of `type` = `spam`, according to the logistic GLM from part c., exceeds 1/2. Obtain the confusion matrix based on the test data in `transfSpamTest` and estimate the misclassification rate.

e. Repeat part d. using the fit from `cv.gamsel()`. Is one rule significantly better than the other?
f. Suppose that e-mail messages classified as spam are likely to be discarded before being read and that you are particularly concerned about failing to read messages that are not spam. In this case, which rule would you prefer? (Find the proportion of nonspam messages that are misclassified by each rule and compare.)
g. The results in parts d. to f. depend on the random split of the dataset into training and test data. Repeat ten times the random splitting of the dataset, the calculation of misclassification rates, and the calculation of the proportion of nonspam messages that are misclassified by each rule. How consistent are the results?

9. Ensure that the packages `Ecdat` (Croissant 2016) and `polspline` (Kooperberg 2015) are installed in your R environment.

a. Consider the problem of selecting a GAM fit to the California schools data analyzed in Sects. 3.3.1 and 3.4 with `mathscr` as the response variable and the following candidate predictors: `calwpct`, `mealpct`, `compstu`,`expnstu`, `str`, `log.avginc`, and `elpct`. Issue the following command to select the model using the `polymars()` function in the `polspline` package:
```
> library(polspline) ; library(Ecdat) ; data(Caschool)
> Caschool$log.avginc <- log(Caschool$avginc)
> y <- Caschool$mathscr
> X <- Caschool[,c(7,8,11,12,13,18)]
> fit <- polymars(y,X,knots = 15,additive = TRUE)
```
The `polymars()` fit is selected from the set of all possible models with an overall intercept and each predictor being modeled according to a linear combination of the predictor and 15 truncated line functions of the predictor. The command:
```
> print(fit$model)
```
leads to the output:
```
  pred1    knot1 pred2 knot2       coefs         SE
1     0       NA     0    NA 671.1796944 7.88319588
2     2       NA     0    NA  -0.4853888 0.02870374
3     6       NA     0    NA   0.8854078 2.68204476
4     6 2.905972     0    NA  23.3997860 4.29266210
```
which means that the selected model is:

$$671.1796944 - 0.4853888\, x_2 + 0.8854078\, x_6 + 23.3997860\, (x_6 - 2.905972)_+$$

where x_2 is the second predictor (`mealpct`) and x_6 is the sixth predictor (`log.avginc`).

b. Issue the following commands to obtain plots of the fitted functions of `mealpct` and `log.avginc` with the other predictor set to its average:

```
> f2hat <- predict(fit,cbind(0,X[,2],0,0,0,mean(X[,6])))
> f6hat <- predict(fit,cbind(0,mean(X[,2]),0,0,0,X[,6]))
> par(mfrow=c(1,2))
> plot(X[,2],f2hat,
+      xlab = "percent qualifying for reduced-price lunch",
+      ylab = "mean average mathematics score",
+      ylim = range(c(f2hat,f6hat)))
> plot(X[,6],f6hat,xlab = "log average income",
+      ylab="mean average mathematics score",
+      ylim = range(c(f2hat,f6hat)))
```

c. The gcv argument in the function polymars() controls the parsimony of the selected model and is set to 4 by default. Obtain and graphically display the models selected by polymars() with

   i. gcv set to 3,
   ii. gcv set to 2,
   iii. gcv set to 1.

d. Repeat the tasks of Exercise 7 for classification of the spam data from the kernlab package, but with polymars() used to select the additive model. Set classify = TRUE in the call to polymars().

10. Recall that, in Sect. 3.3.3, a factor-by-curve extension of a GAM was fit to the Boston mortgages data via the code:

```
> library(mgcv) ; library(HRW) ; data(BostonMortgages)
> fitFacByCurvBostMort <- gam(deny ~ black
                          + factor(self)
+                         + s(dir,by = factor(self))
+                         + s(lvr,by = factor(self)) + pbcr
+                         + self + single + s(ccs,k = 4),
+                         family = binomial,data = BostonMortgages)
```

a. Fit a similar model, but with factor-by-curve interactions for the factor black, which equals yes if the mortgage applicant has black ethnicity and no otherwise.
b. Using predict() function with type = "response" obtain a plot showing, for both black and non-black applicants, the estimated probability of denial as a function of dir (debt payments to income ratio) ranging between 0 and 1, lvr set to its average value, pbrc, self, and single all set to "yes" and ccs set to its average value.
c. Do the same as in part b. but as a function of lvr (loan size to property value ratio) ranging between 0 and 1 and dir set to its average value.
d. Make some interpretational remarks based on the plots in parts b. and c.

11. a. Issue the following commands to load and describe the data frame ragweed:

```
> library(HRW) ; data(ragweed) ; help(ragweed)
```

Of interest are regression models that explain pollenCount, which is a count variable.

b. Fit a Poisson GAM, with a factor-by-curve extension, using the following commands:

```
> fit1 <- gam(pollenCount ~ factor(year)
+        + s(dayInSeason,k = 27,by = factor(year))
+        + temperatureResidual + rain + s(windSpeed,k = 27),
+          data = ragweed,family = poisson)
```

Then issue the following commands to obtain residual plots for the fitted model:

```
> gam.check(fit1)
> qqnorm(residuals(fit1,type = "scaled.pearson"))
> abline(0,1,col = "red")
```

The second residual plot, which uses *scaled Pearson* residuals, rather than the unscaled residuals supported by `gam.check()`, to better assess approximate standard normality.

c. Obtain alternative fit objects `fit2` and `fit3` as follows:

   i. using the same response, `pollenCount`, as for `fit1` but with `family = quasipoisson` corresponding to the quasi-likelihood extension of the Poisson response GAM,
   ii. using the response `sqrt(pollenCount)`, corresponding to the square-root transformation applied to `pollenCount`, and with the specification `family = gaussian` in the call to `gam()`.

d. Obtain residual plots, analogous to those obtained for `fit1` in part b. for each of `fit2` and `fit3`.

e. Choose among the three models according to best agreement between residual plots and model assumptions. Make numerical and graphical summaries of chosen model fit.

f. For the selected model, use the `predict()` function to make a plot showing the four penalized spline fits for the effect of `dayInSeason` on the mean response together on the same set of axes.

Chapter 4
Semiparametric Regression Analysis of Grouped Data

4.1 Introduction

Grouped data arise in several diverse contexts in statistical design and analysis. Examples include medical studies in which patients are followed over time and measurements on them recorded repeatedly, educational studies in which students grouped into classrooms and schools are scored on examinations, and sample surveys in which the respondents to questionnaires are grouped within geographical districts. Major areas of Statistics such as *longitudinal data analysis*, *panel data analysis*, *multilevel models*, and *small area estimation* are concerned with analysis of grouped data.

Books concerned with analyzing grouped data include Baltagi (2013), Diggle et al. (2002), Fitzmaurice et al. (2008), Frees (2004), Gelman and Hill (2007), Goldstein (2010), and Rao and Molina (2015). Longitudinal data analysis is the branch of grouped data analysis that has had the biggest interplay with semiparametric regression. One of the five parts of Fitzmaurice et al. (2008) is titled *Non-Parametric and Semi-Parametric Methods for Longitudinal Data* and contains five chapters on various aspects of semiparametric longitudinal data analysis.

Mixed models are central to the analysis of grouped data, with random effects used to account for within-group dependence. Penalized splines with mixed model representations allow for natural extension from parametric to semiparametric grouped data models. The mixed model functionality of R can be used for fitting quite elaborate semiparametric models to grouped data. Relevant packages are `mgcv` (Wood 2017), which directly supports a wide range of semiparametric regression models for grouped data, and `nlme` (Pinheiro et al. 2017) which allows more elaborate models to be fit using user-specified design matrices. However, not all models of interest can be handled using mixed model packages such as `nlme`, and occasionally we need to call upon the package `rstan` (Guo et al. 2017) for MCMC-based fitting and inference.

J. Harezlak et al., *Semiparametric Regression with R*, Use R!,
https://doi.org/10.1007/978-1-4939-8853-2_4

4.2 Additive Mixed Models

Additive mixed models have emerged as a popular vehicle for semiparametric grouped data analysis. An early reference is Donnelly et al. (1995). Here we follow the approach described in Sect. 9.2 of Ruppert et al. (2003).

The ideas of additive mixed models are nicely illustrated using the *spinal bone mineral density* data. These data are from Bachrach et al. (1999) and involve longitudinal measurements for US-based female and male cohorts with recruitment during 1992–1996. We will restrict attention to the female data, which are stored in the data frame `femSBMD` in the package `HRW`. The data can be visualized using graphics from the package `lattice` (Sarkar 2017) via the commands:

```
> library(lattice) ; library(HRW); data(femSBMD)
> femSBMDvis <- xyplot(spnbmd ~ age|factor(ethnicity),
+                 group = idnum,as.table = TRUE,
+                 data = femSBMD,
+                 strip = strip.custom(par.strip.text
+                                   = list(cex = 1.5)),
+                 par.settings = list(layout.heights
+                                   =list(strip=1.6)),
+                 scales = list(cex = 1.25),
+                 xlab = list("age (years)",cex = 1.5),
+                 ylab = list(expression(paste(
+                 "spinal bone mineral density (g/c",m^2,")")),
+                 cex = 1.5),
+                 panel = function(x,y,subscripts,groups)
+                 {
+                    panel.grid()
+                    panel.superpose(x,y,subscripts,groups,
+                              type = "b",pch = 16,lwd = 2)
+                 })
> plot(femSBMDvis)
```

and produce the plot shown in Fig. 4.1. Here `idnum` is an array having the same length as the response vector `femSBMD$spnbmd`. It contains the subject identification number corresponding to each response measurement. We will not delve deeply into the specifics of the `xyplot()` within `lattice` and refer the reader to the detailed documentation provided by the R commands:

```
> library(lattice)  ;   help(xyplot)
```

and the book Sarkar (2008) dedicated to `lattice` graphics. Briefly, however, we note that

```
spnbmd ~ age|factor(ethnicity)
```

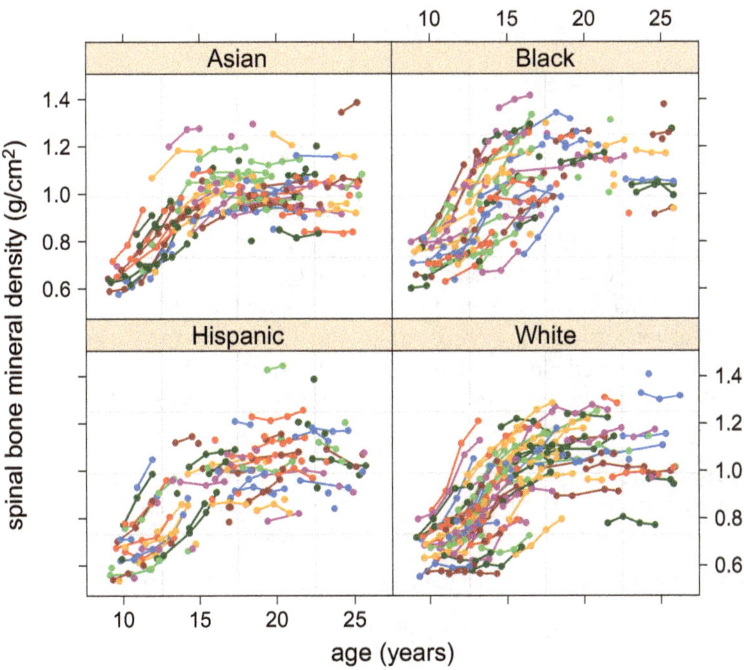

Fig. 4.1 Visualization of the female subset of the spinal bone mineral density data, broken down by ethnicity. Points for the same subject are connected by lines.

specifies plotting spinal bone mineral density values against age with a separate panel for each ethnicity category. The `panel` argument of `xyplot()` permits panel-level specifications. As we will see in this chapter, `lattice` is very useful for displaying grouped data and fits from semiparametric regression analyses. A more recent package, `ggplot2` (Wickham and Chang 2016), has similar functionality.

The number of subjects is $m = 230$. Let n_i, $1 \le i \le m$, denote the number of measurements for the ith subject. One question of interest concerns differences in mean spinal bone mineral density among the four ethnic groups, Asian, Black, Hispanic, and White, after accounting for age. Define the indicator variable:

$$\texttt{black}_i \equiv \begin{cases} 1 & \text{if the } i\text{th subject is black,} \\ 0 & \text{otherwise} \end{cases}$$

and let $\texttt{hispanic}_i$ and $\texttt{white}_i$ be defined analogously. An appropriate *additive mixed model* is

$$\begin{aligned}\texttt{spnbmd}_{ij} &= U_i + f(\texttt{age}_{ij}) + \beta_1\,\texttt{black}_i + \beta_2\,\texttt{hispanic}_i\\ &\quad +\beta_3\,\texttt{white}_i + \varepsilon_{ij}, \quad 1 \le j \le n_i, \quad 1 \le i \le m, \qquad (4.1)\\ U_i &\overset{\text{ind.}}{\sim} N(0, \sigma_U^2),\ \varepsilon_{ij} \overset{\text{ind.}}{\sim} N(0, \sigma_\varepsilon^2).\end{aligned}$$

where, for example, $\texttt{spnbmd}_{ij}$ is jth the spinal bone mineral density measurement on the ith subject. With this formulation the Asian subjects comprise the reference group and β_1, β_2, and β_3 represent mean differences in spinal bone mineral density between the other ethnic groups and Asians. The additivity assumption in (4.1) is somewhat tenuous and a more flexible model that allows for a factor-by-curve interaction between ethnicity and age is considered in Exercise 1.

Model (4.1) can be fitted in R using the function gamm() within the package mgcv (Wood 2017) as follows:

```
> library(mgcv)
> fit <- gamm(spnbmd ~ s(age) + black + hispanic + white,
+             random = list(idnum = ~1),data = femSBMD)
```

The fitted penalized spline for the age effect can be viewed using:

```
> plot(fit$gam,shade = TRUE,shade.col = "palegreen",bty = "l")
```

The resulting plot is shown in Fig. 4.2. The shaded region corresponds to pointwise approximate 95% confidence intervals. Note that default plotting of the estimate of $f(\text{age})$ in mgcv involves vertical centering about zero.

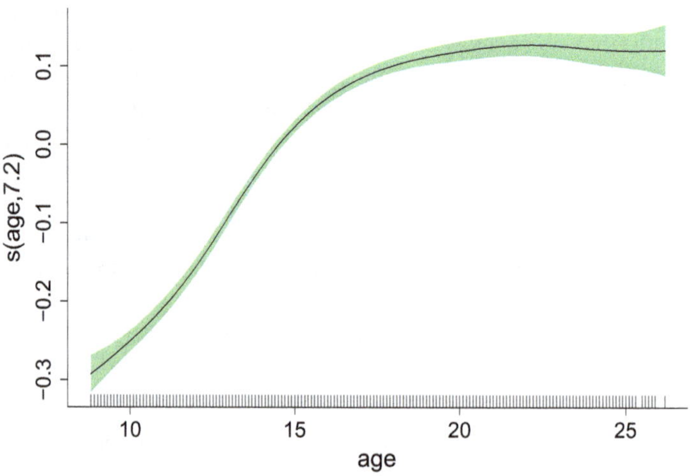

Fig. 4.2 The fitted penalized spline for the age effect in the fit of the additive mixed model (4.1) to the female spinal bone mineral density data. The function estimate is vertically centered about zero. The shaded region corresponds to pointwise approximate 95% confidence intervals.

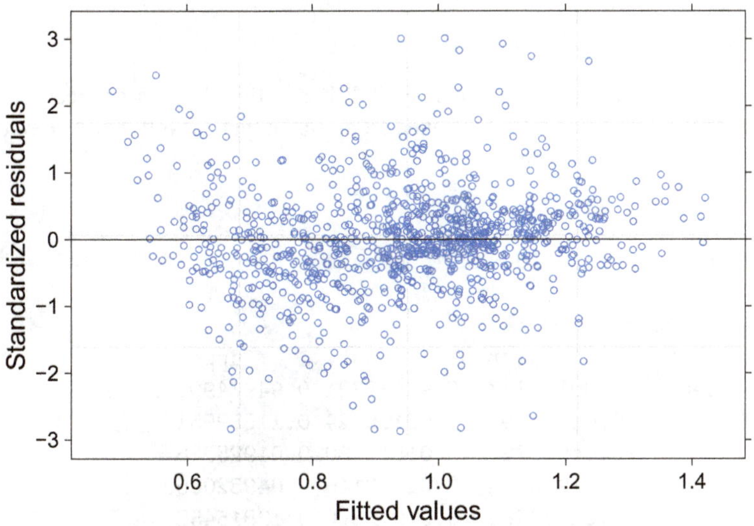

Fig. 4.3 Standardized residuals versus fitted values for the fit of the additive mixed model (4.1) to the female spinal bone mineral density data.

A standardized residual plot is produced from:

```
> plot(fit$lme)
```

The residual plot, shown in Fig. 4.3, shows reasonable accordance with the model assumptions.

A summary of the additive model aspects of the fit is produced using:

```
> summary(fit$gam)

Family: gaussian
Link function: identity

Formula:
spnbmd ~ s(age) + black + hispanic + white

Parametric coefficients:
            Estimate Std. Error t value Pr(>|t|)
(Intercept)  0.92538    0.01243  74.444  < 2e-16
black        0.08191    0.01718   4.769 2.13e-06
hispanic    -0.01516    0.01754  -0.864    0.388
white        0.01503    0.01748   0.860    0.390

Approximate significance of smooth terms:
         edf Ref.df     F p-value
s(age) 7.201  7.201 225.6  <2e-16
```

```
R-sq.(adj) =  0.519
  Scale est. = 0.0013551   n = 1003
```

Note that the fitted age effect involves 7.201 effective degrees of freedom. The effects of other variables can be gleaned from this output as well, but a more direct assessment is available from:

```
> intervals(fit$lme)

Approximate 95% confidence intervals

 Fixed effects:
                      lower         est.      upper
X(Intercept)  0.90101757  0.92538331 0.94974906
Xblack        0.04821097  0.08190524 0.11559951
Xhispanic    -0.04956727 -0.01515680 0.01925366
Xwhite       -0.01925714  0.01503146 0.04932006
Xs(age)Fx1    0.02267288  0.07541370 0.12815452
attr(,"label")
[1] "Fixed effects:"

 Random Effects:
  Level: g
                   lower       est.     upper
sd(Xr - 1) 0.01371071 0.04113408 0.1241885
  Level: idnum
                      lower      est.     upper
sd((Intercept)) 0.1138351 0.1222052 0.1311908

 Within-group standard error:
      lower        est.       upper
0.03474367 0.03681208 0.03900364
```

This output shows that an approximate 95% confidence interval for β_1 in (4.1) is (0.0482,0.116), which indicates a statistically significant difference between the Asian and Black females in terms of mean spinal bone mineral density. However, there is no significant difference between Hispanic or White females and Asian females. An approximate 95% confidence interval for σ_U is $(0.114, 0.131)$, which implies significant within-subject correlation. The confidence interval for σ_ε is $(0.0347, 0.0390)$. The output also contains a confidence interval for the standard deviation of the penalized spline coefficients, which is less interpretable.

We can also perform omnibus tests for multiparameter hypotheses concerning ethnicity. For example, a test for the overall effect of ethnicity, corresponding to

$$H_0 : \beta_1 = \beta_2 = \beta_3 = 0, \tag{4.2}$$

can be performed using the code:

```
> fitReduced <- gamm(spnbmd ~ s(age),random = list(idnum = ~1),
+                   data = femSBMD)
> logLRstat <- 2*(fit$lme$logLik - fitReduced$lme$logLik)
> pValue <- 1 - pchisq(logLRstat,df = 3) ; print(pValue)
```

```
[1] 6.14458e-08
```

The approximate p-value is 0.0000000614, and based on twice the log-likelihood ratio having an approximate χ^2 distribution with three degrees of freedom, since (4.2) involves three parameters (e.g. McCulloch et al. 2008). Note `anova()` function used in Chap. 3 to compare models fit via `gam()` is not applicable to `gamm()` fit objects.

4.2.1 Bayesian Approach

An alternative route for additive mixed model analysis in R involves adopting a Bayesian approach and using MCMC for approximate inference. As discussed in Sect. 2.10, MCMC-based inference is supported in R by the package `rstan` (Guo et al. 2017). In this section we explain how to fit a Bayesian version of (4.1). Unless you are a hardcore Bayesian there is a sense in which this analysis is redundant, since we have just shown that `mgcv` supports the additive mixed model needed here. However, the Bayesian analysis takes into account the uncertainty in the smoothing parameters, which for a non-Bayesian would requiring bootstrapping. In addition, it is useful to see how to fit such models in `rstan` because it is the only option for R implementation of more elaborate semiparametric grouped data models given later in this chapter.

The Bayesian model we consider here is:

$$
\begin{gathered}
\texttt{spnbmd}_{ij}|\boldsymbol{\beta}, U_i, u_1, \ldots, u_K, \sigma_U, \sigma_u, \sigma_\varepsilon \stackrel{\text{ind.}}{\sim} N\Big(\beta_0 + U_i + \beta_{\text{age}}\texttt{age}_{ij} \\
+ \textstyle\sum_{k=1}^{K} u_k\, z_k(\texttt{age}_{ij}) + \beta_1\,\texttt{black}_i + \beta_2\,\texttt{hispanic}_i + \beta_3\,\texttt{white}_i, \sigma_\varepsilon^2\Big), \\
U_i|\sigma_U \stackrel{\text{ind.}}{\sim} N(0, \sigma_U^2), \quad u_k|\sigma_u \stackrel{\text{ind.}}{\sim} N(0, \sigma_u^2), \qquad (4.3) \\
\beta_0, \beta_{\text{age}}, \beta_1, \beta_2, \beta_3 \stackrel{\text{ind.}}{\sim} N(0, \sigma_\beta^2), \quad \sigma_U \sim \text{Half-Cauchy}(A_U), \\
\sigma_u \sim \text{Half-Cauchy}(A_u), \quad \sigma_\varepsilon \sim \text{Half-Cauchy}(A_\varepsilon).
\end{gathered}
$$

where, as in Sect. 2.10, the notation $\sigma \sim \text{Half-Cauchy}(A)$ means that σ has the prior density function $p(\sigma) = 2A/\{\pi(\sigma^2 + A^2)\}$ for $\sigma > 0$. We impose noninformative priors via the hyperparameter settings $\sigma_\beta = A_U = A_u = A_\varepsilon = 10^5$. However, we work with the *standardized* spnbmd and age values in our Bayesian fitting and revert to the original units at the very end of the analysis. This guarantees that the hyperparameter settings are scale invariant. The assumption that $u_k|\sigma_u \overset{\text{ind.}}{\sim} N(0, \sigma_u^2)$ is appropriate if, for example, we use O'Sullivan splines. Such a spline basis is used here.

The R script femSBMDbayes.R facilitates fitting of (4.3) via MCMC and rstan. The Stan code is stored in the object addMixModModel which is assigned inside the script femSBMDbayes.R as shown below. It is a computer code representation of model (4.3).

```
> addMixModModel <-
+ 'data
+ {
+     int<lower=1> numObs;                int<lower=1> numGrp;
+     int<lower=1> ncX;                   int<lower=1> ncZ;
+     real<lower=0> sigmaBeta;            real<lower=0> AU;
+     real<lower=0> Aeps;                 real<lower=0> Au;
+     vector[numObs] y;                   int<lower=1> idnum[numObs];
+     matrix[numObs,ncX] X;               matrix[numObs,ncZ] Zspl;
+ }
+ parameters
+ {
+     vector[ncX] beta;                   vector[numGrp] U;
+     vector[ncZ] u;                      real<lower=0> sigmaU;
+     real<lower=0> sigmau;               real<lower=0> sigmaEps;
+ }
+ model
+ {
+     y ~ normal(X*beta + U[idnum] + Zspl*u,sigmaEps);
+     U ~ normal(0,sigmaU);                  u ~ normal(0,sigmau);
+     beta  ~ normal(0,sigmaBeta) ;          sigmaEps ~ cauchy(0,Aeps);
+     sigmaU ~ cauchy(0,AU) ;                sigmau ~ cauchy(0,Au);
+ }'
```

To run this script issue:

```
> library(HRW) ; demo(femSBMDbayes,package = "HRW")
```

Its location for possible copying and modifying is determined by:

```
> system.file("demo","femSBMDbayes.R",package = "HRW")
```

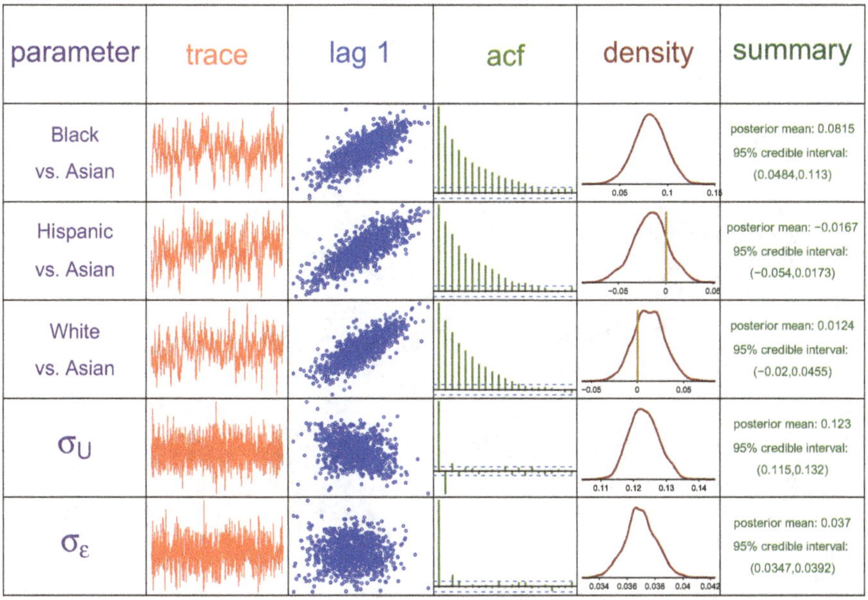

Fig. 4.4 Summary of MCMC-based inference for parameters in the fitted Bayesian model for the spinal bone mineral density data. The columns are: parameter, trace plot of MCMC sample, plot of sample against 1-lagged sample, sample autocorrelation function, kernel estimates of posterior density, and basic numerical summaries.

Figure 4.4 summarizes the MCMC output for the model parameters. The difference between the Black and Asian subjects is apparent from the estimated posterior densities and 95% credible sets. Some of the MCMC samples are "sticky" in that they have high autocorrelation. Nevertheless, the trace plots are suggestive of convergence to a reasonable degree.

Figure 4.5 shows the fitted curves and pointwise 95% credible sets for the age effect.

4.2.2 *Serial Correlation Extension*

Additive mixed models such as (4.1) assume that the errors are independent, as conveyed by the

$$\varepsilon_{ij} \overset{\text{ind.}}{\sim} N(0, \sigma_\varepsilon^2).$$

Often such an assumption is not reasonable and it is more appropriate to assume that the errors are *serially correlated.* One of the simplest types of serial correlation is order 1 autoregression, usually denoted by the abbreviation AR(1):

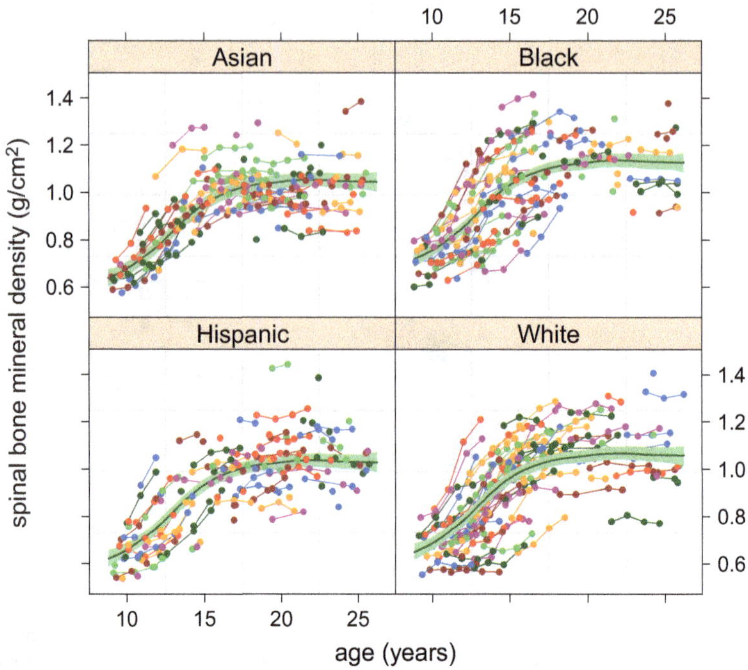

Fig. 4.5 The female subset of the spinal bone mineral density data with the Bayes estimates of the mean function for each ethnicity group overlaid. The Bayes estimates are based on MCMC via the `rstan` package. The shaded regions correspond to pointwise 95% credible sets.

$$\varepsilon_{ij} = \rho \varepsilon_{i,j-1} + \xi_{ij}$$

where $|\rho| < 1$ and the ξ_{ij} are independent. Illustration of the serial correlation extension is provided by Exercise 9.

4.3 Models with Group-Specific Curves

Figure 4.6 shows data on adolescent somatic growth obtained from a study of the mechanisms of human hypertension development conducted at the Indiana University School of Medicine, Indianapolis, Indiana, USA Pratt et al. (1989) contains a full description of the study. In Fig. 4.6 we restrict attention to the black males in the study. The full dataset, which is stored in the data frame `growthIndiana` in the `HRW` package, includes adolescents of both genders and categorized according to being black or white. The R code that produced Fig. 4.6 is:

```
> library(lattice)   ;   library(HRW)
> data(growthIndiana)
```

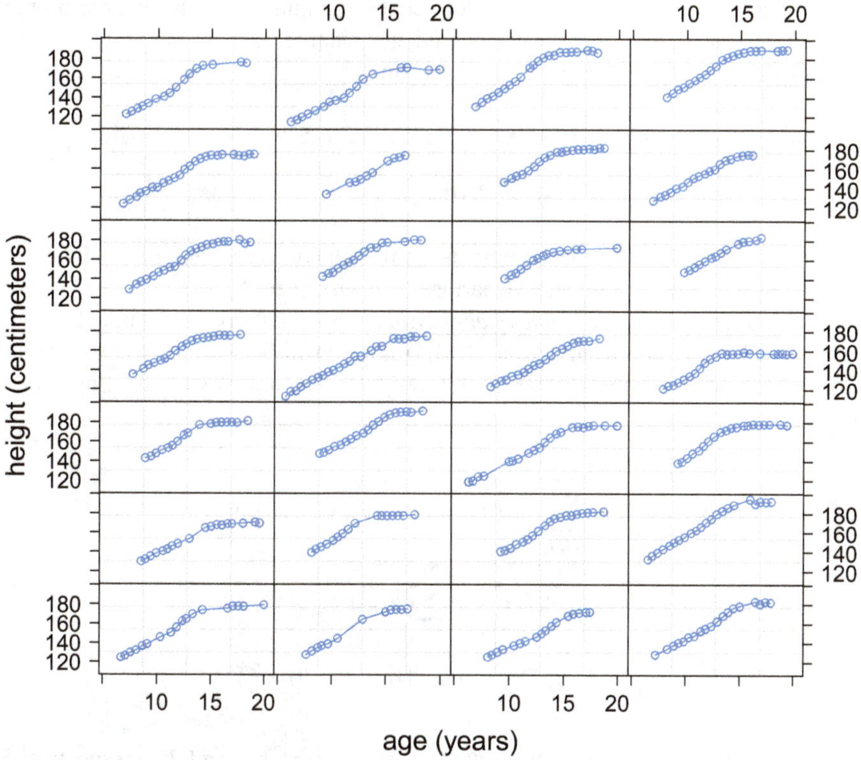

Fig. 4.6 Indiana adolescent growth data for black male adolescents subset. The panels show height versus age for each of the 28 subjects.

```
> growthINblackMales <- growthIndiana[(growthIndiana$male == 1)
+                                     &(growthIndiana$black == 1),]
> figBlkMalRaw <- xyplot(height ~ age|idnum,groups=idnum,
+                        data = growthINblackMales,
+                        layout = c(4,7),
+                        strip = FALSE,scales = list(cex = 1.25),
+                        xlab = list("age (years)",cex = 1.5),
+                        ylab = list("height (centimeters)",
+                        cex = 1.5),as.table = TRUE,
+                        panel = function(x,y,subscripts,groups)
+                        {
+                           panel.grid()
+                           panel.superpose(x,y,subscripts,groups,
+                                  col = "dodgerblue",type = "b")
+                        })
> plot(figBlkMalRaw)
```

The shapes of the curves for each adolescent differ quite markedly and the simple additive mixed models of Sect. 4.2 would not capture such behavior very well. Instead we should consider models of the form

$$\begin{gathered}\texttt{height}_{ij} = f(\texttt{age}_{ij}) + g_i(\texttt{age}_{ij}) + \varepsilon_{ij}, \quad 1 \le j \le n_i, \quad 1 \le i \le 28, \\ \varepsilon_{ij} \overset{\text{ind.}}{\sim} N(0, \sigma_\varepsilon^2)\end{gathered} \tag{4.4}$$

where n_i is the number of measurements for the ith adolescent and g_i is a function that represents that adolescent's departure from the overall mean function f. We call (4.4) a *group-specific curves semiparametric mixed model*. A mixed-model based penalized spline formulation of (4.4) is given in Sect. 9.3 of Ruppert et al. (2003) and embellished in Durbán et al. (2005). Briefly, it involves modeling f and the g_i according to

$$\begin{gathered}f(x) = \beta_0 + \beta_1 x + \sum_{k=1}^{K_{\text{gbl}}} u_{\text{gbl},k} z_{\text{gbl},k}(x), \quad u_{\text{gbl},k} \,|\, \sigma_{\text{gbl}} \overset{\text{ind.}}{\sim} N(0, \sigma_{\text{gbl}}^2), \\ g_i(x) = U_{0i} + U_{1i}\, x + \sum_{k=1}^{K_{\text{grp}}} u_{\text{grp},ik}\, z_{\text{grp},k}(x), \\ \begin{bmatrix} U_{0i} \\ U_{1i} \end{bmatrix} \overset{\text{ind.}}{\sim} N(\mathbf{0}, \mathbf{\Sigma}), \quad u_{\text{grp},ik} \,|\, \sigma_{\text{grp}} \overset{\text{ind.}}{\sim} N(0, \sigma_{\text{grp}}^2)\end{gathered} \tag{4.5}$$

where $z_{\text{gbl},k}$ and $z_{\text{grp},k}$ are suitable spline bases of sizes K_{gbl} and K_{grp} respectively. In the examples we use canonical O'Sullivan splines as described in Sect. 2.2. Typically, K_{grp} is smaller than K_{gbl} since fewer basis functions are needed to handle group-specific deviations.

Model (4.5) can be fit in R using the function `lme()` from the package `nlme`. However, before the call to `lme()`, a fair amount of setting up is required. First, we need to extract the main variables from the `growthINblackMales` data frame and then create an array with identification numbers matching the i subscript of (4.5):

```
> library(nlme)
> age <- growthINblackMales$age
> height <- growthINblackMales$height
> idnum <-  growthINblackMales$idnum
> idnumBM <- rep(NA,length(idnum))
> uqID <- unique(idnum)
> for (i in 1:length(uqID))
+         idnumBM[idnum == uqID[i]] <- i
> growthINblackMales$idnum <- idnumBM
```

Next, set up design matrices $\mathbf{Z}_{\text{gbl}}$, containing the $z_{\text{gbl},k}$, and $\mathbf{Z}_{\text{grp}}$, containing the $z_{\text{grp},k}$:

```
> numObs <- length(height)
> numGrp <- length(unique(idnum))
> numIntKnotsGbl <- 20
> intKnotsGbl <- quantile(unique(age),
+    seq(0,1,length=numIntKnotsGbl+2))[-c(1,numIntKnotsGbl+2)]
> range.age <- c(5.5,20)
> Zgbl <- ZOSull(age,range.x=range.age,intKnots=intKnotsGbl)
> numIntKnotsGrp <- 10
> intKnotsGrp <- quantile(unique(age),
+  seq(0,1,length=numIntKnotsGrp+2))[-c(1,numIntKnotsGrp+2)]
> Zgrp <- ZOSull(age,range.x=range.age,intKnots=intKnotsGrp)
```

The next chunk of code sets up the random effect structure for the call to lme(), corresponding to (4.5):

```
> dummyId <- factor(rep(1,numObs))
> Zblock <- list(dummyId = pdIdent( ~ -1 + Zgbl),
+                idnumBM = pdSymm( ~ age),
+                idnumBM = pdIdent( ~ -1 + Zgrp))
```

Note that we need the dummy identification variable dummyID, an array of length numObs, the total number of observations, with all entries equal to one to trick lme() into accommodating the global penalized spline component. The list entry dummyId = pdIdent(~-1+Zgbl) invokes the multiple of identity matrix structure $\boldsymbol{u}_{\text{gbl}} \sim N(\mathbf{0}, \sigma^2_{\text{gbl}} \boldsymbol{I})$ across the entire dataset regardless of within-subject grouping. The list item idnumBM = pdSymm(~age) invokes the block-diagonal unstructured 2×2 covariance matrix form on the $[U_{0i}\ U_{1i}]^T$, $1 \le i \le 28$, as required by (4.5). Similarly pdIdent(~-1+Zgrp) accommodates $u_{\text{grp},ik} \,|\, \sigma_{\text{grp}} \overset{\text{ind.}}{\sim} N(0, \sigma^2_{\text{grp}})$.

We are now ready to call lme() with the random argument set to Zblock:

```
> blkMalGD <- groupedData(height ~ age|rep(1,length = numObs),
+                data = data.frame(height,age,Zgbl,Zgrp,idnumBM))
> fit <- lme(height ~ age,data = blkMalGD,random = Zblock)
```

We are not aware of any R packages that handle group-specific curves directly. This is the reason for using lme() with explicit construction of spline basis functions.

The code for the lattice graphics display of the fits, shown in Fig. 4.7, is:

```
> ng <- 101
> ageg <- seq(range.age[1],range.age[2],length = ng)
> Xg <- cbind(rep(1,ng),ageg)
> Zgblg <- ZOSull(ageg,range.x = range.age,
+                 intKnots = intKnotsGbl)
> Zgrpg <- ZOSull(ageg,range.x = range.age,
```

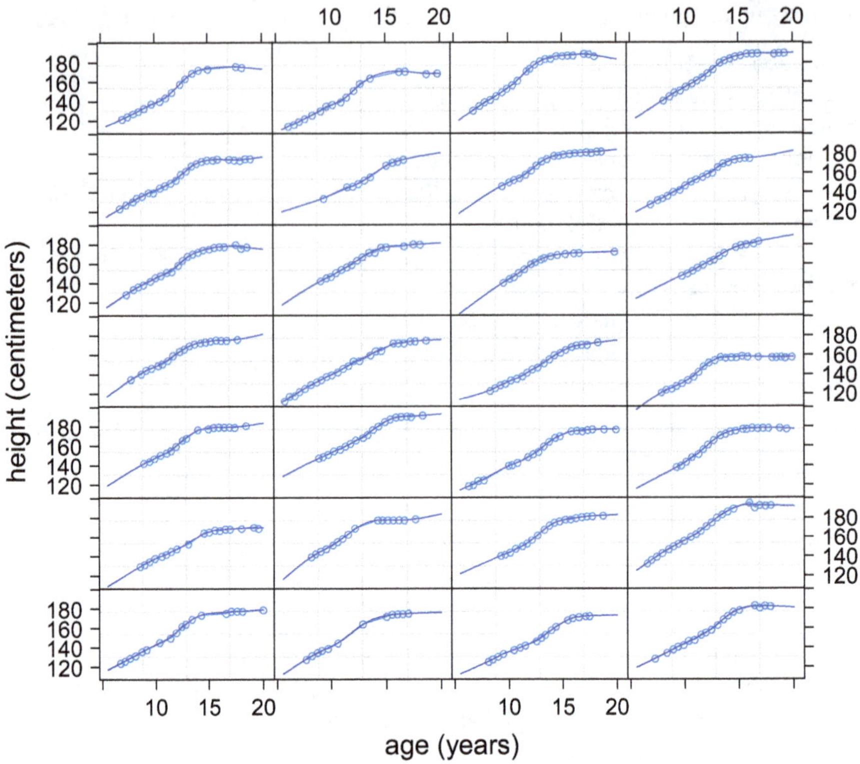

Fig. 4.7 Fitted group-specific curves for the data shown in Fig. 4.6 and model (4.5). The fits are obtained via `lme()` as described in the text.

```
+                    intKnots = intKnotsGrp)
> betaHat <- as.vector(fit$coef$fixed)
> uHat <- as.vector(fit$coef$random[[1]])
> fHatg <- as.vector(Xg%*%betaHat + Zgblg%*%uHat)
> curvEsts <- vector("list",numGrp)
> for (i in 1:numGrp)
+ {
+    uLinHati <- as.vector(fit$coef$random[[2]][i,])
+    uSplHati <- as.vector(fit$coef$random[[3]][i,])
+    ghati <- Xg%*%uLinHati + Zgrpg%*%uSplHati
+    curvEsts[[i]] <- fHatg + ghati
+ }
> figBlkMalFit <- xyplot(height ~ age|idnumBM,
                     groups = idnumBM,
+                    data = growthINblackMales,
+                    strip = FALSE,scales = list(cex = 1.25),
+                    xlab = list("age (years)",cex = 1.5),
```

```
+                   ylab = list("height (centimeters)",cex = 1.5),
+                   as.table = TRUE,layout = c(4,7),
+                   panel = function(x,y,subscripts,groups)
+                   {
+                      panel.grid()
+                      adolNum <- idnumBM[subscripts][1]
+                      panel.superpose(x,y,subscripts,groups,
+                                col = "dodgerblue",type = "b")
+                      panel.xyplot(ageg,curvEsts[[adolNum]],
+                                   col = "blue",type = "l")
+                   })
> plot(figBlkMalFit)
```

Finally, we can check the standardized residuals from the fit shown in Fig. 4.7 simply by issuing the command:

```
> print(plot(fit))
```

Result is shown in Fig. 4.8. It indicates no distinct patterns or outliers.

Figure 4.9 is similar to Fig. 4.6, but now has the growth data for 88 white male adolescents in addition to those for the black males.

Of primary interest is the *contrast function*:

$$c(\text{age}) \equiv f_B(\text{age}) - f_W(\text{age})$$

where f_B is the global mean function for black adolescents and f_W is its counterpart for white adolescents. We now consider the group-specific curves extension of the factor-by-curve interaction models discussed in Sect. 3.6. Coull et al. (2001)

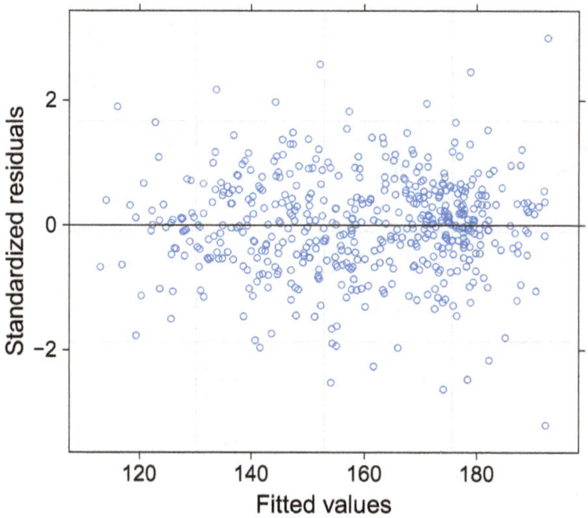

Fig. 4.8 Standardized residuals for the `lme()`-based group-specific curves fit shown in Fig. 4.7.

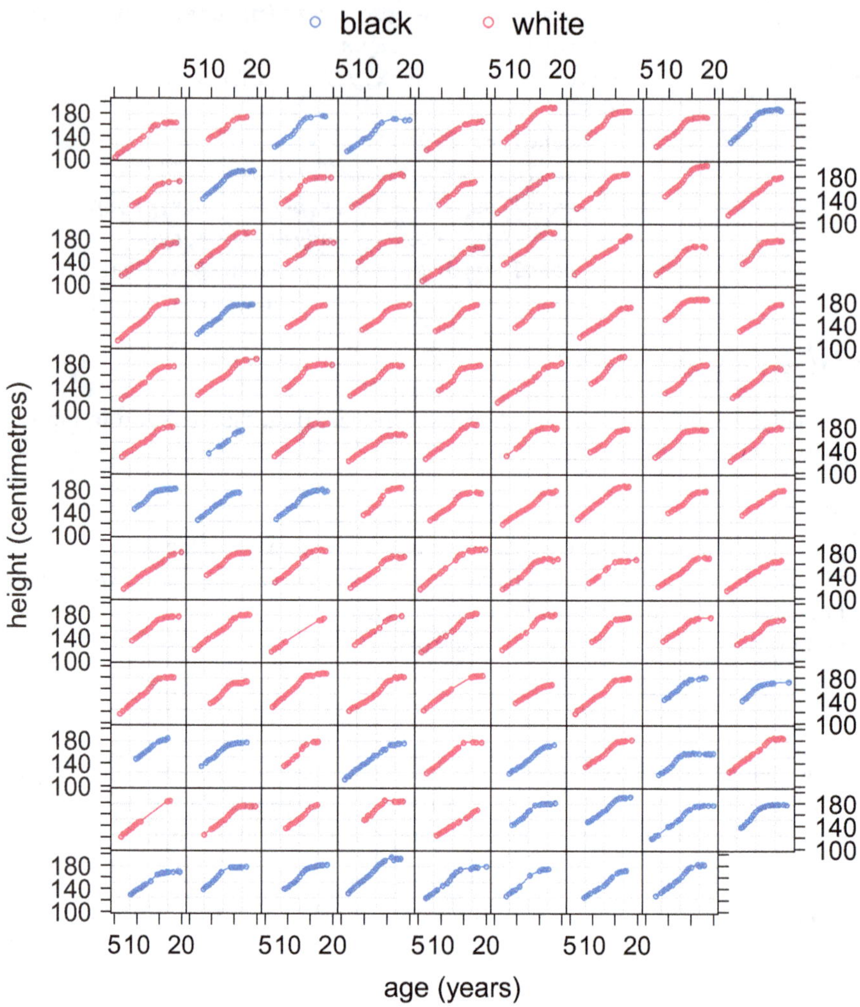

Fig. 4.9 Indiana adolescent growth data for the male adolescents subset, categorized according to black and white ethnicity. The panels show height versus age for each of the 116 subjects.

advocate symmetry in the random effects-based spline models for f_B and f_W as follows:

$$f_W(x) = \beta_0^W + \beta_1^W x + \sum_{k=1}^{K_{\text{gbl}}} u_{\text{gbl},k}^W z_{\text{gbl},k}(x) \text{ and}$$

$$f_B(x) = \beta_0^W + \beta_0^{\text{BvsW}} + (\beta_1^W + \beta_1^{\text{BvsW}})\, x + \sum_{k=1}^{K_{\text{gbl}}} u_{\text{gbl},k}^B z_{\text{gbl},k}(x),$$

$$\text{where} \quad u^W_{\text{gbl},k} \,|\, \sigma^W_{\text{gbl}} \overset{\text{ind.}}{\sim} N(0, (\sigma^W_{\text{gbl}})^2), \qquad u^B_{\text{gbl},k} \,|\, \sigma^B_{\text{gbl}} \overset{\text{ind.}}{\sim} N(0, (\sigma^B_{\text{gbl}})^2),$$

leading to the contrast function

$$c(x) \equiv f_B(x) - f_W(x) = \beta_0^{\text{BvsW}} + \beta_1^{\text{BvsW}}\, x + \sum_{k=1}^{K_{\text{gbl}}} (u^B_{\text{gbl},k} - u^W_{\text{gbl},k}) z_{\text{gbl},k}(x). \tag{4.6}$$

An alternative model, that is *asymmetric* in f_B and f_W, differs from (4.6) in that

$$f_B(x) = \beta_0^W + \beta_0^{\text{BvsW}} + (\beta_1^W + \beta_1^{\text{BvsW}})\, x + \sum_{k=1}^{K_{\text{gbl}}} (u^B_{\text{gbl},k} + u^{\text{BvsW}}_{\text{gbl},k}) z_{\text{gbl},k}(x),$$

$$\text{where} \quad u^W_{\text{gbl},k} \,|\, \sigma^W_{\text{gbl}} \overset{\text{ind.}}{\sim} N(0, (\sigma^W_{\text{gbl}})^2), \quad u^{\text{BvsW}}_{\text{gbl},k} \,|\, \sigma^{\text{BvsW}}_{\text{gbl}} \overset{\text{ind.}}{\sim} N(0, (\sigma^{\text{BvsW}}_{\text{gbl}})^2),$$

and induces the contrast function model

$$c(x) \equiv f_B(x) - f_W(x) = \beta_0^{\text{BvsW}} + \beta_1^{\text{BvsW}}\, x + \sum_{k=1}^{K_{\text{gbl}}} u^{\text{BvsW}}_{\text{gbl},k} z_{\text{gbl},k}(x). \tag{4.7}$$

In model (4.6) the effective degrees of freedom for the estimates f_W and f_B are an increasing function of the estimates of

$$(\sigma^W_{\text{gbl}})^2/\sigma_\varepsilon^2 \quad \text{and} \quad (\sigma^B_{\text{gbl}})^2/\sigma_\varepsilon^2.$$

However, in model (4.7) they are an increasing function of the estimates of

$$(\sigma^W_{\text{gbl}})^2/\sigma_\varepsilon^2 \quad \text{and} \quad \{(\sigma^W_{\text{gbl}})^2 + (\sigma^{\text{BvsW}}_{\text{gbl}})^2\}/\sigma_\varepsilon^2.$$

This implies that the estimate of f_W must have fewer effective degrees of freedom than that of f_B regardless of their relative curviness. Therefore, model (4.6) is preferable since it does not impose such a ranking on the effective degrees of freedom values although in practice this will often not matter a great deal. Model (4.7) has the advantage that fitting and standard error estimation is easier via `lme()` and the script `maleGrowthIndiananlme.R` in the `HRW` package contains the required code. To run it type:

```
> library(HRW) ; demo(maleGrowthIndiananlme,package = "HRW")
```

Its location is determined by the command:

```
> system.file("demo","maleGrowthIndiananlme.R",package = "HRW")
```

The resulting fits are shown in Fig. 4.10 and are seen to be very good.

Figure 4.11 shows the estimated contrast function produced by the R script `maleGrowthIndiananlme.R` and corresponding approximate pointwise 95% confidence intervals. It should be noted that the `lme()` fit object does not provide the

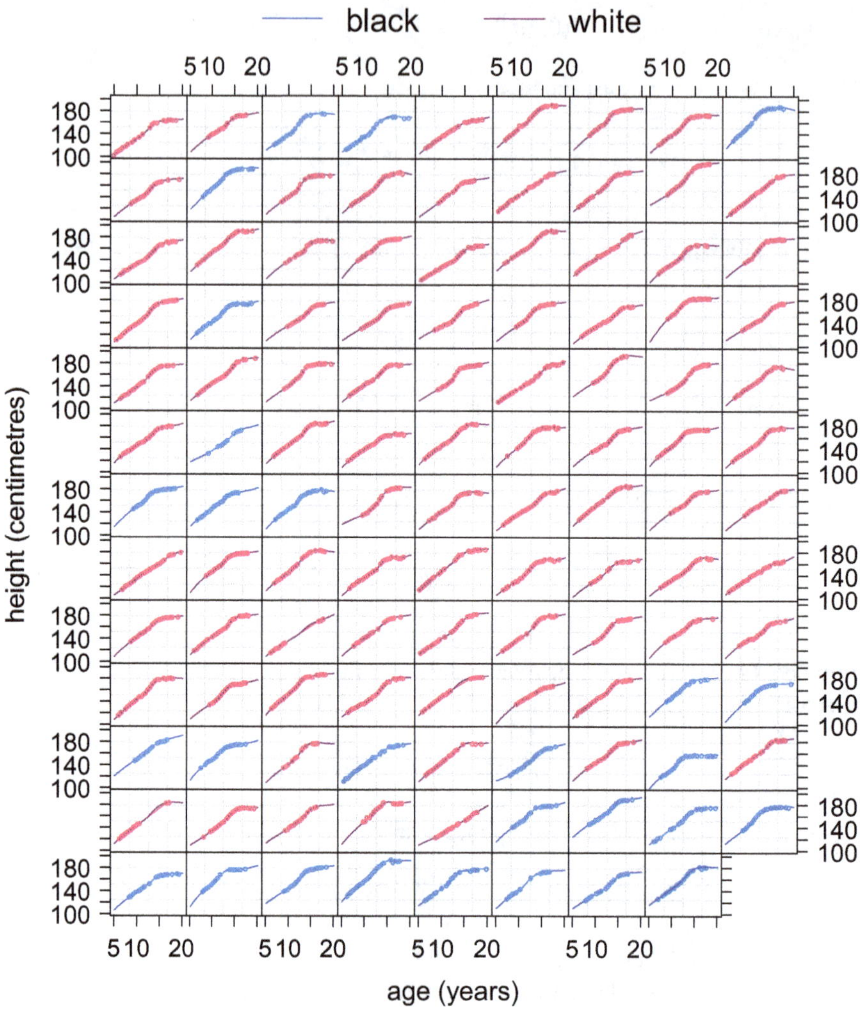

Fig. 4.10 Fitted group-specific curves for the data shown in Fig. 4.9 and model (4.7). The fits are obtained via `lme()` within the script `maleGrowthIndiananlme.R`.

covariance matrix of the estimated random effects and, hence, the approximate confidence interval computations have to be programmed in R using theory laid out in Chap. 4 of (Ruppert et al. 2003). The required code is given in `maleGrowthIndiananlme.R`.

An alternative approach that (a) allows the theoretically more appealing symmetric group-specific curves model (4.6) to be fitted and (b) involves considerably less R programming is to adopt a Bayesian approach and use an MCMC-based package such as `rstan`, albeit at the cost of increased computing time. The script `maleGrowthIndianaBayes.R` in the `HRW` achieves this and produces the fitted

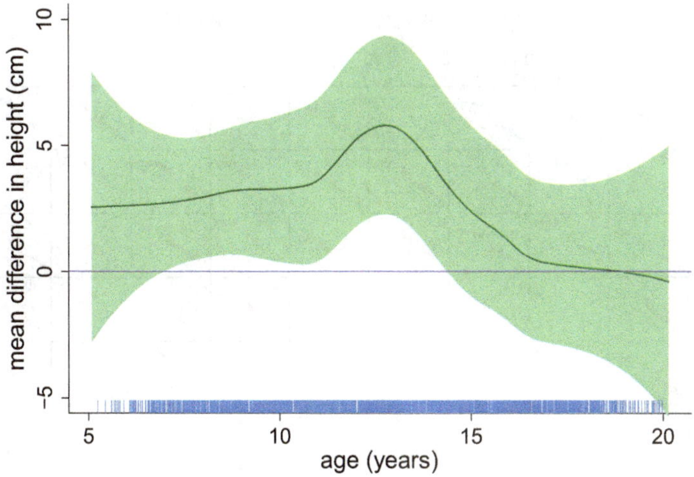

Fig. 4.11 Estimated contrast function and approximate pointwise 95% confidence intervals for the mean height of black adolescents minus mean height of white adolescents for the Indiana adolescent growth data. The estimate is obtained using `lme()` as described in the text.

curves shown in Fig. 4.12 and the estimated contrast function shown in Fig. 4.13. In each of these figures pointwise 95% credible sets are added to the curve estimates. Unlike the `lme()` approach, obtaining such variability bands in `rstan` is quite easy. The script can be run using:

```
> library(HRW) ; demo(maleGrowthIndianaBayes,package = "HRW")
```

and located on the computer on which `HRW` resides using:

```
> system.file("demo","maleGrowthIndianaBayes.R",package = "HRW")
```

Visual inspection of Figs. 4.10 and 4.13 shows that inference for the contrast function based on model (4.7) and `lme()` and that based on model (4.6) and `rstan` produce similar results. A direct comparison is left as an exercise (Exercise 3). Both inferences show that black adolescents have a significantly higher mean height between 7 and 14 years of age with a peak difference at about 13 years of age. This is evidence of black males experiencing an earlier pubertal spurt compared with white males. The difference then decreases and becomes nonsignificant for 14 years old and older.

We finish up with a check of the MCMC output in Fig. 4.14. It shows that good MCMC convergence is achieved for the error standard deviation and three vertical slices of the contrast function fit.

With simplicity in mind, we have focused on the male subset of the Indiana adolescent growth data throughout this section. Exercise 8 is concerned with analysis of the full dataset.

The group-specific curves analyses performed in this section are based on the low-rank penalized spline mixed model approach developed in Durbán et al. (2005).

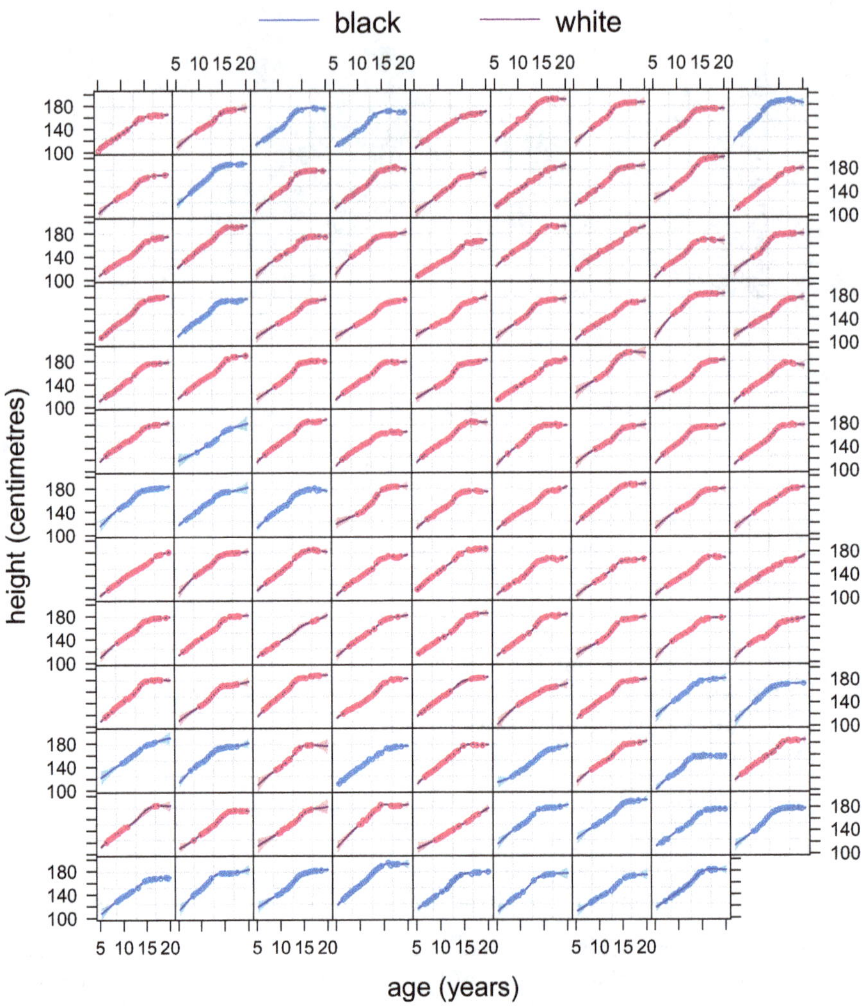

Fig. 4.12 Fitted group-specific curves for the data shown in Fig. 4.9 and model (4.6). The shading corresponds to pointwise 95% credible sets. The fits are obtained via `rstan` within the script `maleGrowthIndianaBayes.R`.

There are some earlier articles that develop similar models such as Donnelly et al. (1995), Brumback and Rice (1998), Zhang et al. (1998), Wang (1998), Verbyla et al. (1999), and Guo (2002). The approach in Durbán et al. (2005), which includes two authors of this book, was developed to allow direct implementation of group-specific curves in R as well as the mixed model capabilities of the SAS environment. In this section we have demonstrated that the same applies to the MCMC functionality of R via the `rstan` package. In Sect. 6.9.1 we show how `rstan` can handle group-specific curves models in the more challenging binary response situation.

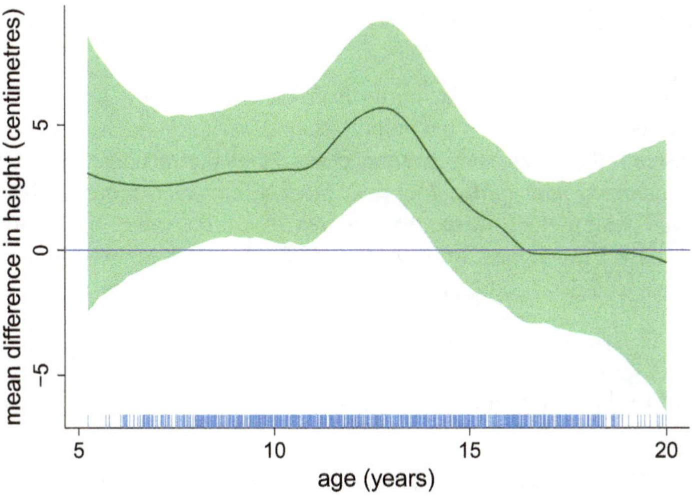

Fig. 4.13 Estimated contrast function and approximate pointwise 95% credible sets for the mean height of black adolescents minus mean height of white adolescents for the Indiana adolescent growth data. The inference is based on MCMC via the `rstan` package.

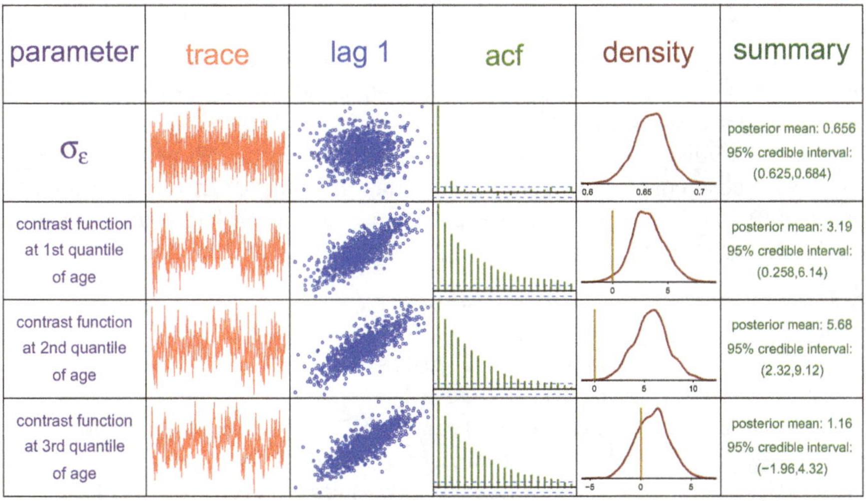

Fig. 4.14 Summary of MCMC-based inference for parameters in the fitted Bayesian group-specific curves model (4.6) for the male subset of the Indiana adolescent growth data. The columns are: parameter, trace plot of MCMC sample, plot of sample against 1-lagged sample, sample autocorrelation function, kernel estimates of posterior density, and basic numerical summaries.

4.4 Marginal Models

Marginal models for grouped data differ from the models in the preceding sections in that there is no modeling of the within-group covariance structure using random effects. Instead, the mean and covariance matrix of the response vector for each group are modeled *marginally*. Marginal models are much more straightforward for *balanced* designs, i.e., where the group sizes are the same. In the notation of Sects. 4.2 and 4.3 this means we have $n_i = n$ for $1 \le i \le m$. We will assume the design is balanced throughout this section.

4.4.1 Marginal Nonparametric Regression

Suppose we observe the two-level grouped regression dataset (x_{ij}, y_{ij}), $1 \le i \le m$, $1 \le j \le n$ where n is small compared with m. Let $\boldsymbol{y}_i$ be the vector of responses for the ith group.

The *marginal nonparametric regression* model is

$$y_{ij} = f(x_{ij}) + \varepsilon_{ij}, \quad \text{Cov}(\boldsymbol{\varepsilon}_i) = \boldsymbol{\Sigma}, \quad 1 \le i \le m, \ 1 \le j \le n \tag{4.8}$$

where $\boldsymbol{\varepsilon}_i \equiv [\varepsilon_{i1}, \ldots, \varepsilon_{in}]^T$ is the vector of residuals for the ith group. The mean function f is assumed to be smooth, while the covariance matrix $\boldsymbol{\Sigma}$ is an unstructured $n \times n$ covariance matrix.

Figure 4.15 displays longitudinal data from the Observing Protein and Energy Nutrition study (conducted in the USA during 1999–2000) that is amenable to model (4.8). The data are in the dataset `protein` in the `HRW` package and their source is Kipnis et al. (2003). The code needed to produce Fig. 4.15 is:

```
> library(HRW)   ;  library(lattice)
> data(protein)
> femInds <- (1:nrow(protein))[protein$female == 1]
> femProtein <- protein[femInds,]
> rawFemProtein <- xyplot(proteinBioM ~ BMI,groups = idnum,
+                    data = femProtein,type = "b",
+                    xlab = list("body mass index", cex = 1.5),
+                    ylab = list("log(protein biomarker)",
+                           cex = 1.5),
+                    scales = list(x = list(cex = 1.5),
+                                  y = list(cex = 1.5)))
> plot(rawFemProtein)
```

To illustrate (4.8) we use body mass index (x) as a predictor of a logarithmically transformed protein biomarker (y) for the females in the study. The sample sizes are $m = 130$ and $n = 2$ with $j = 1, 2$ corresponding to visit number. Also, note that

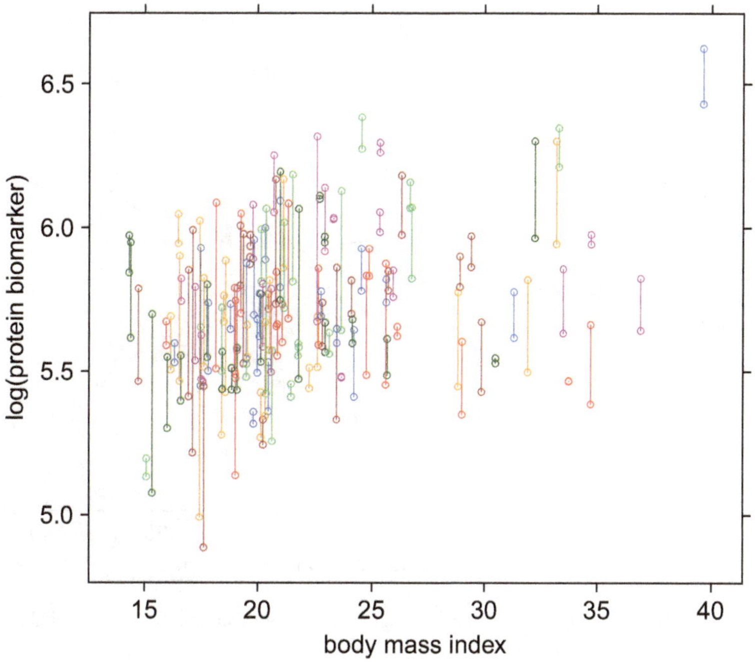

Fig. 4.15 Protein biomarker versus body mass index for females in data from a nutritional epidemiology study (source: Kipnis et al. 2003). The vertical lines join measurements on the same subject.

the body mass index measurement does not vary between visits so, in the notation of (4.8), $x_{i1} = x_{i2}$.

In Fig. 4.16 we have assessed the trend in the data by fitting a five degrees of freedom smoothing spline. The code to produce this figure is:

```
> BMI <- femProtein$BMI
> proteinBioM <- femProtein$proteinBioM
> fitSS <- smooth.spline(BMI,proteinBioM,df = 5)
> BMIg <- seq(min(BMI),max(BMI),length = 101)
> fHatg <- predict(fitSS,BMIg)$y
> smooBMI <- xyplot(proteinBioM ~ BMI,groups = idnum,
+             data = femProtein,type = "b",
+             subscripts = TRUE,
+             panel=function(x,y,subscripts,groups)
+             {
+                panel.superpose(x,y,subscripts,groups,type="b")
+                panel.xyplot(BMIg,fHatg,lwd = 3,
+                             type = "l",col = "darkgreen")
+             }
```

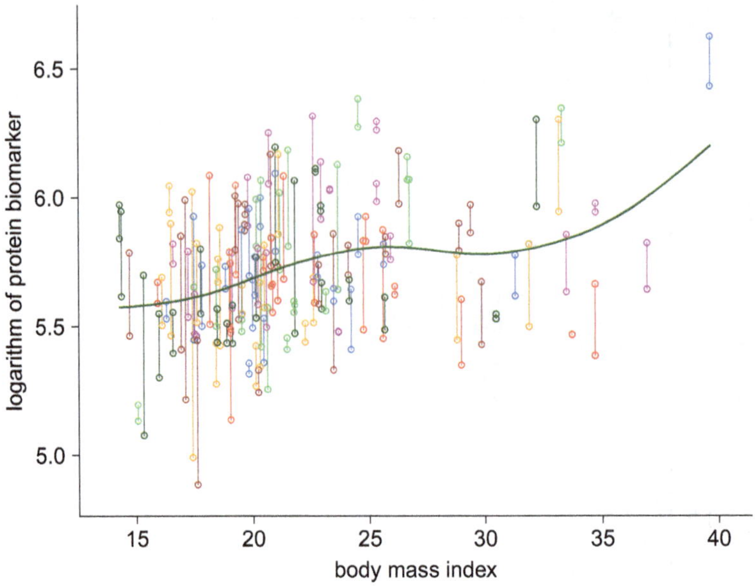

Fig. 4.16 The data from Fig. 4.15 with a five degrees of freedom smoothing spline added to assess trend.

```
+               )
> plot(smooBMI)
```

A nonlinear trend is apparent in Fig. 4.16.

The residuals can then be extracted using:

```
> residsSS <- residuals(fitSS)
```

Figure 4.17 displays the residuals for each subject as follows:

$$\widehat{\varepsilon}_{i2} \quad \text{versus} \quad \widehat{\varepsilon}_{i1}, \qquad 1 \le i \le 130.$$

The scatterplot in Fig. 4.17 allows for an appreciation of $\boldsymbol{\Sigma} = \text{Cov}(\boldsymbol{\varepsilon}_i)$. In particular, it shows a substantial positive correlation between measurements on the same subject. Therefore, model (4.8) is worth entertaining for these data.

4.4.1.1 Comparison with Random Intercept Model

In Sect. 4.2 we considered the random intercept model.

$$y_{ij} = U_i + f(x_{ij}) + \varepsilon_{ij} \tag{4.9}$$

$$U_i \stackrel{\text{ind.}}{\sim} N(0, \sigma_U^2) \quad \text{independent of} \quad \varepsilon_{ij} \stackrel{\text{ind.}}{\sim} N(0, \sigma_\varepsilon^2).$$

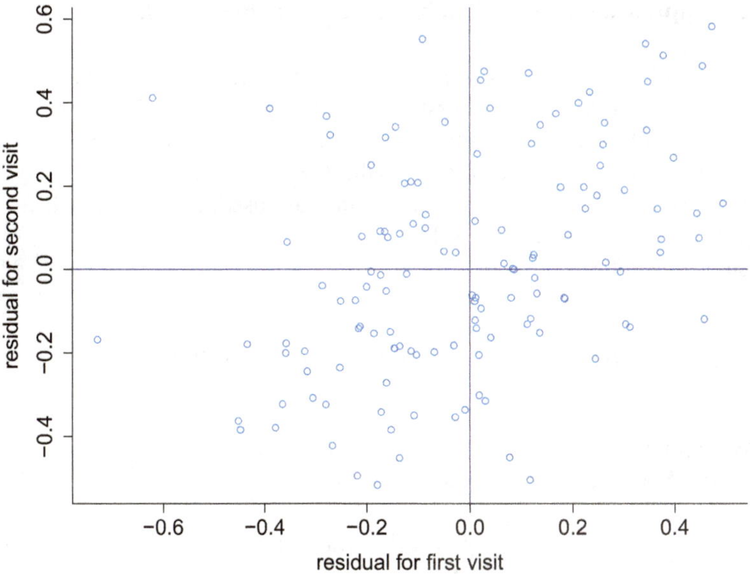

Fig. 4.17 The residual for the second visit versus that of the first visit for each subject.

If $\boldsymbol{y}_i \equiv [y_{i1}, y_{i2}]^T$ is the vector of responses for the ith subject then, under model (4.9),

$$\text{Cov}(\boldsymbol{y}_i) = \begin{bmatrix} \sigma_U^2 + \sigma_\varepsilon^2 & \sigma_U^2 \\ \sigma_U^2 & \sigma_U^2 + \sigma_\varepsilon^2 \end{bmatrix}.$$

It follows that model (4.9) imposes the restriction

$$\text{Cov}(\boldsymbol{y}_i) \in \left\{ \begin{bmatrix} \sigma^2 & \rho\sigma^2 \\ \rho\sigma^2 & \sigma^2 \end{bmatrix} : \sigma^2 > 0, 0 < \rho < 1 \right\}.$$

Compare this with the marginal model (4.8) where $\text{Cov}(\boldsymbol{y}_i)$ is unrestricted:

$$\text{Cov}(\boldsymbol{y}_i) \in \left\{ \begin{bmatrix} \sigma_1^2 & \rho\sigma_1\sigma_2 \\ \rho\sigma_1\sigma_2 & \sigma_2^2 \end{bmatrix} : \sigma_1^2, \sigma_2^2 > 0, -1 < \rho < 1 \right\}.$$

For this $n = 2$ case marginal model allows for (a) non-equal variances between visits, and (b) negative within-group correlation. For larger n the marginal model offers considerably more flexibility.

4.4.1.2 Approaches to Marginal Nonparametric Regression

Estimation and inference for model (4.8) have been the subject of intense research since about 2000, although it was first studied by Zeger and Diggle (1994). A survey of most of the literature on the topic is given in Sect. 3.9 of Ruppert et al. (2009). The approach that is most in keeping with this book is that of Al Kadiri et al. (2010), in which mixed model-based penalized splines are used. Hence, we will focus on that approach for the remainder of the section.

Assume that f takes the form

$$f(x) = \beta_0 + \beta_1 x + \sum_{k=1}^{K} u_k z_k(x), \quad u_k|\sigma \overset{\text{ind.}}{\sim} N(0, \sigma^2). \tag{4.10}$$

This assumption is appropriate if we use O'Sullivan splines.

Then the model can be written as

$$\boldsymbol{y} = \boldsymbol{X\beta} + \boldsymbol{Zu} + \boldsymbol{\varepsilon} \tag{4.11}$$

where

$$\boldsymbol{y} = \begin{bmatrix} \boldsymbol{y}_1 \\ \vdots \\ \boldsymbol{y}_m \end{bmatrix}, \quad \boldsymbol{X} = \begin{bmatrix} \boldsymbol{1} & \boldsymbol{x}_1 \\ \vdots & \vdots \\ \boldsymbol{1} & \boldsymbol{x}_m \end{bmatrix}, \quad \boldsymbol{Z} = \begin{bmatrix} z_1(\boldsymbol{x}_1) & \cdots & z_K(\boldsymbol{x}_1) \\ \vdots & \ddots & \vdots \\ z_1(\boldsymbol{x}_m) & \cdots & z_K(\boldsymbol{x}_m) \end{bmatrix}, \quad \boldsymbol{\varepsilon} = \begin{bmatrix} \boldsymbol{\varepsilon}_1 \\ \vdots \\ \boldsymbol{\varepsilon}_m \end{bmatrix},$$

$$\boldsymbol{\beta} = \begin{bmatrix} \beta_0 \\ \beta_1 \end{bmatrix} \quad \text{and} \quad \boldsymbol{u} = \begin{bmatrix} u_1 \\ \vdots \\ u_K \end{bmatrix}.$$

The random vectors on the right-hand side of (4.11) have mean zero and covariance matrix:

$$\text{Cov}\begin{bmatrix} \boldsymbol{u} \\ \boldsymbol{\varepsilon} \end{bmatrix} = \begin{bmatrix} \sigma^2 \boldsymbol{I} & \boldsymbol{0} \\ \boldsymbol{0} & \boldsymbol{I}_m \otimes \boldsymbol{\Sigma} \end{bmatrix}. \tag{4.12}$$

In principle, we could fit (4.10–4.12) using (restricted) maximum likelihood (REML) and best linear unbiased predictor (e.g., McCulloch, Searle, and Neuhaus). Al Kadiri et al. (2010) provide some of the details of this approach. However, implementation in `lme()` is not straightforward and we are yet to succeed in this regard. On the other hand, a Bayesian version of the model can be fit in Stan. The following function fits the Bayesian version of (4.10)–(4.12) with priors:

$$\boldsymbol{\beta} \sim N(\boldsymbol{0}, 10^{10}), \quad \sigma \sim \text{Half-Cauchy}(10^5)$$

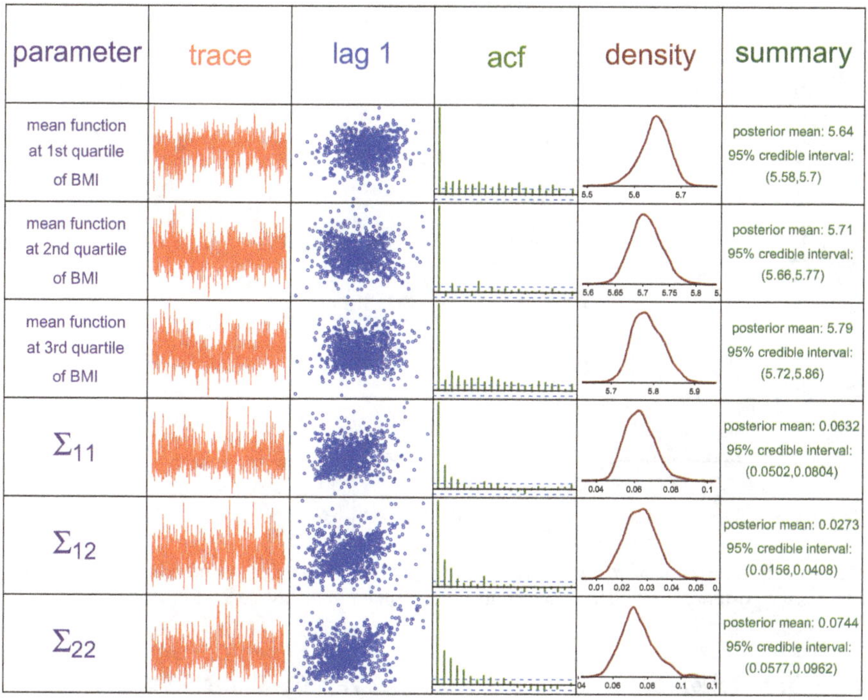

Fig. 4.18 Summary of MCMC-based inference for the mean function at the quartiles of body mass index (BMI) and entries of $\boldsymbol{\Sigma}$ in the fitted marginal nonparametric regression model (4.10)–(4.12). The columns are: parameter, trace plot of MCMC sample, plot of sample against 1-lagged sample, sample autocorrelation function, kernel estimates of posterior density, and basic numerical summaries.

and $\boldsymbol{\Sigma}$ has a marginally noninformative prior of the type described in Huang and Wand (2013). Specifically,

$$\boldsymbol{\Sigma}|a_1, a_1 \sim \text{Inverse-Wishart}(3, 4\,\text{diag}(a_1, a_2)^{-1}),$$
$$\text{where} \quad a_1, a_2 \overset{\text{ind.}}{\sim} \text{Inverse-Gamma}(\tfrac{1}{2}, 10^{-10})$$

with support from the package `rstan` (Guo et al. 2017). The required Stan code is contained in the object `margNPRegnModel`, listed below, within the script `proteinMargNPRegn.R` in the package `HRW`. Note that the Stan code stores the predictor and response data in $m \times n$ matrices named `xMat` and `yMat`. The code in `proteinMargNPRegn.R` produces Figs. 4.18 and 4.19.

```
> margNPRegnModel <-
+ 'data
+ {
+     int<lower=1> m;              int<lower=1> n;
```

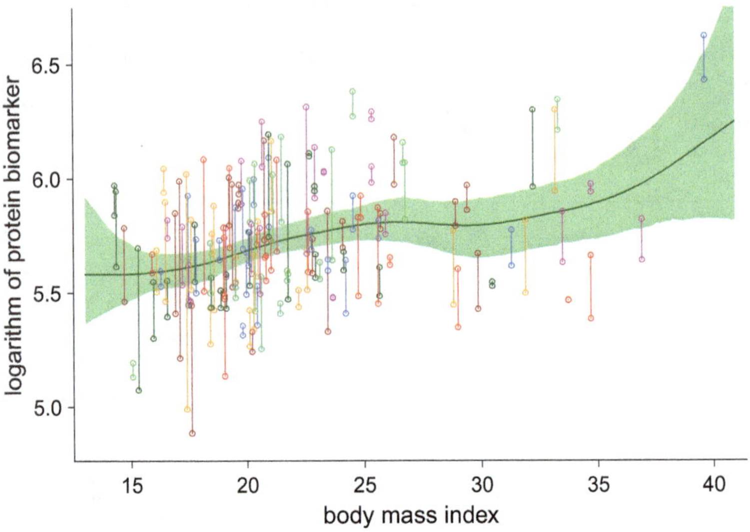

Fig. 4.19 Fitted Bayesian marginal parametric regression model for the protein data. The solid curve is the posterior mean. The dashed curves correspond to pointwise 95% credible sets.

```
+     int<lower=1> ncZ;            real<lower=0> sigmaBeta;
+     real<lower=0> Au;            real<lower=0> ASigma;
+     matrix[m,n] xMat;            matrix[m,n] yMat;
+     matrix[m*n,ncZ] Z;
+ }
+ parameters
+ {
+     real beta0;                  real beta1;
+     vector[ncZ] u;               real<lower=0> sigma;
+     cov_matrix[n] Sigma;         vector[n] a;
+ }
+ transformed parameters
+ {
+     matrix[m,n] meanFunc;
+     for (i in 1:m)
+        for (j in 1:n)
+           meanFunc[i,j] = beta0 + beta1*xMat[i,j]
+                                          + dot_product(u,Z[(i-1)*n+j]);
+ }
+ model
+ {
+     vector[n] scaleSigma;
+
+     for (i in 1:m)
```

```
+       yMat[i] ~ multi_normal(meanFuncFull[i],Sigma);
+
+    u ~ normal(0,sigma);
+
+    a ~ inv_gamma(0.5,pow(ASigma,-2));
+    for (j in 1:n) scaleSigma[j] = 4/a[j];
+    Sigma ~ inv_wishart(n+1,diag_matrix(scaleSigma));
+
+    beta0  ~ normal(0,sigmaBeta) ; beta1 ~ normal(0,sigmaBeta);
+    sigma ~ cauchy(0,Au);
+ }'
```

4.4.2 Additive Model Extension

Now suppose that two predictor values are observed for each value of the response. Denote these by x_{1ij} and x_{2ij} for $1 \leq i \leq m$ and $1 \leq j \leq n$. Then the *marginal additive model for grouped data* is

$$E(y_{ij}) = \beta_0 + f_1(x_{1ij}) + f_2(x_{2ij}), \qquad \text{Cov}\{\boldsymbol{y}_i | f_1(\boldsymbol{x}_{1i}), f_2(\boldsymbol{x}_{2i})\} = \boldsymbol{\Sigma}. \qquad (4.13)$$

The smooth functions f_1 and f_2 can be estimated using the penalized spline models:

$$f_1(x) = \beta_{11}\, x + \sum_{k=1}^{K_1} u_{1k} z_{1k}(x), \quad u_{1k}|\sigma_1 \stackrel{\text{ind.}}{\sim} N(0, \sigma_1^2)$$

$$f_2(x) = \beta_{21}\, x + \sum_{k=1}^{K_2} u_{2k} z_{2k}(x), \quad u_{2k}|\sigma_2 \stackrel{\text{ind.}}{\sim} N(0, \sigma_2^2)$$

where $\{z_{1k}(\cdot) : 1 \leq k \leq K_1\}$ and $\{z_{2k}(\cdot) : 1 \leq k \leq K_2\}$ are spline basis functions over the range of the x_1 and x_2 values, respectively.

The script `margAddMod.R` illustrates fitting of (4.13) for some simulated data, generated according to

$$f_1(x) = \sin\{2\pi(x^2 - 0.1)\} \quad \text{and} \quad f_2(x) = \sin\{3\pi(0.05 - x)\}$$

and

$$\boldsymbol{\Sigma} = \begin{bmatrix} 0.97 & 0.83 & 0.77 & 0.41 & 0.18 \\ 0.83 & 0.97 & 0.83 & 0.77 & 0.41 \\ 0.77 & 0.83 & 0.97 & 0.83 & 0.77 \\ 0.41 & 0.77 & 0.83 & 0.97 & 0.83 \\ 0.18 & 0.41 & 0.77 & 0.83 & 0.97 \end{bmatrix}.$$

The `margAddMod.R` can be run using:

```
> library(HRW) ; demo(margAddMod,package = "HRW")
```

and copied and modified from the location given by:

```
> system.file("demo","margAddMod.R",package = "HRW")
```

The extension of (4.13) to more than two predictors is straightforward.

4.4.3 Incorporation of Interactions

Now suppose that we also observe values on a *categorical* predictor x_3. Let x_{3ij} denote the observation partnering y_{ij}. Then the model

$$E(y_{ij}|x_{i1},x_{i2},x_{i3}) = \beta_0 + f_{x_{3ij}}(x_{1ij}) + f_2(x_{2ij}),\ \text{Cov}(\boldsymbol{y}_i|x_{i1},x_{i2},x_{i3}) = \boldsymbol{\Sigma},$$
$$1 \le i \le m,\ 1 \le j \le n \tag{4.14}$$

is an extension of (4.13) that allows for an interaction between x_1 and x_3. The notation $f_{x_{3ij}}(x_{1ij})$ signifies that we have a separate function for each value of x_{3ij}. For example, if the x_{3ij} are restricted to the set {female, male}, then we would have two functions: f_{female} and f_{male}. Model (4.14) is an instance of a factor-by-curve interaction, which is the focus of Sect. 3.6.

We now illustrate R fitting of (4.14) for a fuller version of the nutritional epidemiology data (source: Kipnis et al. 2003). The sample sizes are $m = 294$ and $n = 2$ and each penalized spline fit involved $K = 25$ basis functions. In the notation of (4.14), the variables are

$$\begin{aligned} y &= \text{logarithm of intake of protein as measured by the biomarker urinary} \\ &\quad\ \text{nitrogen,} \\ x_1 &= \text{body mass index,} \\ x_2 &= \text{logarithm of intake of protein as measured by a 24-hour recall} \\ &\quad\ \text{instrument,} \\ x_3 &= \text{gender.} \end{aligned}$$

The script `proteinMargAddInt.R` contains R code for the full additive/interaction model analysis. Figure 4.20 summarizes the MCMC-based inference for f_{female}, f_{male} and f_2 at the median of the relevant predictor, and the unique entries of $\boldsymbol{\Sigma}$. Excellent chain mixing is observed. The within-group covariance is significantly positive, as indicated by the 95% credible set for $\boldsymbol{\Sigma}_{12}$.

Figure 4.21 shows the estimates of f_{male}, f_{female} and f_2 and corresponding 95% pointwise credible sets.

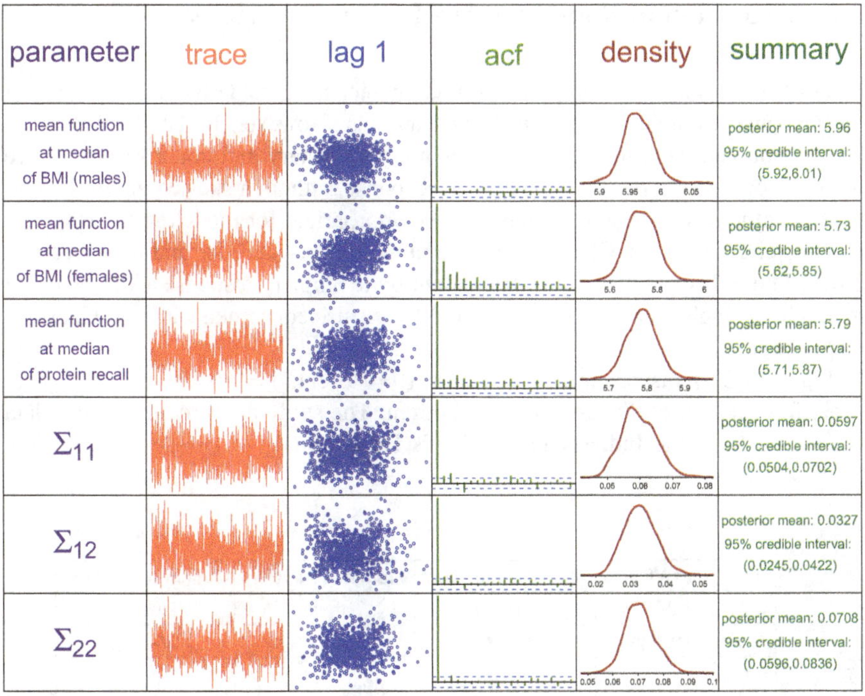

Fig. 4.20 Summary of MCMC-based inference for parameters in the fitted additive/interaction model for the nutritional epidemiology data. The columns are: parameter, trace plot of MCMC sample, plot of sample against 1-lagged sample, sample autocorrelation function, kernel estimates of posterior density, and basic numerical summaries.

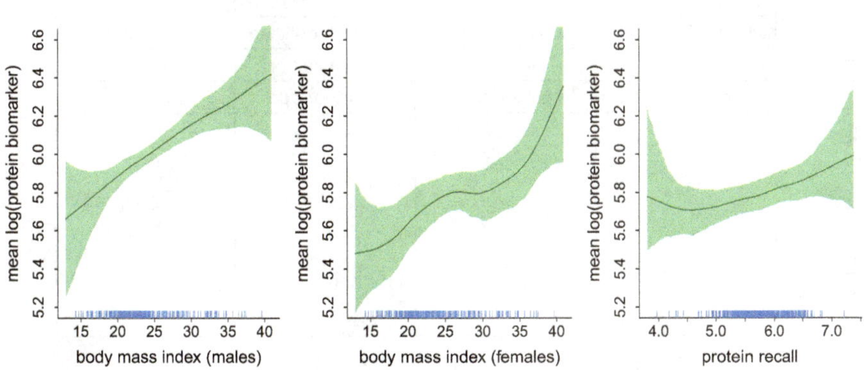

Fig. 4.21 Estimated functions for the additive/interaction model (4.14) for the protein data described in the text. The dashed curves correspond to 95% pointwise credible sets.

4.5 Extension to Non-Gaussian Response Variables

Each of the examples given so far in this chapter have made the assumption that the response variable is approximately Gaussian. However, longitudinal datasets having non-Gaussian response variables abound. Binary and count responses are the most common type of non-Gaussian response, although skewed and/or heavy-tailed continuous distributions may also be appropriate for modeling the response in some circumstances. Clearly, there are numerous possibilities and not all can be covered here. Instead, we will restrict attention to a binary response additive mixed model for a single dataset. Exercises 2 and 10 are also concerned with non-Gaussian response situations.

Figure 4.22 plots respiratory infection (0 = absent, 1 = present) versus age in years for a cohort of 275 Indonesian children. The study that produced these data is referred to as the Indonesian Children's Health Study in Diggle et al. (2002).

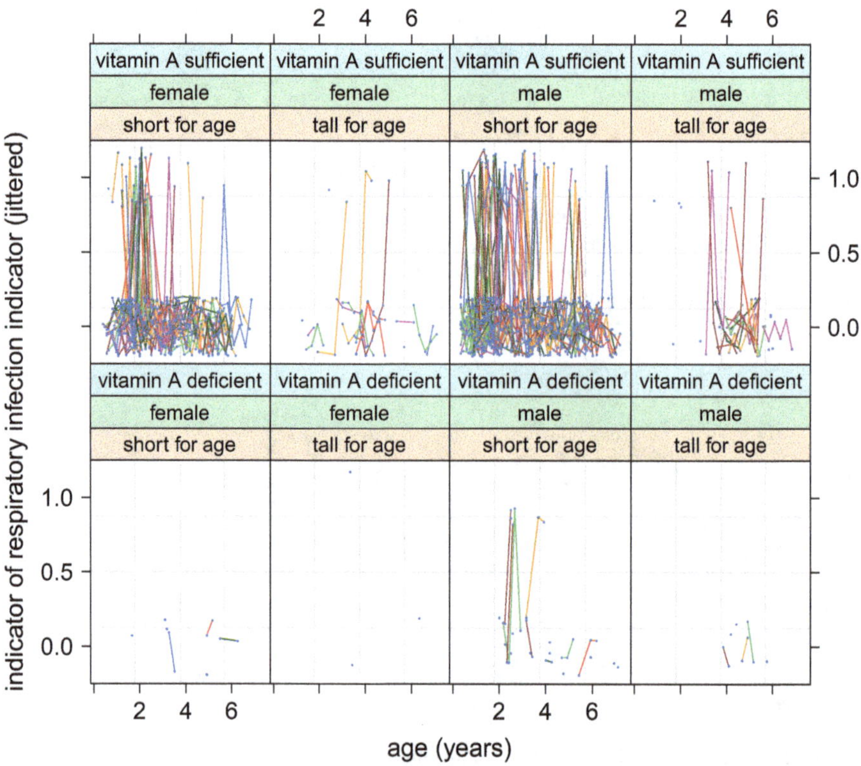

Fig. 4.22 Respiratory infection indicator (0 = absent, 1 = present) versus age in years for the Indonesian Children's Health Study data. The indicator is jittered to aid visualization. Repeated measures on same child are connected by lines. Each panel corresponds to a different combination of three predictors: vitamin A status, gender, and height for age.

The original source of the data is Sommer (1982). The data frame `indonRespir` in the `HRW` package contain the Indonesian Children's Health Study data. Besides age, data are available on height, vitamin A status (sufficient or deficient), gender, stunted (indicator of short for age), and number of visits for each child. The panels correspond to the three binary variables: vitamin A status, gender, and stunted. In our analyses we work with indicators for visit number. Five such indicators are required since there can be as many as six visits. Let $\boldsymbol{x}_{ij}$ be the 9×1 vector containing values of these predictors, which will be modeled as having linear effects. Here $1 \leq i \leq 275$ indexes child and $1 \leq j \leq n_i$ indexes the repeated measures for each child. Then a generalized additive mixed model of interest is:

$$\begin{aligned}\text{logit}\{P(\text{respir. infec.}_{ij} = 1 | U_i, \boldsymbol{x}_{ij}, \text{age}_{ij})\} &= U_i + \boldsymbol{\beta}^T \boldsymbol{x}_{ij} + f(\text{age}_{ij}),\\ U_i | \sigma^2_{\text{grp}} &\overset{\text{ind.}}{\sim} N(0, \sigma^2_{\text{grp}}).\end{aligned} \tag{4.15}$$

If a mixed model O'Sullivan penalized spline is used for the $f(\text{age}_{ij})$ component:

$$f(x) = \beta_{\text{age}}\, x + \sum_{k=1}^{K} u_k z_k(x), \quad u_k \overset{\text{ind.}}{\sim} N(0, \sigma^2_{\text{spl}}) \tag{4.16}$$

then (4.15) reduces to a generalized linear mixed model. The difficulty of fitting and inference for such non-Gaussian response models is well documented and approximations need to be used. Chapter 10 of McCulloch et al. (2008) provides access to some of this approximation literature, although it continues to grow.

The following subsections describe two different ways of fitting variants of (4.15) in R, via:

1. *penalized quasi-likelihood* (e.g. Breslow and Clayton 1993) via the package `mgcv` (Wood 2017),
2. MCMC via the package `rstan` (Guo et al. 2017).

Penalized quasi-likelihood is less computationally demanding, but is often criticized (e.g. McCulloch et al. 2008) for being too crude. MCMC can be made quite accurate by using a higher number of MCMC samples and becomes exact in the limit as the number of samples increases to infinity, but can be quite slow. It is also the most extendible to other non-Gaussian responses.

4.5.1 Penalized Quasi-Likelihood Analysis

The function `gamm()` in the package `mgcv` (Wood 2017) uses penalized quasi-likelihood (PQL) for binary response models. Note that `gamm()` actually calls the generic generalized linear mixed model function `glmmPQL()`, of the package `MASS` (Ripley et al. 2015), to carry out the PQL computations.

To fit model (4.15) via gamm() we issue the commands:

```
> library(mgcv)   ;   library(HRW)   ;   data(indonRespir)
> fit <- gamm(respirInfec ~ s(age) + vitAdefic + female + height
+                          + stunted + visit2 + visit3 + visit4
+                          + visit5 + visit6,
+                          random = list(idnum = ~1),
+                          family = binomial,data=indonRespir)
```

```
 Maximum number of PQL iterations:  20
```

A summary of the additive model aspects of the fit is produced using:

```
> summary(fit$gam)
```

```
Family: binomial
Link function: logit

Formula:
respirInfec ~ s(age) + vitAdefic + female + height + stunted +
    visit2 + visit3 + visit4 + visit5 + visit6

Parametric coefficients:
            Estimate Std. Error t value Pr(>|t|)
(Intercept) -2.16519    0.26765  -8.090 1.46e-15
vitAdefic    0.66979    0.47263   1.417  0.15670
female      -0.46434    0.24454  -1.899  0.05783
height      -0.03091    0.02544  -1.215  0.22462
stunted      0.40793    0.43277   0.943  0.34608
visit2      -1.06311    0.38877  -2.735  0.00634
visit3      -0.55626    0.36500  -1.524  0.12777
visit4      -1.23521    0.44846  -2.754  0.00597
visit5       0.43218    0.30730   1.406  0.15987
visit6      -0.01073    0.33938  -0.032  0.97477

Approximate significance of smooth terms:
         edf Ref.df     F  p-value
s(age) 2.302  2.302 12.38 4.66e-06

R-sq.(adj) =  0.0585
  Scale est. = 1         n = 1200
```

Note that the fitted age effect involves 2.302 effective degrees of freedom. Differences among the categories can be obtained from this. A more direct assessment is available from:

```
> intervals(fit$lme,which="fixed")
```

```
Approximate 95% confidence intervals

Fixed effects:
                   lower        est.       upper
X(Intercept) -2.68826653 -2.16519109 -1.64211566
XvitAdefic   -0.25389718  0.66979454  1.59348626
Xfemale      -0.94375879 -0.46434375  0.01507128
Xheight      -0.08063775 -0.03091272  0.01881231
Xstunted     -0.43786332  0.40792745  1.25371821
Xvisit2      -1.82290211 -1.06310891 -0.30331572
Xvisit3      -1.26959339 -0.55625846  0.15707647
Xvisit4      -2.11166707 -1.23520952 -0.35875196
Xvisit5      -0.16839380  0.43217841  1.03275061
Xvisit6      -0.67399803 -0.01073369  0.65253066
Xs(age)Fx1   -1.53419404 -0.49355253  0.54708899
attr(,"label")
[1] "Fixed effects:"
```

This output shows that an approximate 95% confidence interval for the coefficient of `vitAdefic` in (4.15) is the insignificant (−0.254, 1.593), while that for `visit2` is the significant (−1.823, −0.303) although these need to be interpreted conditionally on the U_i value for each subject. Marginal interpretation requires adjustments such as that developed by Zeger et al. (1988). The row labeled `Xs(age)Fx1` corresponds to the coefficient of the linear component of the penalized spline model for age, but the values do not have a simple interpretation.

A plot of the estimated $f(\text{age})$ curve can be obtained via:

```
> plot(fit$gam,shade = TRUE,shade.col = "palegreen")
```

The vertical axis of the resulting plot is on the logit scale. The following R code produces a plot of the age effect on the *probability* scale, and with each of the other predictors set at their average values:

```
> ng <- 101
> ageg <- seq(min(indonRespir$age),max(indonRespir$age),
+             length = ng)
> aveOthers <- apply(indonRespir[,4:12],2,mean)
> newDataDF <- as.data.frame(cbind(ageg,
+              matrix(rep(aveOthers,each = ng),
+                         ng,length(aveOthers))))
> names(newDataDF) <- names(indonRespir)[-c(1,2)]
> newDataList <- as.list(newDataDF)
> predObj <- predict(fit$gam,newdata = newDataList,se = TRUE)
> muHatg <- 1/(1+exp(-predObj$fit))
> stdErrg <- predObj$se*muHatg*(1-muHatg)
> lowerg <- muHatg - 2*stdErrg ; upperg <- muHatg + 2*stdErrg
> ylimVal <- range(c(lowerg,upperg))
```

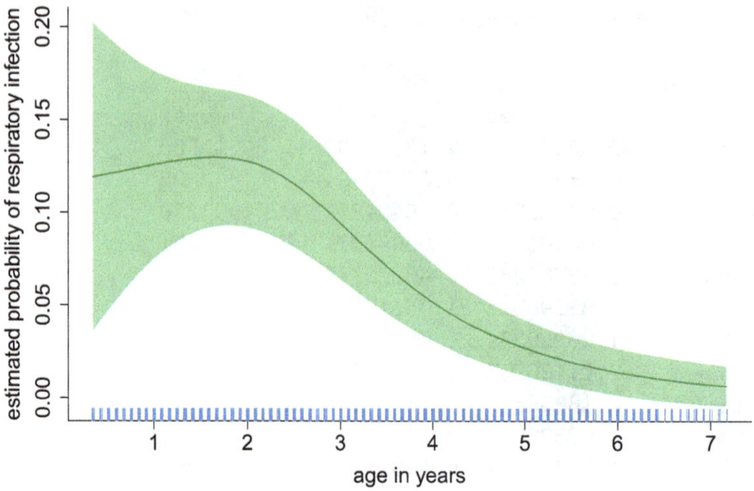

Fig. 4.23 The fitted penalized spline for the age effect in the fit of the gamm()-based logistic additive mixed model (4.15) to the Indonesian children's health study data. The shaded region corresponds to pointwise approximate 95% confidence intervals.

```
> plot(ageg,muHatg,type="n",,ylim=ylimVal,xlab="age in years",
+      ylab="estimated probability of respiratory infection",
+      cex.lab = 1.5,cex.axis = 1.5,bty = "l")
> polygon(c(ageg,rev(ageg)),c(lowerg,rev(upperg)),
+         col = "palegreen",border = FALSE)
> lines(ageg,muHatg,lwd = 2,col = "forestgreen")
> rug(jitter(indonRespir$age),col = "dodgerblue")
```

The result is shown in Fig. 4.23. The code `stdErrg <- predObj$se *muHatg*(1-muHatg)` computes a delta-method standard error and uses the result that $L'(x) = L(x)\{1 - L(x)\}$ where $L(x) \equiv \{1 + \exp(-x)\}^{-1}$.

4.5.2 *Markov Chain Monte Carlo Analysis*

Our final analysis of the Indonesian respiratory data involves MCMC fitting via the package rstan. This involves adopting a Bayesian approach. We supplement (4.15) and (4.16) with the prior specifications:

$$\sigma_{\text{grp}}, \sigma_{\text{spl}} \overset{\text{ind.}}{\sim} \text{Half-Cauchy}(10^5), \quad \beta_{\text{age}} \sim N(0, 10^{10}) \quad \text{and} \quad \boldsymbol{\beta} \sim N(\mathbf{0}, 10^{10}\,\boldsymbol{I}).$$

The IndonRespirBayes.R carries out the full analysis using rstan. The Stan code in this script is contained in:

```
> logistAddMixModModel <-
+ 'data
```

```
+ {
+     int<lower=1> numObs;         int<lower=1> numGrp;
+     int<lower=1> ncX;            int<lower=1> ncZspl;
+     real<lower=0> sigmaBeta;     real<lower=0> Agrp;
+     real<lower=0> Aspl;          matrix[numObs,ncX] X;
+     int<lower=1> idnum[numObs]; matrix[numObs,ncZspl] Zspl;
+     int<lower=0,upper=1> y[numObs];
+ }
+ parameters
+ {
+     vector[ncX] beta;             vector[numGrp] U;
+     vector[ncZspl] u;             real<lower=0> sigmaGrp;
+     real<lower=0> sigmaSpl;
+ }
+ model
+ {
+     y ~ bernoulli_logit(X*beta + U[idnum] + Zspl*u);
+     U ~ normal(0,sigmaGrp);           u ~ normal(0,sigmaSpl);
+     beta  ~ normal(0,sigmaBeta) ;  sigmaGrp  ~ cauchy(0,Agrp);
+     sigmaSpl ~ cauchy(0,Aspl);
+ }'
```

Figure 4.24 summarizes the MCMC output for the parametric components of (4.15). Figures 4.25 and 4.26 show the fitted curve and MCMC summaries for the nonparametric function of age. The findings are similar to those in the previous analyses in this section. Zhao et al. (2006) give further details on MCMC-based analysis for these data.

4.6 Further Readings

Part Two of Fitzmaurice et al. (2008) provides access to the wider literature on semiparametric grouped data analysis in the context of longitudinal data. Many of the models in this chapter are described in Chap. 9 of Ruppert et al. (2003). Similar models are treated by Wu and Zhang (2006). Pinheiro and Bates (2000) provide details on early versions of the package `nlme`. Wood (2006a) performs a similar service for the package `mgcv`.

Background material on longitudinal data analysis may be found in the books by Diggle et al. (2002), Fahrmeir and Kneib (2011), Fitzmaurice et al. (2004), and Verbeke and Molenberghs (2000).

Gelman and Hill (2007) and Goldstein (2010) are major references for multilevel models. Chapter 15 of the latter reference combines nonparametric regression with multilevel model concepts.

Panel data analysis is very similar to longitudinal and multilevel data analysis, but has been mainly developed within the Econometrics literature. Summaries are provided by Baltagi (2013) and Frees (2004). Small area estimation has evolved

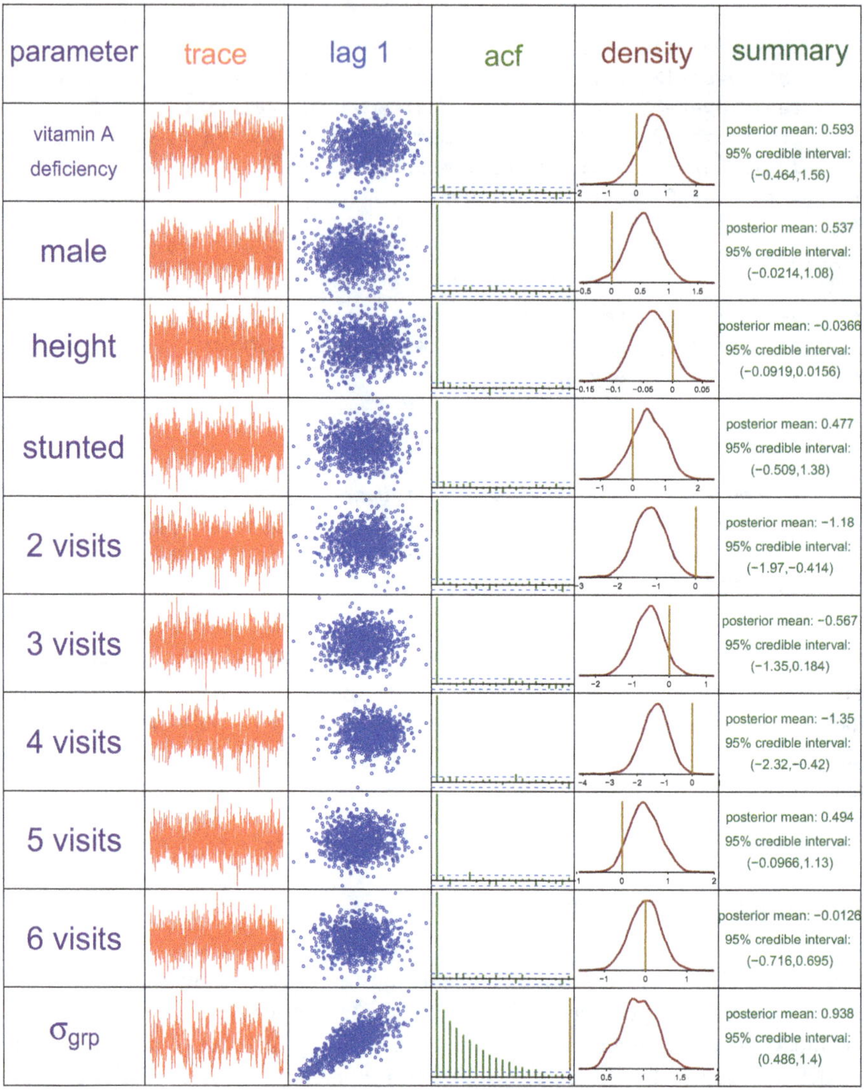

Fig. 4.24 Summary of MCMC-based inference for parameters in the fitted Bayesian model for the Indonesian children's respiratory data. The columns are: parameter, trace plot of MCMC sample, plot of sample against 1-lagged sample, sample autocorrelation function, kernel estimates of posterior density, and basic numerical summaries.

within the sample survey literature but also involves similar concepts and models. References include Longford (2005) and Rao and Molina (2015). Even though the examples and exercises are geared towards longitudinal and multilevel data analysis, the concepts described in this chapter apply equally well to panel data analysis and small area estimation.

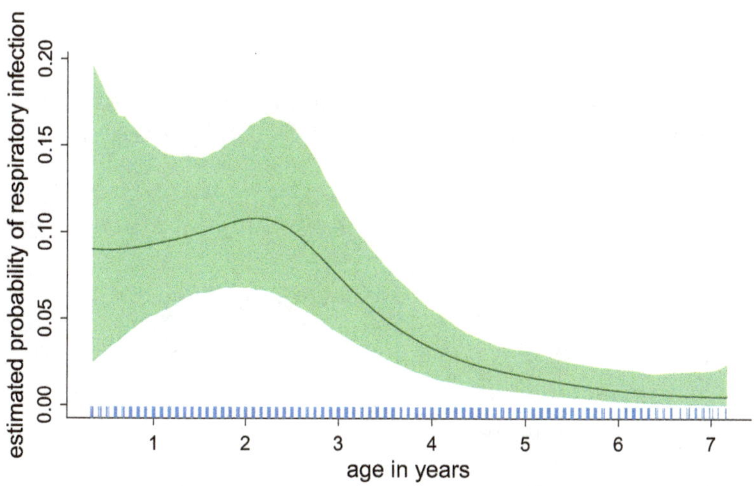

Fig. 4.25 Estimated age effect on probability scale, with shaded region corresponding to pointwise 95% credible sets.

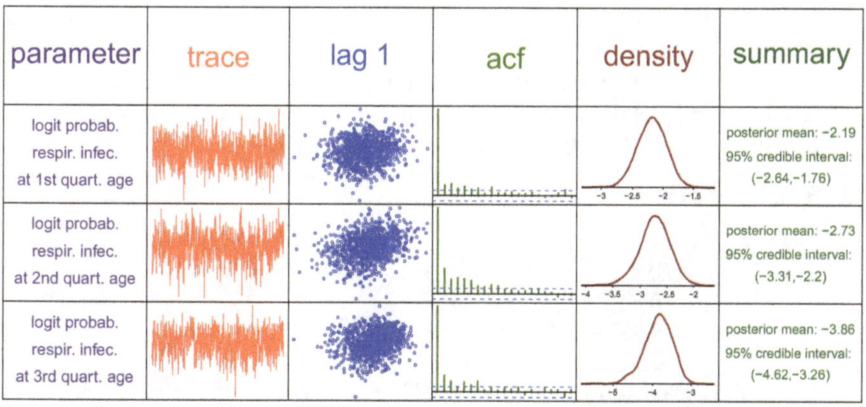

Fig. 4.26 Summary of MCMC-based inference for parameters in the fitted Bayesian logistic additive mixed model for the Indonesian children's health study data. The columns are: parameter, trace plot of MCMC sample, plot of sample against 1-lagged sample, sample autocorrelation function, kernel estimates of posterior density, and basic numerical summaries.

Many of the mixed model theoretical results that underpin grouped data analysis may be found in McCulloch et al. (2008).

4.7 Exercises

1. Model (4.1) for the female spinal bone mineral density data assumes that the mean functions for each ethnicity category differ only by vertical shifts. A more flexible model is

$$\begin{gathered}\texttt{spnbmd}_{ij} = U_i + f_{\texttt{ethnicity}_i}(\texttt{age}_{ij}) + \varepsilon_{ij}, \quad 1 \le j \le n_i, \quad 1 \le i \le m, \\ U_i \stackrel{\text{ind.}}{\sim} N(0, \sigma_U^2), \quad \varepsilon_{ij} \stackrel{\text{ind.}}{\sim} N(0, \sigma_\varepsilon^2),\end{gathered} \tag{4.17}$$

   where

$$\texttt{ethnicity}_i \in \{\texttt{asian}, \texttt{black}, \texttt{hispanic}, \texttt{white}\}, \quad 1 \le i \le m,$$

   is a categorical variable that signifies the ethnicity of the ith subject. Write an R script that uses Stan, via the package `rstan`, to:

   a. Fit a Bayesian version of (4.17) with each of $f_{\texttt{asian}}$, $f_{\texttt{black}}$, $f_{\texttt{hispanic}}$, and $f_{\texttt{white}}$ modeled using a separate Bayesian mixed-model based penalized spline as described in Sect. 2.10. The script should plot the Bayes estimates of each mean function and corresponding 95% pointwise credible sets.
   b. Obtain and plot the estimated contrast functions for each ethnicity pair along with corresponding 95% pointwise credible sets.

2. Simulate data according to the model

$$\begin{gathered}y_{ij}|U_i \stackrel{\text{ind.}}{\sim} \text{Poisson}[\exp\{U_i + f(s_{ij}) + \beta_x x_{ij}\}] \\ U_i \stackrel{\text{ind.}}{\sim} N(0, \sigma^2), \ 1 \le i \le m, \ 1 \le j \le n,\end{gathered} \tag{4.18}$$

   where $f(s) = \cos(4\pi s) - s^2$, $\beta_x = 0.83$, $\sigma = 0.37$, $m = 50$, $n = 10$. Generate the s_{ij} to be uniformly distributed on $(0, 1)$ and the x_{ij} to take 0 or 1 with equal probabilities.

   a. Fit the data using the function `gamm()` in the package `mgcv` and compare the results with the true parameters and function from which the data were generated.
   b. Write an R script that uses Stan, via the `rstan` package, to fit a Bayesian version of (4.18). For estimation of f, use a mixed model-based penalized spline model of the form

$$f(s) = \beta_0 + \beta_s\, s + \sum_{k=1}^{25} u_k z_k(s), \quad u_k|\sigma \stackrel{\text{ind.}}{\sim} N(0, \sigma^2),$$

where the z_k, $1 \leq k \leq 25$, are O'Sullivan splines. Use noninformative prior distributions for β_0, β_s and β_x and σ.

c. Compare the results produced by gamm() with those produced by rstan.

3. In Sect. 4.3 we pointed out that the first group-specific curves analysis of the male Indiana adolescent growth data has an unattractive asymmetry in the mean functions for each ethnicity group. As discussed there, this asymmetry is convenient when using lme() in the package nlme. The R script maleGrowthIndianaBayes.R in the HRW package performs a Bayesian analysis with symmetry between the ethnicity groups using MCMC via Stan and rstan. Compare the estimated contrast function and the pointwise 95% confidence intervals from the lme() analysis with the estimated contrast function and the pointwise 95% credible sets obtained from the MCMC-based Bayesian analysis.
4. The data frame Dialyzer in the package nlme contains measurements on 20 high-flux hemodialyzers to assess their *in vivo* ultrafiltration characteristics. The ultrafiltration rates (milliliters per hour) were measured at seven different transmembrane pressures (decimeters of mercury). The evaluation of the dialyzers used two different bovine flow rates: 200 deciliters per minute and 300 deciliters per minute. Further details and references can be obtained from the R commands library(nlme) ; help(Dialyzer).

   a. Plot the data by issuing the R commands:

```
> library(nlme)   ;   data(Dialyzer)
> x <- Dialyzer$pressure   ;   y <- Dialyzer$rate
> idnum <- as.numeric(as.character(Dialyzer$Subject))
> typeIsB <- as.numeric(Dialyzer$QB == "300")
> plot(x,y,type = "n",bty = "l",
+       xlab = "transmembrane pressure (dmHg)",
+       ylab = "ultrafiltration rate (ml/hr)")
> for (i in 1:20)
+ {
+    currInds <- (1:length(x))[idnum == i]
+    xCurr <- x[currInds] ; yCurr <- y[currInds]
+    if (typeIsB[currInds[1]] == 0) colCurr <- "blue"
+    if (typeIsB[currInds[1]] == 1) colCurr <- "deeppink"
+    lines(xCurr[order(xCurr)],yCurr[order(xCurr)],
+          type = "b",col = colCurr)
+ }
> legend("bottomright",lty = rep(1,2),
+        legend = c("low bovine blood flow rate",
+                   "high bovine blood flow rate"),
+        col = c("blue","deeppink"))
```

b. The graphic produced in a. suggests that a good model for these data is

$$y_{ij} = (1 - I_i^B)\, f^A(x_{ij}) + I_i^B\, f^B(x_{ij}) + U_{0i} + U_{1i}\, x_{ij} + \varepsilon_{ij},$$
$$\begin{bmatrix} U_{0i} \\ U_{1i} \end{bmatrix} \overset{\text{ind.}}{\sim} N(0, \boldsymbol{\Sigma}) \quad \text{independently of} \quad \varepsilon_{ij} \overset{\text{ind.}}{\sim} N(0, \sigma_\varepsilon^2), \tag{4.19}$$

where (x_{ij}, y_{ij}) is jth ultrafiltration rate/transmembrane pressure pair for the ith hemodialyzer ($1 \leq i \leq 20$, $1 \leq j \leq 7$) and I_i^B is an indicator of measurements for the ith hemodialyzer being at a bovine flow rate of 300 deciliters per minute. Write an R script to fit a Bayesian version of (4.19) with penalized spline modeling similar to that used in the R script `maleGrowthIndianaBayes.R` from Exercise 3. The number of O'Sullivan spline basis functions should be reduced to about 5 or 6 since there are only seven unique predictor values. The script should produce a plot showing the fitted curves for each hemodialyzer and an estimate of the contrast function $c(x) \equiv f^B(x) - f^A(x)$. Pointwise 95% credible sets should accompany each curve estimate.

c. Assess the statistical significance of the diagonal entries of $\boldsymbol{\Sigma}$, which correspond to the variability in the random intercepts U_{0i} and slopes U_{1i}.

5. The data frame `Milk` in the package `nlme` contains measurements of the protein content of milk versus time since calving for each of 79 cows. The cows are divided into three diet groups: barley, barley+lupin, and lupin. Perform a group-specific curves analysis of these data. In particular, obtain contrast functions for (a) barley versus barley+lupins, (b) barley+lupins versus lupins, and (c) barley versus lupins.

6. Consider the frequentist approach to group-specific curve fitting described in Sect. 4.3, and its application to the analysis of the Indiana adolescent growth data via the function `lme()` in the `nlme` package. These data are available in the data frame `growthIndiana` within the `HRW` package.

   a. Estimate the speed of growth for both black and white adolescents as well as their difference.
   b. Estimate the group-specific speed for each individual growth curve.
   c. Is there evidence for a difference in the timing of the pubertal growth spurt between black and white adolescents?

7. The data frame `schoolResults` in the `HRW` package contains assessment results for 1905 school children in 73 different schools in a region of the United Kingdom (source: Creswell 1991). Consider a group-specific curves model with groups corresponding to schools, the predictor variable corresponding to the scores from teacher-evaluated coursework, and the response variable corresponding to the scores on a written paper. The model should have separate global mean curves for boys and girls. Using code similar to that contained in the `maleGrowthIndianaBayes.R` (see Exercise 3) fit a Bayesian version of the model using Stan via `rstan`. The script should produce a plot showing the fitted curves for each school and an estimate of the contrast function of

mean written paper score for boys versus girls, conditional on the score for teacher-evaluated coursework. Pointwise 95% credible sets should accompany each curve estimate.

8. In Exercise 3, the `maleGrowthIndianaBayes.R` fits a Bayesian group-specific curves model data for the male subjects within the data frame titled `growthIndiana` in the `HRW` package. Create a new R script that analyses both genders together using a model similar to that fit in the script `maleGrowthIndianaBayes.R`. The model should have separate global mean curves for black females, black males, white females and white males. The script should produce a plot or plots showing the fitted curves for each subject and an estimated contrast functions for comparison of

   a. black females versus white females,
   b. black males versus white males,
   c. black males versus black females, and
   d. white males versus white females.

9. The data frame `UtahPEF` in the `HRW` contains data from a study on respiratory health involving 41 Utah Valley, USA, schoolchildren as described in Pope et al. (1991). Data on peak expiratory flow (PEF), amount of particulate matter with an aerodynamic diameter less than or equal to 10 micrometers (PM_{10}), and weather were recorded for 109 consecutive days. The variable `devPEF` in `UtahPEF` corresponds to

$$\Delta\text{PEF}_{ij} \equiv \text{deviation of the } i\text{th child's PEF on day } j \text{ from the child's average PEF}, \quad 1 \leq i \leq 41,\ 1 \leq j \leq 109.$$

   The variable `PM10withMA5` is the 5-day moving average value of PM_{10} for each day and `lowTemp` is the lowest temperature recorded on that day. Use the function `gamm()` in the `mgcv` package to conduct an additive mixed model analyses of the data in `UtahPEF` with `devPEF` as the response variable. Allow for the possibility of serial correlated errors, as described in Sect. 4.2.2.

10. The data frame `BanglaContrac` in the `HRW` package contains data from the 1988 Bangladesh Fertility survey (source: Huq and Cleland 1990). The data comprise information on contraception use, family size, age, and rural/urban residential status for 1934 women grouped in 60 districts of Bangladesh. The age data have had the average age subtracted from them.

   a. Use the `gamm()` in the R package `mgcv` to fit the logistic additive mixed model

$$\begin{aligned}\text{logit}\{P(\texttt{useContraception}_{ij} = 1)\} &= U_i + \beta_1\texttt{oneChild}_{ij} \\ &+\beta_2\texttt{twoChildren}_{ij} + \beta_3\,\texttt{moreTwoChildren}_{ij} \\ &+\beta_4\texttt{isUrban}_{ij} + f(\texttt{ageMinusAve}_{ij}), \qquad U_i \stackrel{\text{ind.}}{\sim} N(0, \sigma_U^2),\end{aligned} \tag{4.20}$$

where, for the jth woman in the ith district, $\texttt{useContraception}_{ij}$ is an indicator of contraception use and $\texttt{ageMinusAve}_{ij}$ is her age in years minus the average age. The other variables appearing in (4.20) are indicators of the woman having one child, two children, more than two children, and living in an urban area. Your answer should include estimates of β_k, $1 \le k \le 4$, and corresponding standard errors and a plot of the estimate of f with pointwise approximate 95% confidence intervals.

b. Use `rstan` to fit a Bayesian version of (4.20) with suitable noninformative priors placed on the model parameters. Your answer should include MCMC-based Bayes estimates and 95% credible intervals for β_j, $1 \le j \le 4$, σ_U, and f. It should also include some convergence diagnostic checks of the MCMC.

c. The *odds ratio* for the effect of dwelling in an urban area, conditional on the district's random effect, corresponds to the parameter $\exp(\beta_4)$. Use the MCMC output to approximate the Bayes estimate of $\exp(\beta_4)$ along with a 95% credible set. Do the same for the other binary predictor variables in (4.20).

Chapter 5
Bivariate Function Extensions

5.1 Introduction

We now focus on models for the joint effect of two continuous predictor variables. Additive models are convenient, but there is no reason to assume that they are always adequate. In the general bivariate models studied in this chapter, the joint effect of the two variables is a smooth, but otherwise unrestricted, function of these variables. Therefore, these models allow interactions so that the effect of one predictor depends upon the value of the other predictor.

In keeping with other parts of this book, our approach to handling bivariate functional effects involves bivariate extensions of penalized splines. We pay particular attention to the impressive support for bivariate penalized splines provided by the `mgcv` package (Wood 2017). This package also allows bivariate functional effects to be incorporated into generalized additive models via the function `gam()` and generalized additive mixed models via the function `gamm()`. However, bivariate penalized splines are just one of number of approaches in the literature to fitting bivariate surfaces and the field known generally as *geostatistics* also deals with problems of this type. Instead of penalized splines, geostatistical approaches are based upon notions such as *spatial processes* and *kriging*. Prominent geostatistics references include Cressie (2015) and Diggle and Ribeiro (2007). Commonalities and differences between penalized spline and geostatistical approaches are discussed in Chap. 13 of Ruppert et al. (2003). This multitude of approaches to bivariate smoothing is also reflected in R. For example, multivariate penalized splines are supported by the function `Tps()` in the `fields` package (Nychka et al. 2017) and alternative geostatistical approaches are supported by the function `ksline()` in the `geoR` package (Ribeiro and Diggle 2016). *Gaussian processes* (e.g. Rasmussen and Williams 2006) is another name for kriging. *Kernel machines*, which are treated in detail in Sect. 6.7, are yet another way to approach bivariate smoothing.

The related topics of geoadditive models, varying-coefficient models, and covariance function estimation are also covered in this chapter.

J. Harezlak et al., *Semiparametric Regression with R*, Use R!,
https://doi.org/10.1007/978-1-4939-8853-2_5

5.2 Bivariate Nonparametric Regression

The bivariate nonparametric regression model is

$$y_i = f(x_{1i}, x_{2i}) + \varepsilon_i,$$

where, for $1 \le i \le n$, y_i is the response for the ith case and (x_{1i}, x_{2i}) are the corresponding predictors. The only assumption about the regression function f is that it is smooth.

There are two commonly used spline bases for estimating f, thin plate splines and tensor product splines. In principle, both can be readily extended to more than two dimensions, but for two reasons we will only discuss the bivariate case. The first reason is simplicity; it is difficult to understand and visualize functions of three or more arguments. The second is the so-called *curse of dimensionality* which dictates that large amounts of data are required for penalized splines in three or more dimensions to be feasible. In the case of very large sample sizes, penalized splines in moderate dimensions higher than two may be feasible although the methodological and applied literature is relatively scant. See, for example, O'Connell and Wolfinger (1997) and Wood et al. (2017) for some analyses involving trivariate penalized splines. For moderate sample sizes, when a smooth function is fit to more than two variables it is typically additive with each component a function of only one or two of these variables. For example, a model for three variables might be

$$y_i = f_1(x_{1i}) + f_2(x_{2i}) + f_3(x_{3i}) + \varepsilon_i$$

or

$$y = f_1(x_{1i}) + f_{23}(x_{2i}, x_{3i}) + \varepsilon_i.$$

Neither model requires one to model or visualize a function of more than two variables. With the second model, for example, we would plot f_1 and f_{23}, and f_1 would be a univariate spline while f_{23} would be a bivariate spline of the type we introduce in this chapter.

A bivariate thin plate spline with a second derivative penalty is an analog of a univariate cubic regression spline and minimizes

$$\sum_{i=1}^{n}\{y_i - f(x_{1i}, x_{2i})\}^2 + \lambda \int_{-\infty}^{\infty}\int_{-\infty}^{\infty}\left\{\left(\frac{\partial^2 f(x_1,x_2)}{\partial x_1^2}\right)^2 + 2\left(\frac{\partial^2 f(x_1,x_2)}{\partial x_1 \partial x_2}\right)^2 + \left(\frac{\partial^2 f(x_1,x_2)}{\partial x_2^2}\right)^2\right\} dx_1 dx_2.$$

where $\lambda \ge 0$ is a smoothing parameter. It can be shown (e.g. Green and Silverman 1994) that the thin plate spline is a linear combination of the basis functions $1, x_1, x_2,$

and $r(\|[x_1\ x_2]^T - [x_{1i}\ x_{2i}]^T\|)$ where $r(x) \equiv x^2 \log(x)$ and $\boldsymbol{x}_i \equiv [x_{1i}\ x_{2i}]^T$ ranges over the unique values of the predictor vector. It is shown in Ruppert et al. (2003) and in Wood (2003, 2006a) that the spline coefficients satisfy three constraints, so that the number of independent parameters is equal to the number of unique $\boldsymbol{x}_i$.

Computations can become infeasible when the number of unique $\boldsymbol{x}_i$ values is large. To alleviate this problem, low-rank approximations to thin plate splines are used. Ruppert et al. (2003) use $r(\|\boldsymbol{x} - \boldsymbol{\kappa}_k^*\|)$ where $\boldsymbol{\kappa}_k,\ k = 1, \ldots, K$, is a selected subset of the unique $\boldsymbol{x}_i$. Wood (2003, 2006a) uses an eigendecomposition of $\boldsymbol{E}$ where $E_{ij} = r(\|\boldsymbol{x}_i - \boldsymbol{x}_j\|)$. Then the K eigenvectors corresponding to the K largest eigenvalues are used as basis functions along with 1, x_1, and x_2. Wood calls this approximation a thin plate regression spline and they are used by the `gam()` function in the `mgcv` package.

A bivariate tensor product spline uses a univariate spline basis in each variable and takes all possible products. Specifically, the basis is

$$\{z_{1k}(x_1)z_{2k'}(x_2) :\ k = 1, \ldots, K_1,\ k' = 1, \ldots, K_2\}$$

where $\{z_{1k} :\ k = 1, \ldots, K_1\}$ is a basis for the x_1 variable and $\{z_{2k'} :\ k' = 1, \ldots, K_2\}$ is a basis for the x_2 variable.

Thin plate splines have a single penalty parameter, smooth to the same degree in all directions, and are invariant to rotation of the coordinate axes. This is not true of tensor product splines. Tensor product splines have a separate smoothing parameter for each variable and so can smooth one variable more than the other. A tensor product spline fit depends on the choice of the coordinate axes and is not invariant to rotations.

Often the two predictors are comparable or, at least, on the same scale. This is true of geographical or image data where the predictors are locations along two orthogonal coordinates. In such cases, thin plate and tensor product spline fits are often similar. For data where the variables are on different scales, a thin plate spline may not be appropriate, although rescaling may remedy this problem. The same is true of geographical data in cases where the north-south and east-west directions are not comparable.

Tensor product splines are generally faster to compute than thin plate splines (Wood 2006b). For data on regular grids, there are very fast tensor product algorithms; see Sects. 5.6 and 5.8.

Finally, note that the `mgcv` package supports other types of bivariate basis functions such as, for example, the Gaussian process and soap film smooths families. The R command `help(smooth.terms)` provides details.

5.2.1 *Example: Ozone Levels in Midwest USA*

As an example of bivariate nonparametric regression, we will use the `ozoneSub` dataset in the `HRW` package. This dataset is a subset of the `ozone2` dataset in the fields package (Nychka et al. 2017). There are three variables, longitude, latitude, and ozone, the 8-hour average ozone concentration; corresponding to 143 monitoring sites in the Midwest region of the USA. We fit the data using the gam() function in the mgcv package (Wood 2017). A penalized thin plate spline is specified by bs = "tp" and REML is used to select the amount of smoothing.

```
> library(HRW) ; data(ozoneSub) ; library(mgcv)
> fitOzoneThinPlate <-
+         gam(ozone ~ s(longitude,latitude,bs = "tp"),
+                          data = ozoneSub,method = "REML")
```

Alternatively, one can use tensor product splines by replacing s() by te() as follows.

```
> fitOzoneTensProd <- gam(ozone ~ te(longitude,latitude),
+                           data = ozoneSub,method = "REML")
```

The default for the univariate bases of a tensor product is cr (cubic regression spline), but other bases can be specified. For example, te(longitude, latitude, bs = c("ps","cr")) specifies a tensor product spline with a P-spline basis for the first variable and a cubic regression spline basis for the second variable.

There are many options for plotting a bivariate gam() fit. For details see help(plot.gam). Here we use plot() with scheme = 2 and hcolors = terrain.colors(1000) to produce an image plot with the terrain palette via the code:

```
> plot(fitOzoneThinPlate,scheme = 2,
+       hcolors = terrain.colors(1000),
+       main="",bty="l", cex.lab = 2,cex.axis = 2,
+       xlab = "degrees longitude",
+       ylab = "degrees latitude")
```

We also add blue circles to show the sampling locations using:

```
> points(ozoneSub$longitude,ozoneSub$latitude,
+         col = "dodgerblue")
```

The US() function in the fields package (Nychka et al. 2017) adds US state borders:

```
> library(fields) ; US(add = TRUE,lwd = 2)
```

Finally, the names and locations of major cities are added using:

```
> cityNames <- c("Chicago","Indianapolis","Milwaukee",
+                "St Louis")
> cityLonLatMat <- rbind(c(-87.6298,41.8781),
+                        c(-86.1581,39.7694),
```

```
+                             c(-87.9065,43.0389),
+                             c(-90.1994,38.6270))
> for (iCity in 1:length(cityNames))
+ {
+    points(cityLonLatMat[iCity,1],cityLonLatMat[iCity,2],
+           col = "navy",pch = 16)
+    text(cityLonLatMat[iCity,1] + 0.15,cityLonLatMat[iCity,2],
+              cityNames[iCity],adj = 0,cex = 1.8)
+ }
```

The resulting plot is in Fig. 5.1 and is interpreted as a terrain map so that the mean ozone concentration increases as the color changes from green to yellow to brown to white. The white "snow-capped peaks" show that the region of highest ozone concentration is near the city of Milwaukee.

To compare the estimates based on thin plate and tensor product splines, we can examine a scatterplot of the two sets of fitted values—the blue points in Fig. 5.2. We can see that the thin plate and tensor product fitted values are reasonably similar. One reason that they are not even more similar is that, as will be discussed next, the default numbers of thin plate and tensor product basis functions used here are too small. Increasing the sizes of these bases beyond the default values results in fits that are more similar—shown by the orange points in Fig. 5.2.

We now apply the function `gam.check()` discussed in Chap. 2 to produce some diagnostic output and plots. The following code produces Fig. 5.3 and the diagnostics that follow the code.

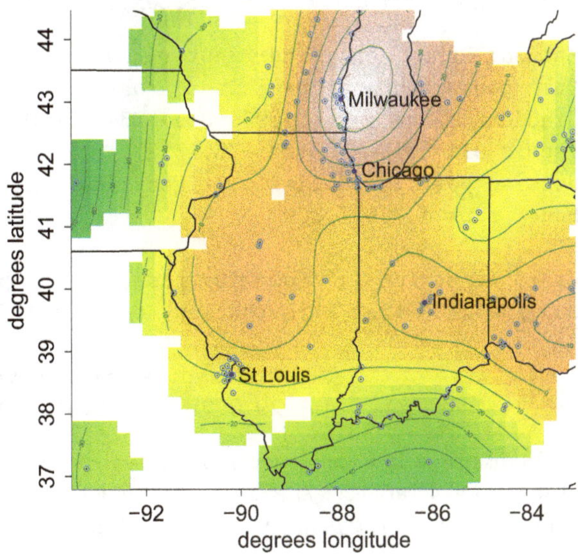

Fig. 5.1 Bivariate penalized thin plate spline estimate of mean ozone concentration as a function of degrees longitude and degrees latitude based on data in the `ozoneSub` data frame from the `HRW` package. The blue circles mark the locations where the ozone concentration was measured, and corresponds to the Midwest region of the USA. The shading indicates the height of the estimate using the `terrain.colors()` color scheme.

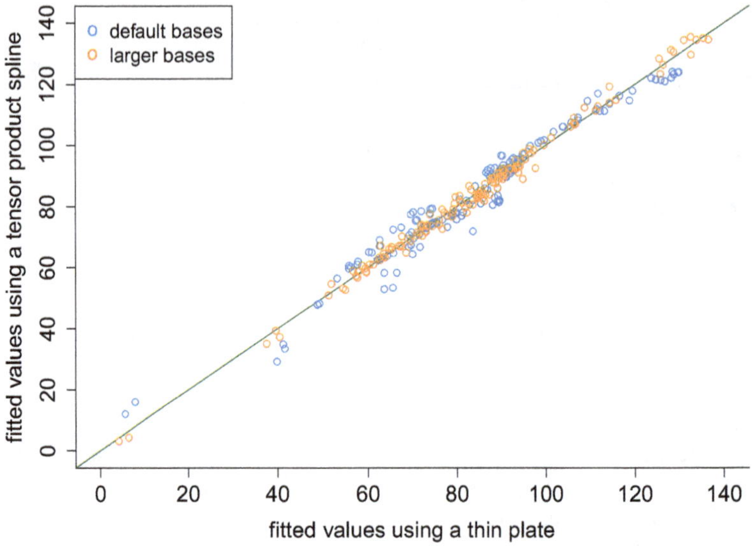

Fig. 5.2 Fitted values using the thin plate penalized spline and tensor product estimates of ozone concentration in the U.S. Midwest with default bases and larger bases. The default bases use `k = 30` for the thin plate spline and `k = c(5,5)` for the tensor product spline. The larger bases use `k = 60` for the thin plate spline and `k = c(8,8)` for the tensor product spline. The 1 : 1 line is plotted to aid comparison.

```
> library(mgcv)
> gam.check(fitOzoneThinPlate,cex.lab = 1.5,cex.main = 1.5)

Method: REML   Optimizer: outer newton
full convergence after 5 iterations.
Gradient range [-2.581699e-07,4.734945e-09]
(score 611.5318 & scale 178.1712).
Hessian positive definite, eigenvalue range [6.558447,73.64495].
Model rank =  30 / 30

Basis dimension (k) checking results. Low p-value (k-index<1)
may indicate that k is too low, especially if edf is close to k'.

                          k'  edf k-index p-value
s(longitude,latitude) 29.0 22.8    0.91   0.075
```

As discussed in Chap. 2, the `k-index` diagnostic is the ratio of two estimates of the residual variance. A `k-index` that is less than 1 is an indication that `k` is not large enough. The *p*-value is of a test of thenull hypothesis that `k` is adequate,

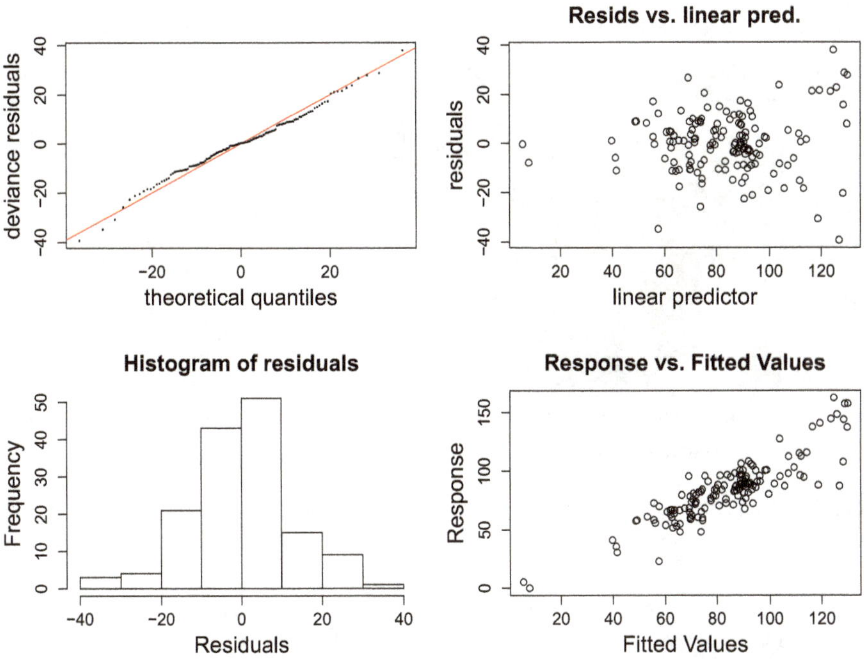

Fig. 5.3 Diagnostic plots for the penalized thin plate spline fit to ozone data based on the default number of basis functions used by the `gam()` in the `mgcv` package. These plots are obtained using the function `gam.check()`.

meaning that expected value of the `k-index` is 1. The *p*-value is calculated by randomly permuting the data to find the null distribution of the `k-index`. Note that here the `k-index` is less than 1 and the *p*-value is small. These diagnostics suggest that the default choice of `k` for the thin plate spline, which is 30 here, might be too low. An application of `gam.check()` to the tensor product fit also indicates that the default choice of the number of basis functions (25, 5 for each variable) is also too low. It is useful to compare the number of basis functions `k` with the effective degrees of freedom, `edf`. Note that `k` is an upper bound for `edf`, and `edf` is generally much smaller than `k` because of the shrinkage due to the penalty. The output from `gam.check()` suggests comparing `edf` to `k'` which is `k` -1. Although the difference between `k'` and `edf` is 6.2 and is reasonably large, an even larger difference between them would add confidence that `k` is large enough.

Both estimates were recomputed with larger bases and stored as the objects `fitOzoneThinPlate2` and `fitOzoneTensProd2`. The penalized thin plate spline has 60 basis functions. The univariate bases of the tensor product basis each have eight basis functions, so the size of the tensor product basis is 64. Increasing the number of basis functions changes the fits somewhat, and with the larger bases the

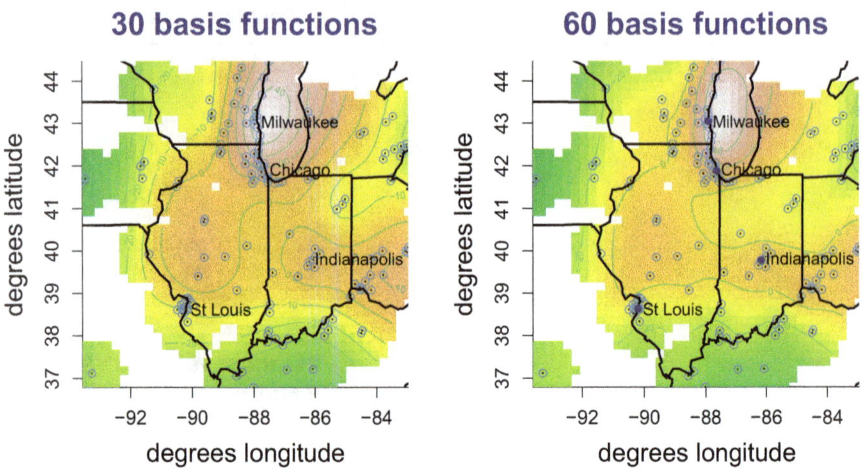

Fig. 5.4 Comparison of penalized thin plate fits to the ozone data with bases consisting of 30 and 60 basis functions.

thin spline and tensor product fits agree more than before, as can be seen in Fig. 5.2. Figure 5.4 compares the thin-plate fits with 30 and 60 basis functions. The plot of the left side of Fig. 5.4 is the same as in Fig. 5.1. The two fits in Fig. 5.4 are similar, but their details differ.

```
> gam.check(fitOzoneThinPlate2)

Method: REML   Optimizer: outer newton
full convergence after 4 iterations.
Gradient range [-8.108796e-06,1.892707e-06]
(score 606.6715 & scale 138.1293).
Hessian positive definite, eigenvalue range [5.656503,76.2219].
Model rank =  60 / 60

Basis dimension (k) checking results. Low p-value (k-index<1)
may indicate that k is too low, especially if edf is close
to k'.

                          k'  edf k-index p-value
s(longitude,latitude) 59.0 35.5     1.1    0.94

> gam.check(fitOzoneTensProd2)

Method: GCV   Optimizer: magic
Smoothing parameter selection converged after 6 iterations.
The RMS GCV score gradient at convergence was 0.00312892.
The Hessian was positive definite.
```

```
Model rank =  64 / 64

Basis dimension (k) checking results. Low p-value (k-index<1)
may indicate that k is too low, especially if edf is close to k'.

                              k'  edf k-index p-value
te(longitude,latitude) 63.0 33.3    1.11    0.92
```

We see from the `gam.check()` results that for both the thin plate and tensor product fits, with larger bases, the `k-index` is greater than 1 and the p-value is large. Also, the `edf` values are greater than those using the previous values of `k`.

With the larger bases, `edf` is considerably smaller than `k`. These results suggest that the default values of `k` used earlier were too small, but that `k` is now adequate for both spline fits.

In Fig. 5.3, the quantile-quantile plot shows that the assumed normal error distribution is reasonable, but the plot of the residuals against the fitted values shows some indication of heteroscedasticity—the residuals are more scattered where the fitted values are larger. The heteroscedasticity is seen more easily if we plot the absolute residuals versus the fitted values. In fact, the correlation between the fitted values and absolute residuals is 0.35. The heteroscedasticity could be reduced by applying a square-root variance-stabilizing transformation to ozone concentration. A variance-stabilizing transformation is advisable if one is constructing prediction intervals, because without a constant residual variance the prediction intervals will be too narrow (wide) where the residual variance is smaller (larger) than typical. See Carroll and Ruppert (1988) for a full discussion of transformations and for methods to back-transform results to the original scale, e.g., here to predict ozone concentration, not its square-root.

The image plots produced by `plot()` for `gam()` fit objects switch off pixels where the spatial data are sparse. If this is not done, then the image plot shades tend to be dominated by extrapolations of the fitted surface beyond the spatial data, resulting in a less meaningful display. Having the pixels confined to a user-specified region is also worthwhile—especially when the spatial data are distributed over a very complicated region. Figure 5.5 is an alternative image plot of the $k = 60$ surface with the switched-on pixels confined to the convex hull of the spatial data. It involves use of the function `chull()` for convex hull determination and the **HRW** package function `pointsInPoly()` for determining which points are inside the convex hull. This display also has a legend that conveys the height of the surface, courtesy of the function `image.plot()` from the **fields** package. Exercise 1

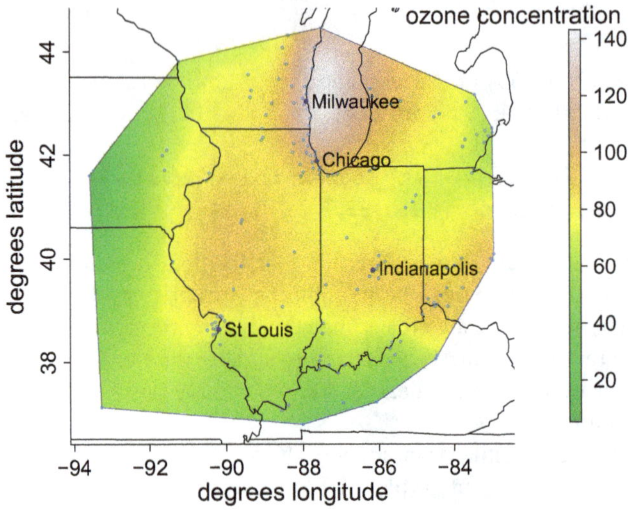

Fig. 5.5 A more sophisticated display of the fit shown in Fig. 5.1 with (a) the image pixels restricted to the convex hull of the geographical locations at which ozone concentrations were recorded, (b) a legend bar showing how the image shades relate to mean ozone concentration, and (c) names and locations of major cities in the region.

provides tuition on `image.plot()`. The code that produced Fig. 5.5 is in the script `ozoneDisplayConvHull.R` and can be run via the following commands:

```
> library(HRW) ; demo(ozoneDisplayConvHull,package = "HRW")
```

To access and, if desired, copy and edit `ozoneDisplayConvHull.R` run the following code to determine its location on the computer on which `HRW` is installed:

```
> system.file("demo","ozoneDisplayConvHull.R",package = "HRW")
```

The convex hull of the spatial data does not always lead to a good region for switching on pixels in image plots. Section 5.3.1 and Exercises 2 and 7 deal with bivariate-component semiparametric regression analyses of the Sydney real estate data analyzed in Sect. 5.3.1. In this example, the spatial data are distributed over region that is far from convex and the convex hull is not a good pixel switch-on boundary. In such circumstances a good boundary needs to be chosen by eye. The `createBoundary()` function in the `HRW` package facilitates manual creation of a polygonal boundary. The following commands:

```
> library(HRW) ; data(ozoneSub)
> myOzoneBdry <- createBoundary(ozoneSub$longitude,
+                               ozoneSub$latitude)
```

lead to the opening of a graphics window with points plotted at the locations of the ozone observations. Using the mouse or touchpad and following the instruction in the R console, one clicks on the points that will be the vertices of the boundary.

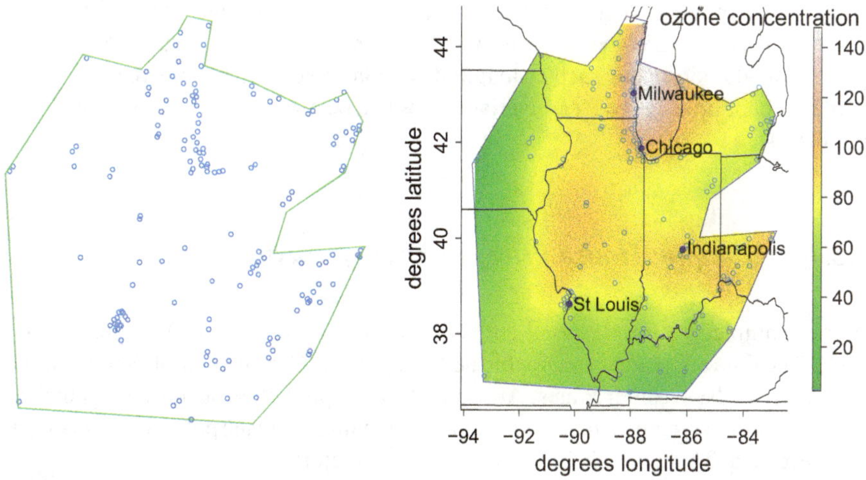

Fig. 5.6 Left panel: A polygonal boundary for the ozone data running example selected with interactive graphics and mouse or touchpad clicks using the function `createBoundary()` in the `HRW` package. Right panel: image plot representation of the bivariate penalized spline surface with pixels confined to selected polygon.

The vertices matrix `myOzoneBdry` can be stored for possible future use in the file `myOzoneBdry.txt` using the command:

```
> write.table(myOzoneBdry,"myOzoneBdry.txt",col.names = FALSE,
+             row.names = FALSE)
```

Figure 5.6 provides an illustration of the `createBoundary()` approach. The left panel shows a polygon obtained using the `createBoundary()` graphics window. The right panel shows the image plot of the surface with switched-on pixels confined to this polygon.

Other options for displaying a fitted bivariate surface include perspective plots using the function `persp()` and three-dimensional spin graphics using the function `rgl.surface()` in the package `rgl` (Adler et al. 2017). These alternative methods of display are illustrated in Exercises 1 and 3.

5.3 Geoadditive Models

Geoadditive models (e.g. Kammann and Wand 2003) are an extension of additive models that allow for bivariate function components. For example, the model

$$y_i = \beta_0 + \beta_1 x_{1i} + \beta_2 x_{2i} + f_3(x_{3i}) + f_4(x_{4i}) + f_{56}(x_{5i}, x_{6i}) + \varepsilon_i, \quad \varepsilon_i \overset{\text{ind.}}{\sim} N(0, \sigma^2),$$

for $1 \leq i \leq n$ is a geoadditive model due to the presence of the bivariate function f_{56} of the predictors x_5 and x_6. Often the bivariate function corresponds to geographical coordinates such as longitude and latitude, which is the reason for the name "geoadditive." However, bivariate terms can be used for any pair of continuous predictors.

5.3.1 *Example: House Prices in Sydney, Australia*

As an example, we use data on houses that were sold in Sydney, Australia, during 2001. The data are part of an unpublished study by A. Chernih and M. Sherris at the University of New South Wales, Australia. Several predictors such as geographical position, socioeconomic status, proximity to amenities, and air pollution levels were also recorded. There are 37,676 cases and 38 predictors.

The following code loads the libraries that will be needed and processes the Sydney housing price data:

```
> library(mgcv) ; library(HRW) ; data(SydneyRealEstate)
> logSalePrice <- SydneyRealEstate$logSalePrice
> longitude <- SydneyRealEstate$longitude
> latitude <- SydneyRealEstate$latitude
> income <- SydneyRealEstate$income
> PM10 <- SydneyRealEstate$PM10
```

The variables used here are:

`logSalePrice`	natural logarithm of sale price in Australian dollars,
`longitude`	degrees longitude of location of house,
`latitude`	degrees latitude of location of house,
`income`	mean weekly household income in Australian dollars of the administrative district where the house is located,
`PM10`	mean daily value of PM_{10} (particulate matter with a diameter less than 10 micrometers).

Next we plot a scatterplot matrix of selected predictors. A random subset of 500 cases is used to improve visualization. The plot is in Fig. 5.7. One can see from the plot that price tends to increase as one moves west to east, as `income` increases, and, to a lesser extent, as `PM10` decreases.

As an illustration, we fit a geoadditive model with only two predictors, `PM10` and `income`, in addition to the spatial coordinates that are fit with a penalized thin plate spline. The response is the logarithm of price. An application of `gam.check()` to a preliminary run showed that the default choice of `k` for `income` and `(longitude,latitude)` might be too small. These defaults are 10 and 30, respectively. Instead, we use 15 and 80.

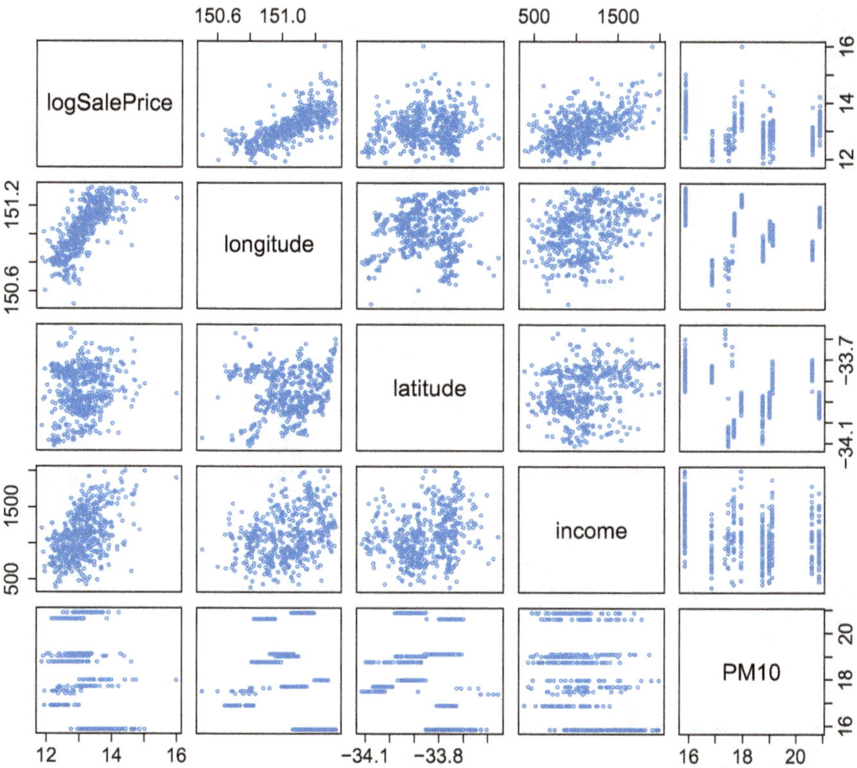

Fig. 5.7 A scatterplot matrix of log price, the geographical coordinates, and two additional variables, PM10 and income from the Sydney real estate data. To aid visualization, this display involves of a random subset of 500 houses.

```
> fitGeoadd <- gam(logSalePrice ~ s(income,k = 15) + s(PM10)
+                   + s(longitude,latitude,bs = "tp", k = 80),
+                     method = "REML")

> summary(fitGeoadd)

Family: gaussian
Link function: identity

Formula:
logSalePrice ~ s(income, k = 15) + s(PM10)
      + s(longitude, latitude,bs = "tp", k = 80)

Parametric coefficients:
             Estimate Std. Error t value Pr(>|t|)
(Intercept) 13.079340   0.001742    7510   <2e-16
```

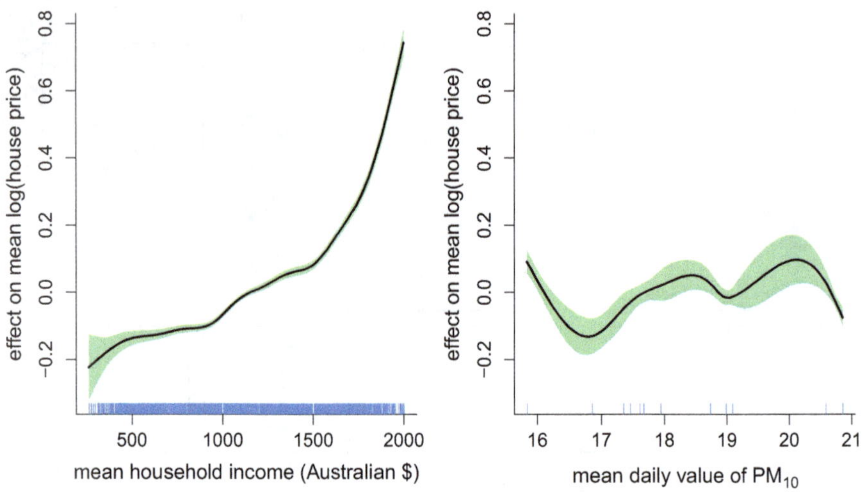

Fig. 5.8 Estimated univariate smooth function components for the geoadditive model fit `fitGeoadd` to the Sydney real estate data using the function `gam()` in the package `mgcv`. The effects are vertically centered around zero. The shaded regions indicate 95% pointwise confidence intervals.

```
Approximate significance of smooth terms:
                             edf Ref.df      F p-value
s(income)                 11.134 12.692 221.42  <2e-16
s(PM10)                    6.842  7.688  12.96  <2e-16
s(longitude,latitude) 76.253 78.789 126.78  <2e-16

R-sq.(adj) =  0.668   Deviance explained = 66.9%
-REML =  12813  Scale est. = 0.11427   n = 37676
```

The results indicate that location, `PM10`, and `income` all are highly significant. Of course, with a sample size of 37,676 it is entirely possible for an effect to be very small and yet highly significant. Therefore, to examine the size of the effects, we plot the component functions for `income`, `PM10`, and location. These plots were produced by setting the argument `select` equal to 1, 2, and 3, respectively, in the `plot()` function. See the `select` argument in `help(plot.gam)` for details. The plot showing the effects of `PM10` and `income` is in Fig. 5.8. The third plot, a contour plot, did not show detail adequately, so it is not shown here and an alternative plot is constructed below.

The estimated effect of `PM10` is not monotonic, which is counter-intuitive. The non-monotonicity might be due to the effects of predictors that have not been put in the model. We see in Fig. 5.7 that `PM10` has a discrete distribution; in fact, it takes only 12 distinct values and these are somewhat irregularly spaced. These features of the distribution of `PM10` explain why the width of the variability bands for `PM10`'s

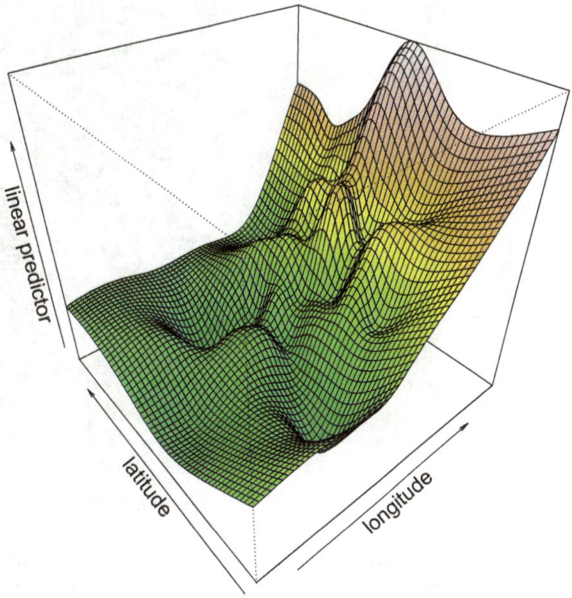

Fig. 5.9 The joint effect of `longitude` and `latitude` on mean `logSalePrice` for the geoadditive model fit to the Sydney real estate data. This plot was produced by the function `vis.gam()` in the package `mgcv`.

effect varies greatly. The difference between the minimum and maximum values of the component function for `PM10` is approximately 0.4. Recall that we are modeling the mean logarithm of price. Since $e^{0.4} \approx 1.5$, the effect of `PM10` could be as much as a 50% net increase in price. No cause-and-effect is implied here. `PM10` might simply be a surrogate for a variable that affects price such as proximity to a major highway or industrial site. Moreover, the minimum and maximum occur where the confidence intervals are widest and probably overestimate the difference between the minimum and maximum. Also, the maximum occurring at a *high* value of `PM10` is somewhat counterintuitive.

The estimated effect of `income` is monotonic and the difference between the minimum and maximum values is approximately 1. Since $e \approx 2.7$, the effect of `income` could be as much as a 170% net increase in price.

As mentioned earlier, the contour plot showing the effect of `(longitude, latitude)` produced by plotting the `gam` object did not show detail adequately. As an alternative, we produce a perspective plot using the function `vis.gam()`.

```
> vis.gam(fitGeoadd,c("longitude","latitude"),
+          color = "terrain",n.grid = 60,theta = 320,phi = 35)
```

The plot is in Fig. 5.9. The value of `n.grid`, the number of grid points on each axis, was increased from its default value of 30–60 to show more detail.

The surface in Fig. 5.9 is over the rectangular region defined by the range of the longitude and latitude values. However, since the geographical data are over a highly irregular region corresponding to Sydney's residential areas, surface plots over the range-defined rectangle can be misleading. In Fig. 5.10 we show the same surface

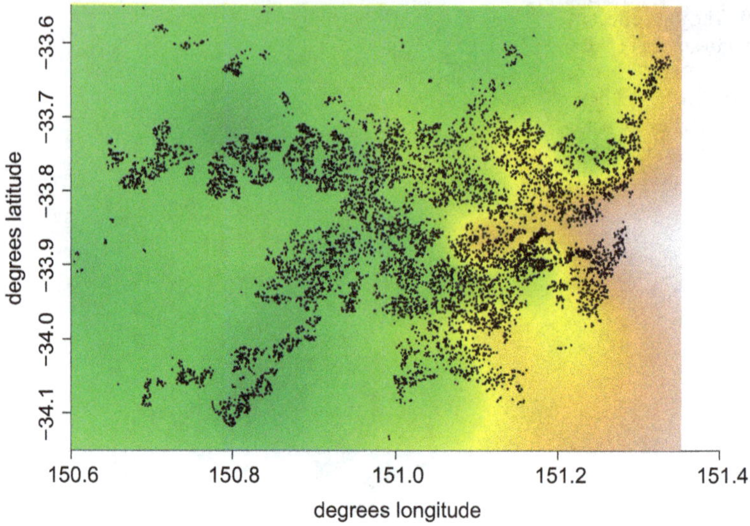

Fig. 5.10 The surface from Fig. 5.9 shown as an image plot. The points show locations of random sample of 10,000 houses from the `SydneyRealEstate` data frame.

as an image plot but also add points corresponding to a random sample of 10,000 houses from the dataset. It is apparent from this that the surface bends upwards substantially to the south-east of the data points corresponding to a section of the Pacific Ocean. Such aberrant extrapolation leads to a misleading display.

As discussed at the end of Sect. 5.2, confining the switched-on pixels to a region defined by the spatial data leads to a more meaningful image plot. We used the `createBoundary()` function in the `HRW` package to create a suitable polygonal boundary and stored the vertices in the data frame `SydneyRealEstateBdry`.

Figure 5.11 improves the Fig. 5.10 image plot by restricting the switched-on pixels to the `SydneyRealEstateBdry` polygon. The `pointsInPoly()` function used to determine the points inside of the Sydney boundary is in the `HRW` package. The locations and names of 15 Sydney suburbs are also plotted and a legend bar is included. The heights in the legend bar correspond to mean log sale price as a function of geographic location with the other predictors, `PM10` and `income`, set to their means. The code that produced Fig. 5.11 is in the script `SydneyDisplaySophis.R` within the `HRW` package and can be run as follows:

```
> library(HRW) ; demo(SydneyDisplaySophis,package = "HRW")
```

To locate, and possibly copy and modify, `SydneyDisplaySophis.R` note that its location can be found using:

```
> system.file("demo","SydneyDisplaySophis.R",package = "HRW")
```

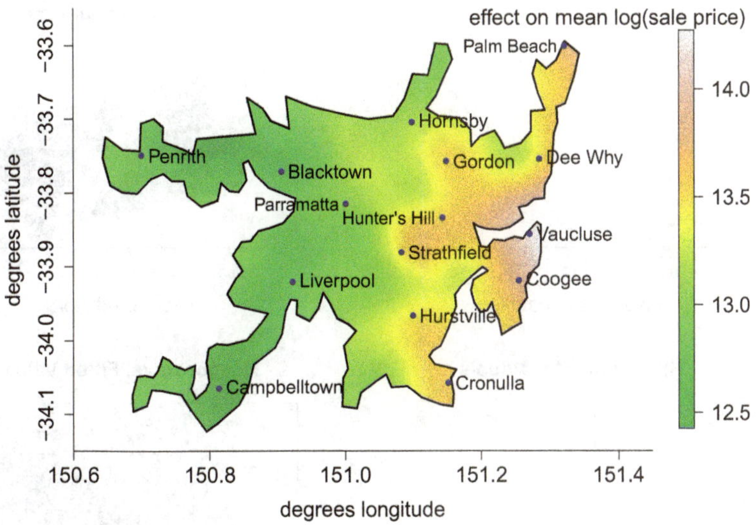

Fig. 5.11 The effect of `longitude` and `latitude` on mean `logSalePrice` of Sydney houses when `PM10` and `income` are fixed at their means. A penalized thin plate spline with 80 basis functions is used for geographic location. The names and locations of 15 Sydney suburbs are also shown.

From Fig. 5.11 we see that mean house prices reach a peak around the suburb of Vaucluse, which is near where Sydney Harbor meets the Pacific Ocean. The prices then drop as one moves from east to west, away from Sydney's ocean beaches. There are smaller effects due to north-to-south changes Some localized nonlinear effects are also apparent.

Next, we run some diagnostics using `gam.check()`. The resulting plots are in Fig. 5.12.

```
> gam.check(fitGeoadd,cex.lab = 1.5,cex.main = 1.5)

Method: REML   Optimizer: outer newton
full convergence after 8 iterations.
Gradient range [-7.427945e-06,3.245751e-06]
(score 12813.26 & scale 0.1142656).
Hessian positive definite, eigenvalue range [1.528962,18835.58].
Model rank =  103 / 103

Basis dimension (k) checking results. Low p-value (k-index<1)
may indicate that k is too low, especially if edf is close to k'.
```

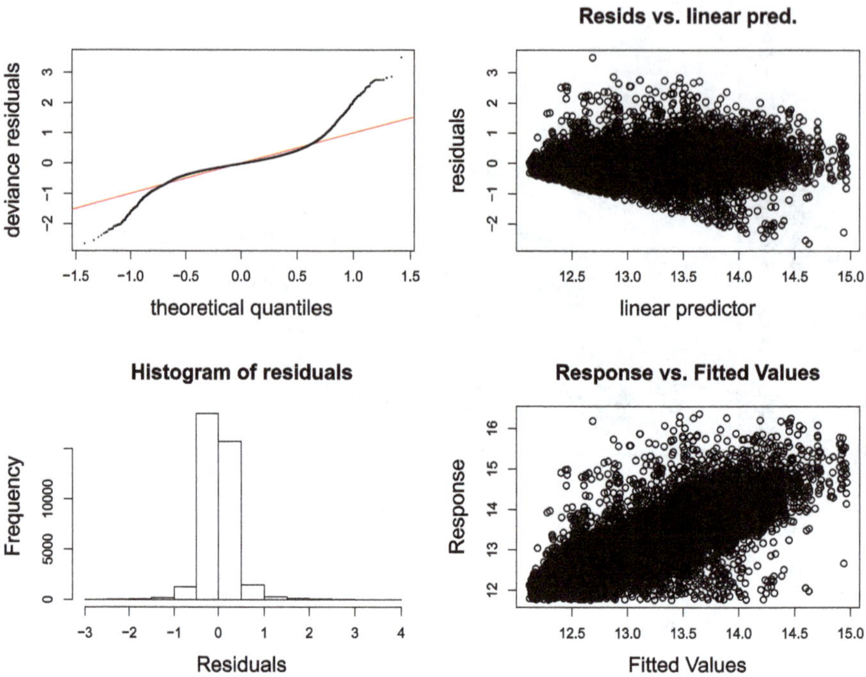

Fig. 5.12 Diagnostic plots for the geoadditive model stored in the fit object `fitGeoadd`.

```
                            k'   edf k-index p-value
s(income)                14.00 11.13    0.97   0.025
s(PM10)                   9.00  6.84    1.01   0.850
s(longitude,latitude) 79.00 76.25    0.84  <2e-16
```

All four plots in Fig. 5.12 show strong evidence of outliers. In particular, the quantile-quantile plot is concave on the left and convex on the right which is a clear sign of heavier than Gaussian tails. (The theoretical quantiles are on the horizontal axis and the residuals are on the vertical axis. If the axes were reversed, as is often done with quantile-quantile plots, then the pattern would have been convex on the left and concave on the right. When examining quantile-quantile plots, it is important to check which variables are on the horizontal and vertical axes.)

There is some evidence of heteroscedasticity. The correlation between the fitted values and the absolute residuals is 0.22. The residuals are somewhat right-skewed and the skewness of the residuals decreases as the fitted values increase. The following code shows that the skewness coefficient of the residuals is 1.87, 1.08, 1.58, 1.01, and 0.05 for fitted values in the first to fifth quintile, respectively. The skewness coefficient for the entire set of residuals is 0.76, which indicates only a small amount of right skewness. Also, the standard deviations of the residuals for these quintiles are 0.24, 0.28, 0.32, 0.35, and 0.46, which reflects the heteroscedasticity just noted. The skewness coefficient is the third central moment divided by the standard deviation raised to the 3/2 power. It is location- and scale-

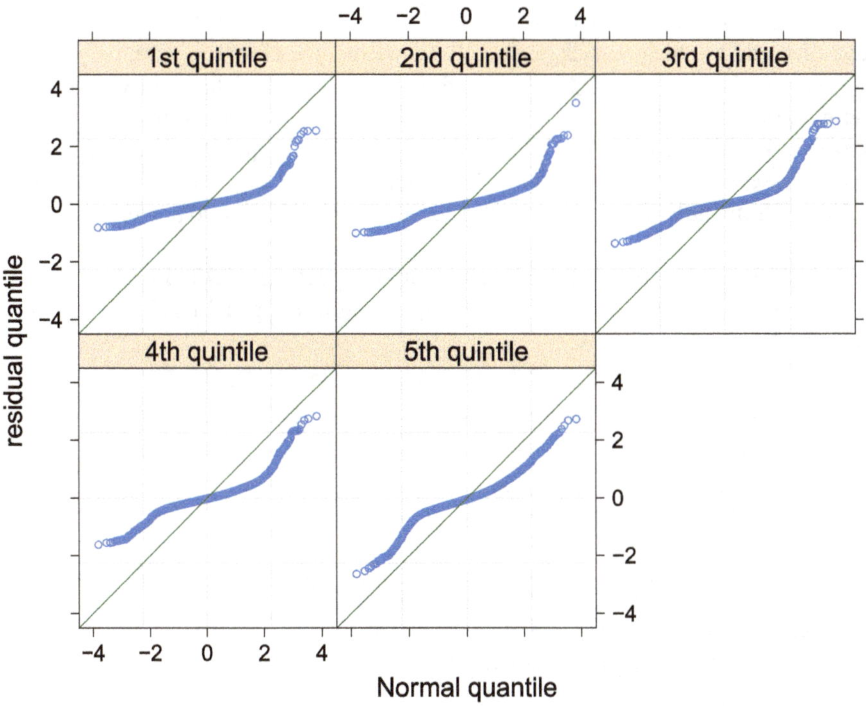

Fig. 5.13 Normal quantile-quantile plots of the residuals from the geoadditive model conditioned upon the quintiles of the fitted values. Notice that the skewness decreases and the spread increases with increasing quintile.

free and a positive (negative) value indicates right (left) skewness with larger magnitudes indicating more skewness. The function `skewness()` is in the `e1071` package. To further illustrate the changes in skewness and heteroscedasticity, the quantile-quantile plots of the residuals for each quintile are shown in Fig. 5.13.

```
> library(e1071) # for the skewness() function
> quintVec <- c(10,quantile(fitted(fitGeoadd),
+                  c(0.2,0.4,0.6,0.8)),20)
> quintInds <- cut(fitted(fitGeoadd),breaks = quintVec,
+                      labels = FALSE)
> print(round(tapply(residuals(fitGeoadd),quintInds,
+              skewness),2))
> print(round(skewness(residuals(fitGeoadd)),2))
> print(round(tapply(residuals(fitGeoadd),quintInds,sd),2))
```

In the code above, `quintInds` indicates the quintile (of the fitted log price) to which a house belongs, so that `quintInds` equals 1 for a house at or below the 20th percentile, equals 2 for a house above the 20th percentile and at or below the

40th percentile, and so forth. Also, tapply() applies the functions skewness() and then sd() to the residuals separately for each value of quintInds.

We can also construct a Normal quantile-quantile plot of the residuals for each quintile of the fitted values. The plots are in Fig. 5.13. The qqmath() function in the lattice package is used so that we can condition on quintInds. One can see that the skewness decreases and the dispersion increases as we move from the first to the fifth quantile. See Sarkar (2008) for a complete description of the lattice package.

We now work through the R code that produced Fig. 5.13. First we obtain the character array quintIndsFac, corresponding to the panel labels in Fig. 5.13, using:

```
> library(lattice)
> quintIndsFac <- rep(NA,length(quintInds))
> quintIndsFac[quintInds==1] <- "1st quintile"
> quintIndsFac[quintInds==2] <- "2nd quintile"
> quintIndsFac[quintInds==3] <- "3rd quintile"
> quintIndsFac[quintInds==4] <- "4th quintile"
> quintIndsFac[quintInds==5] <- "5th quintile"
```

We then apply the function qqmath() within the lattice package to the residuals from the geoadditive model fit using:

```
> library(lattice)
> qqmathObj <- qqmath( ~
+       residuals(fitGeoadd)|factor(quintIndsFac),
+       xlab = list("Normal quantile",cex = 1.5),
+       ylab = list("residual quantile",cex = 1.5),
+       as.table = TRUE,
+       strip = strip.custom(par.strip.text = list(cex = 1.5)),
+       par.settings = list(layout.heights = list(strip = 1.6)),
+       scales = list(cex = 1.25),
+       xlim = c(-4.5,4.5),ylim = c(-4.5,4.5),
+       panel = function(x,...)
+       {
+          panel.grid() ; panel.qqmath(x,...)
+          panel.abline(0,1,col = "darkgreen")
+       })
> print(qqmathObj)
```

The complex distribution of the residuals implies that standard prediction intervals will be inaccurate. A remedy to this problem would be semiparametric quantile regression; see Sect. 6.2.

The residual outliers in Fig. 5.12 might be explained by using additional predictors. The residuals were plotted against all predictors in the dataset (not shown here) and it was seen that lotSize had a moderate positive relationship (correlation equal to 0.26) with the residuals.

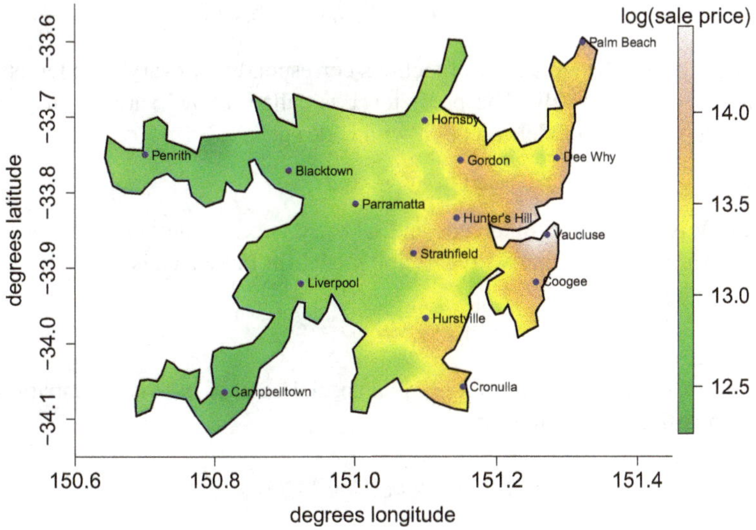

Fig. 5.14 The effect of location on log price when `PM10` and `income` are fixed at their means and a penalized thin plate spline with 200 basis functions is used for `(longitude,latitude)`. Note the increased detail compared to Fig. 5.11.

When the model was enlarged to include `lotSize` it was found that `lotSize` was highly significant and the adjusted R^2 increased slightly from 0.653 to 0.69, but the residual outliers were still abundant.

Also note that `k-index` values are closer to 1 than before (not shown here), but the `k-index` is still rather small for the spatial coordinates. Increasing `k` to 120 (200) for `(longitude,latitude)` further increases the `k-index` but only to 0.902 (0.917). Because of the very large sample size (37,676), it is possible to estimate the effect of location with high resolution by using a very large value of `k`. Whether or not resolving such fine detail is useful depends on the goals of the analysis. The fit in Fig. 5.14 is the same as in Fig. 5.11 except that `k` is 200 instead of 80. It shows more fine detail in the effect of geography on mean log house price.

In this section we have used only `longitude`, `latitude`, and two other predictors to illustrate geoadditive model analysis in R. Geoadditive model analysis involving the full set of candidate predictors is treated in Exercise 7.

5.4 Varying-Coefficient Models

A *varying-coefficient model* has a linear effect for one predictor with the slope and intercept being smooth functions of other predictors. For example if, for $1 \le i \le n$, y_i is the ith response value and (t_i, x_i) is the ith predictor vector, then a varying-coefficient model is

$$y_i = \beta_0(t_i) + \beta_1(t_i)\,x_i + \epsilon_i,$$

where $\beta_0(\cdot)$ and $\beta_1(\cdot)$ are smooth functions corresponding to varying-intercept and varying-slope, respectively. The predictor x is called a "by" variable in `gam()`'s nomenclature and the model can be fit with code such as:

```
> fit <- gam(y ~ s(t) + s(t,by = x))
```

Here `s(t)` is the varying-intercept and `s(t,by = x)` is the varying-slope. The general bivariate nonparametric regression model for these data is

$$y_i = f(t_i, x_i) + \epsilon_i.$$

Hence, the varying-coefficient model is a special case of bivariate nonparametric regression with the restricted form

$$f(t, x) = \beta_0(t) + \beta_1(t)\,x$$

for smooth functions $\beta_0(\cdot)$ and $\beta_1(\cdot)$.

5.4.1 *Example: Daily Stock Returns*

To illustrate varying-coefficient models we use data on daily returns on stock of the General Electric company and the Standard & Poor's 500 (S&P 500) index. The data are in the data frame `capm` in the `HRW` package. The variable `Date` runs from November 1, 1993 to March 31, 2003, so there are over nine years of data. If P_t is the price of a stock on day t, then the log-return on that day is $\log(P_t/P_{t-1})$. The excess log-return is the log-return minus the risk-free interest rate, usually taken to be the short-term Treasury bill rate. The following code computes the excess log-return on both General Electric stock and the S&P 500 index:

```
> library(HRW) ; data(capm) ; n <- dim(capm)[1]
> riskfree <- capm$Close.tbill[2:n]/(100*365)
> elrGE <- diff(log(capm$Close.ge)) - riskfree
> elrSP500 <- diff(log(capm$Close.sp500))  - riskfree
```

Note that `capm$Close.sp500` and `capm$Close.ge` are the daily closing prices and `capm$Close.tbill` is the daily Treasury bill rate expressed as a percentage. In the code, `capm$Close.tbill` is first divided by 100 to convert from a percentage to a fraction and then divided by 365 to convert from a yearly to a daily rate. Hence, `elrGE` and `elrSP500` are the daily excess log-returns on General Electric stock and on the S&P 500 index, respectively.

One of the assumptions of the Capital Asset Pricing Model in finance is that mean excess log-returns on a stock depend linearly on the excess log-returns on the market (e.g. Ruppert and Matteson 2015). The return on the S&P 500 index is often used as a proxy for the return on the market. Usually the simple linear regression model is fit to only recent data, since the slope is expected to change slowly over time. An alternative is to fit a varying-coefficient model to all of the data, with the intercept and slope depending on the date. Doing this would provide evidence as to whether and how fast the slope is changing. Economic theory predicts that the intercept is zero, and so the estimated intercept is expected to be negligible.

First, we fit a simple linear regression model to all of the data using:

```
> library(mgcv) ; fitSLR <- gam(elrGE ~ elrSP500)
> summary(fitSLR)

Family: gaussian
Link function: identity

Formula:
elrGE ~ elrSP500

Parametric coefficients:
             Estimate Std. Error t value Pr(>|t|)
(Intercept) 0.0002971  0.0002628   1.131    0.258
elrSP500    1.2441346  0.0227404  54.710   <2e-16


R-sq.(adj) =  0.559   Deviance explained = 55.9%
GCV = 0.00016324  Scale est. = 0.0001631  n = 2362
```

To check whether a linear model is appropriate, we fit the nonparametric regression model where the excess log-return on General Electric stock is a smooth, time-invariant function of the excess log-return on the S&P 500 index. This model is compared with the simple linear model via an F-test.

```
> fitNPR <- gam(elrGE ~ s(elrSP500),method = "REML")
> anova(fitSLR,fitNPR,test = "F")

Analysis of Deviance Table

Model 1: elrGE ~ elrSP500
Model 2: elrGE ~ s(elrSP500)
  Resid. Df Resid. Dev     Df  Deviance      F    Pr(>F)
1    2360.0    0.38492
2    2352.2    0.37942 7.7743 0.0054985 4.3882 3.497e-05
```

The F-test rejects the simple linear regression model. However, the sample size is quite large since there were 2362 days of returns and, hence, there is substantial

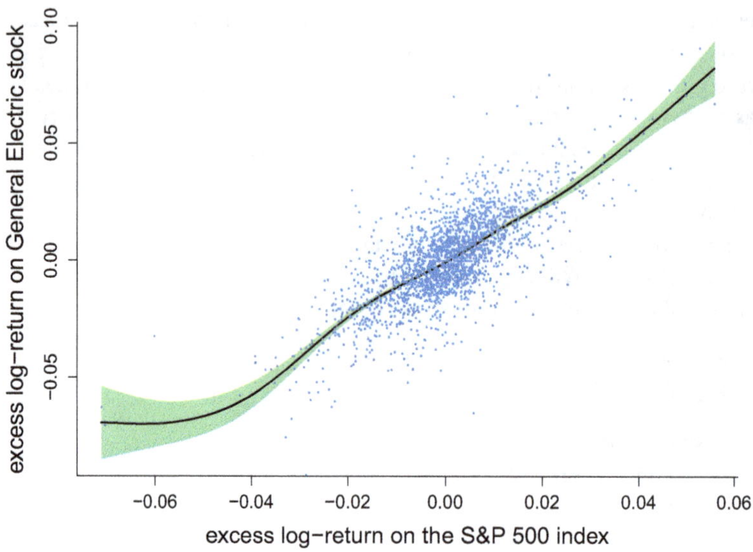

Fig. 5.15 Penalized spline fit of the excess log-returns of General Electric stock on the excess log-returns of the S&P 500 index using the function `gam()` in the package `mgcv`. The pale green shaded region shows 95% pointwise confidence intervals.

power to detect departures from linearity. The plot of the nonparametric regression fit (`fitNPR`) is in Fig. 5.15 and deviates only slightly from a straight line, at least over the range of the bulk of the data. Moreover, the smooth fit has only a slightly larger adjusted R^2, 56.4%, compared with that of the linear model, 55.9%. When we turn to varying-coefficient models, as a working assumption we will assume that the regression of General Electric excess log-returns on market excess log-returns is linear, but with the intercept and slope varying with time.

Now, we fit and plot the varying-coefficient model and compare it the linear, constant-coefficient model by an F-test. In the varying-coefficient model, the intercept and slope of the linear regression of the excess log-return of General Electric stock on the excess log-return of the S&P 500 index are smooth functions of time in years, stored as the variable `t`:

```
> dayNums <- (1:(n-1))/(n-1)
> startTime <- 1993 + 11/12 ; endTime <- 2003 + 3/12
> t <- startTime + (endTime - startTime)*dayNums

> fitVCM <- gam(elrGE ~ s(t) + s(t,by = elrSP500),
+               method = "REML")
> anova(fitSLR,fitVCM,test = "F")

Analysis of Deviance Table
```

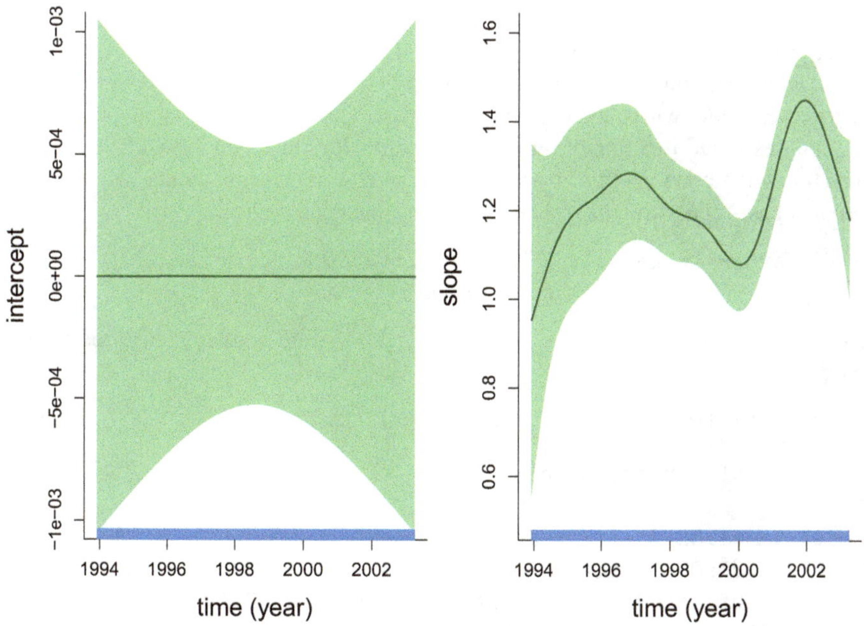

Fig. 5.16 Varying-intercept and varying-slope for the fitted varying-coefficient model of General Electric excess log-returns on S&P 500 excess log-returns. The shaded regions correspond to pointwise 95% confidence intervals.

```
Model 1: elrGE ~ elrSP500
Model 2: elrGE ~ s(t) + s(t, by = elrSP500)
  Resid. Df Resid. Dev     Df  Deviance      F    Pr(>F)
1    2360.0    0.38492
2    2350.9    0.37953 9.0903 0.0053875 3.6744 0.0001309
```

The intercept and slope functions are plotted in Fig. 5.16. The argument `select` of the `gam()` is used to select which component to plot. The code `plot(fitVCM)` without the use of `select` would plot both the intercept and slope but would use the same scale on the y-axis and would not allow us to specify separate values of `ylab` for the two plots. The argument `seWithMean = TRUE` is used to specify that the confidence intervals should include uncertainty about the mean. Basic versions of the plots in Fig. 5.16 are achieved via:

```
> par(mfrow=c(1,2))
> plot(fitVCM,select = 1,ylim = c(-0.001,0.001),
+      seWithMean = TRUE)
> plot(fitVCM,select = 2,ylim = c(0.5,1.6),seWithMean = TRUE)
```

The estimate of the intercept suggests that the intercept was zero, or nearly so, during the entire period, which agrees with economic theory. The estimated slope varies between approximately 1.0 and 1.4 over this period. The simple linear regression model is rejected in favor of the varying-coefficient model, although the latter provides only a small improvement in fit measured by the adjusted R^2.

A generalization of the varying-coefficient model is the bivariate nonparametric regression model where the excess log-return of General Electric stock is modeled as an unrestricted function of time and the excess log-return on the S&P 500 index. The following code fits this bivariate nonparametric regression model with a tensor product basis consisting of 25 basis functions in each direction:

```
> fitBivNPR <- gam(elrGE ~ te(t,elrSP500,k = rep(25,2)),
+                    method = "REML")
```

Next, we compare the varying-coefficient and bivariate nonparametric regression models via an F-test:

```
> anova(fitVCM,fitBivNPR,test = "F")

Analysis of Deviance Table

Model 1: elrGE ~ s(t) + s(t, by = elrSP500)
Model 2: elrGE ~ te(t, elrSP500, k = rep(25, 2))
  Resid. Df Resid. Dev     Df   Deviance     F Pr(>F)
1    2350.9    0.37953
2    2341.9    0.37903 8.9655 0.00050177 0.347 0.9588
```

The F-test indicates that there is little difference between the two models, and this is apparent in Fig. 5.17 which compares the two surfaces via image plots. Here we use the `topo.colors()` palette since it shows the slight differences better than `terrain.colors()`. This color scheme corresponds to topographical maps. For

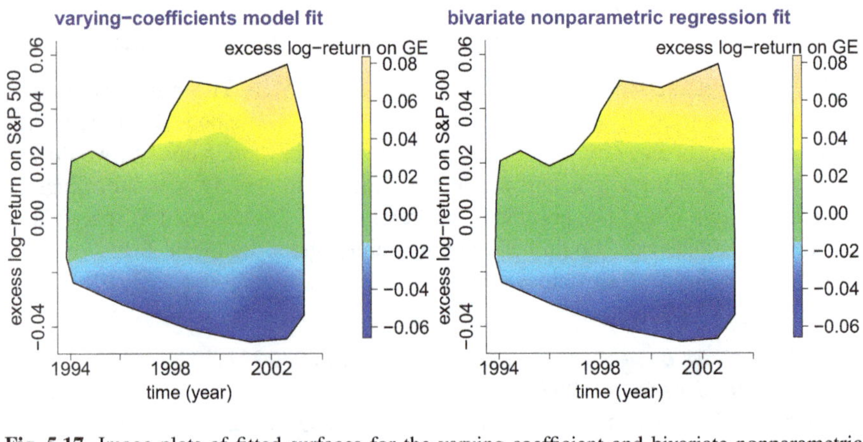

Fig. 5.17 Image plots of fitted surfaces for the varying-coefficient and bivariate nonparametric regression models. In the former, the mean response is linear in the excess log-return of S&P 500 stock with a slope that is a smooth function of time. In the latter, the mean response is a unrestricted bivariate function of excess log-return and time.

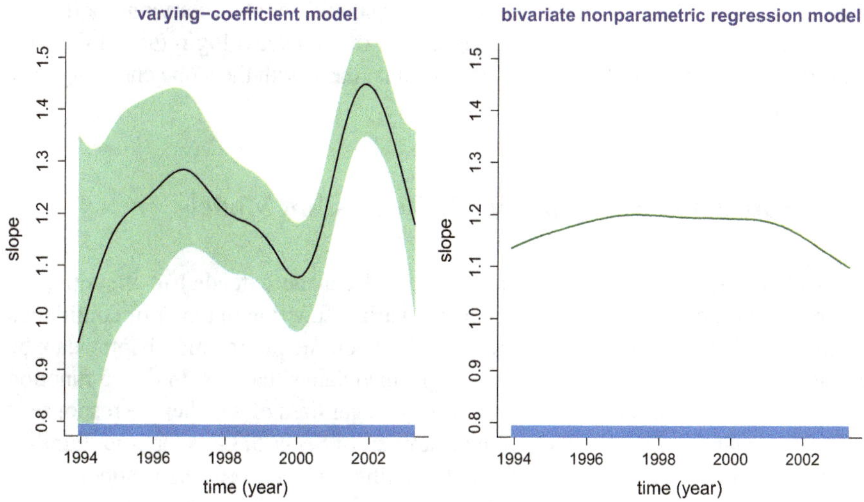

Fig. 5.18 Left panel: the slope of the fitted surface of the varying-coefficient model as a function of time, with corresponding 95% pointwise confidence interval. Right panel: the slope of the fitted surface of the bivariate nonparametric regression model at the slice corresponding to excess log-return of S&P 500 stock equaling zero.

example, lighter blue corresponds to "near sea-level" and darker blue corresponds to "below sea-level."

Both the varying-coefficient and bivariate fits in Fig. 5.17 suggest that the slope of the regression of `elrGE` on `elrSP500` might increase over time, although perhaps not monotonically, but it is difficult to tell for sure from these plots alone. To compare the estimated slopes more clearly, we create additional plots with the code below. Unlike the case of the varying-coefficient model, for the bivariate fit the slope (partial derivative) with respect to `elrSP500` is not constant as `elrSP500` varies with `t` fixed. Therefore, we plot a numerical partial derivative with respect to `elrSP500` at `elrSP500` set to zero, which is approximately the mean of this variable. The plots are in Fig. 5.18; the left panel is the same as the right panel in Fig. 5.16. The bivariate nonparametric regression model suggests that there is a change in the slope over time, but if there is a change at all it seems to be very small. The estimated slope from the varying-coefficient model varies considerably and is non-monotonic. Code for production of Fig. 5.18 is in the script `returnsDrvFig.R` within the `HRW` package. The script can be run and located on the computer where `HRW` is stored by issuing the commands:

```
> library(HRW) ; demo(returnsDrvFig,package = "HRW")
> system.file("demo","returnsDrvFig.R",package = "HRW")
```

Occam's razor states that among competing hypotheses that are consistent with the data, one should choose the least complex. Since the varying-coefficient model is a special case of the bivariate nonparametric regression model and fits the data as well as the latter, Occam's razor implies that we should adopt the varying-coefficient model. That model suggests that the regression of the excess log-return of General Electric stock on the market excess log-return is linear with the slope changing over the time period from `1-Nov-93` to `31-Mar-03`.

5.5 Additional Semiparametric Regression Models

Each of the models covered in Chaps. 3 and 4 can be extended to the situation where one or more of the components is a bivariate function of a pair of continuous predictors. Specifically, each of the models given so far in this chapter can be extended to the generalized response and grouped data situations. In R, the function `gam()` supports bivariate function components regardless of whether the response is in the Gaussian family, as in each of the examples of Sects. 5.2–5.4, or other families such as the Binomial, Poisson, or Gamma families. The `gamm()` also supports non-Gaussian response distributions with bivariate function components for grouped data models. Similar comments apply to approaches based on R's mixed model and Markov chain Monte Carlo software.

5.6 Covariance Function Estimation

Functional data analysis is the field of statistics where the data are functions. We will discuss only two topics in functional data analysis. The first topic, estimation of covariance functions including principal components analysis, is discussed in this section and the next. The second topic is presented in Chap. 6 and is regression where either the responses, the predictors, or both are functions. Examples of functional data include electroencephalography signals which are functions of time, temperatures which can be functions of time of day or date of the year, and spectra which are functions of wavelength. An example of the latter are the near-infrared reflectance spectra of 60 gasoline samples in the `gasoline` dataset of the `refund` package (Goldsmith et al. 2016).

As will be explained shortly, functional data analysis generalizes many techniques of multivariate analysis from vectors to functions. A reader who is not familiar with multivariate analysis would benefit from reading one of the many excellent textbooks on this subject before reading this section, Sect. 5.7, and Sects. 6.3–6.6. Here, we briefly introduce principal components analysis for multivariate data, since this section is concerned with its generalization to functional data. Let $\boldsymbol{y}_i = (y_{1i}, \ldots, y_{di})^T$, $i = 1, \ldots, n$, be an independent sample from a multivariate distribution of dimension d. The sample mean is $\overline{\boldsymbol{y}} = n^{-1} \sum_{i=1}^{n} \boldsymbol{y}_i$ and the $d \times d$ sample covariance matrix is

$$\boldsymbol{\Sigma} = (n-1)^{-1} \sum_{i=1}^{n} (\boldsymbol{y}_i - \overline{\boldsymbol{y}})(\boldsymbol{y}_i - \overline{\boldsymbol{y}})^T .$$

The kth eigenvalue/eigenvector pair, (λ_k, ψ_k), where $k = 1, \ldots, d$, λ_k is a scalar, and ψ_k is a d-dimensional vector, satisfies

$$\boldsymbol{\Sigma}\psi_k = \lambda_k \psi_k. \tag{5.1}$$

We will assume that the eigenvalues are distinct and order them from largest to smallest, that is, $\lambda_1 > \cdots > \lambda_d$. We also assume that the eigenvectors have been normalized to have length one, that is, $\|\psi_k\| = 1$ for $k = 1, \ldots, d$. The first eigenvector, ψ_1, is the direction of maximum variation of the data. More specifically, the variance of $(\boldsymbol{x}^T \boldsymbol{y}_1, \ldots, \boldsymbol{x}^T \boldsymbol{y}_n)$ is maximized over all vectors $\boldsymbol{x}$ of length one by ψ_1. Also, ψ_2 is the direction of maximum variance orthogonal to ψ_1, and, in general, ψ_k is the direction of maximum variance orthogonal to $\psi_1, \ldots, \psi_{k-1}$. The eigendecomposition (5.1) is called *principal components analysis*. Principal components analysis is often used for dimension reduction by retaining only K eigenvalue and eigenvector pairs for some K less than d.

As just mentioned, functional data analysis is in many ways a generalization of multivariate analysis. For example, in functional data analysis, the mean vector and covariance matrix of a random vector are generalized to the mean function and covariance function of a random function. In this section, we apply bivariate smoothing to estimation of covariance functions. Estimation of a covariance function is important for dimension reduction by principal components analysis of functional data and for regression analysis with functional data.

Suppose that $y_1(t), \ldots, y_n(t)$ are independent and identically distributed random functions defined on some interval T. Let $y(t)$ be a generic version of these functions, that is, $y(t)$ has the same distribution as each $y_i(t)$. The mean function of y is $f(t) = E\{y(t)\}$ and can be estimated by using nonparametric regression approaches such as penalized splines discussed in Chap. 2. The covariance function of y is

$$\boldsymbol{\Sigma}(s, t) = E[\{y(s) - f(s)\}\{y(t) - f(t)\}], \quad (s, t) \in T \times T.$$

The standard deviation function is $\sqrt{\boldsymbol{\Sigma}(t, t)}$.

Functional data are necessarily observed at a finite number of points. For example, the near infrared reflectance spectra are only observed at the 401 wavelengths that range from 900 to 1700 nanometers by 2 nanometers.

Suppose for now that for all i, $y_i(t)$ is observed on an equally spaced grid $t_1 < \ldots < t_d$ in T. Given an estimate $\widehat{f}$ of f we define the sample covariance matrix $\widehat{\boldsymbol{\Sigma}}$ to be the $d \times d$ matrix with (j, k)th entry

$$\widehat{\boldsymbol{\Sigma}}_{jk} = \frac{1}{n-1} \sum_{i=1}^{n} \{y_i(t_j) - \widehat{f}(t_j)\}\{y_i(t_k) - \widehat{f}(t_k)\}, \quad j, k = 1, \ldots, d,$$

and $\widehat{\boldsymbol{\Sigma}}_{jk}$ is an estimate of $\boldsymbol{\Sigma}(t_j, t_k)$.

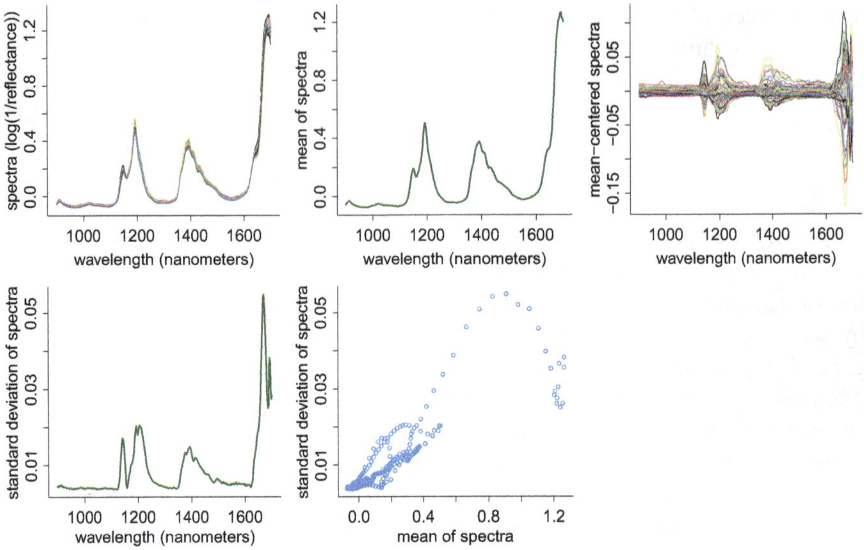

Fig. 5.19 Top left panel: Plots of near infrared reflectance spectra versus wavelength for 60 gasoline samples. These functions are very close to each other and mostly overprint. Top middle panel: The mean function. Top right panel: Plots of mean-centered near infrared reflectance spectra versus wavelength. Bottom left panel: The standard deviation function. Bottom middle panel: The standard deviation plotted against the mean.

The 60 near infrared functions are plotted against wavelength in the top left panel of Fig. 5.19. The other panels of Fig. 5.19 show the mean function, the mean-centered spectra, the standard deviation function, and a plot of the standard deviation versus the mean, respectively. Interestingly, the plot of standard deviation versus mean shows a complex and almost deterministic pattern. Such behavior can be expected when most of the variation in functional data occurs in a low-dimensional space, that spanned by only a few principal components (eigenvectors). As we will see, that is the case here. This phenomenon is also observed in other functional datasets such as that on fat content of meat samples analyzed in Sects. 6.3 and 6.5.

Since they are only observed on a discrete set of points, functional data might be analyzed by the same methods that are applied to multivariate data. However, functional data differ from multivariate data in at least two ways. First, with functional data the dimension d of the observations is often considerably larger than the number n of observations, e.g., $d = 401$ and $n = 60$ for the gasoline spectra. Second, although functional data might be noisy due to random variation in the observed functions as well as possible measurement error, the mean and covariance functions are usually smooth. Therefore, the extension of multivariate techniques such as principal components analysis, canonical correlation analysis, and linear regression to functional data usually involves both dimension reduction and some type of smoothing, for example, using splines.

Although the covariance function $\mathbf{\Sigma}(s, t)$ is usually assumed to be smooth, the sample covariance matrix $\widehat{\mathbf{\Sigma}}$ will be rough because of estimation error which can be large when n is small. The estimation error can be reduced by smoothing $\widehat{\mathbf{\Sigma}}$. Moreover, the sample covariance $\widehat{\mathbf{\Sigma}}$ estimates $\mathbf{\Sigma}(s, t)$ only on the grid $(t_1, \ldots, t_d)^2$ in $T \times T$, but smoothing with, say, a spline extends this estimate to all s and t in $T \times T$.

Although the bivariate smoothers introduced earlier in this chapter are applicable to the smoothing of $\widehat{\mathbf{\Sigma}}$ when d is small, the dimension of $\widehat{\mathbf{\Sigma}}$ is $d \times d$ and d is often so large that these smoothers would take minutes or even hours to compute. Also, one can easily run out of computer memory when attempting to use enough spline basis functions to avoid serious bias. Fortunately, (Xiao et al. 2016) have developed the FAst Covariance Estimator (FACE) that is feasible for d as large as 100,000 and allows a large number of basis functions. FACE is a tensor product spline smoother with a special choice of penalty and an implementation that significantly speeds computations and reduces memory requirements in a number of ways. FACE is implemented by the `fpca.face()` function in the `refund` R package (Goldsmith et al. 2016).

When $d \gg n$, $\widehat{\mathbf{\Sigma}}$ is singular with a relatively low rank that does not exceed $n - 1$. Moreover, a principal components analysis of $\widehat{\mathbf{\Sigma}}$ often shows that most of the variation in the data occurs in the space spanned by a small number, say K, of principal component eigenvectors with $K \ll d$ and often K is between 3 and 10. In this case, $\widehat{\mathbf{\Sigma}}$ can be approximated by a matrix of rank K.

The function `fpca.face()` does not return the entire estimated covariance matrix, which could be too large to fit in memory. Instead, `fpca.face()` returns the first (in decreasing order) K eigenvalues and smooth versions of the corresponding eigenvectors of the estimated covariance matrix. The tuning parameter K is determined by pve, an argument of `fpca.face()`. K is selected so that the estimated fraction of the total variation of the data in the space spanned by the first K eigenvectors is at least pve. In practice, pve is typically in the range 0.95–0.99, but might be even larger.

If λ_k and ψ_k are these eigenvalues and eigenvectors of the smoothed covariance matrix, and if the eigenfunction $\psi_k(t)$ is extension of the eigenvector ψ_k from the grid $(t_1, \ldots, t_d)$ to T, say, by spline interpolation, then the covariance function is estimated by

$$\sum_{k=1}^{K} \lambda_k \psi_k(t) \psi_k^T(t), \ s, t = t_1, \ldots, t_d.$$

In addition to the eigenvalues and eigenfunctions, `fpca.face()` also returns an $n \times K$ matrix of scores, which are the loadings of the observations on the eigenvectors. For simplicity of notation, assume for the remainder of this paragraph that the estimated mean function has been subtracted from $y_i(t)$. The loading of the

ith observation on the kth eigenfunction is denoted by β_{ik} and is defined as the value of β that minimizes $\|y_i(t) - \beta\psi_k(t)\|^2$. Then,

$$\widehat{y_i}(t) = \sum_{k=1}^{K} \beta_{ik}\psi_k(t), \quad t = t_1, \ldots, t_d \tag{5.2}$$

is an approximation to $y_i(t)$. More precisely, it is the projection of y_i onto the space spanned by $\psi_1(t), \ldots, \psi_K(t)$. Using the loadings enables a dimension reduction since the d-dimensional $(y_i(t_1), \ldots, y_i((t_d))$ can be summarized by $(\beta_{i1}, \ldots, \beta_{iK})$ and K is often much smaller than d.

5.6.1 Example: Gasoline Near-Infrared Spectra

The following code computes (1) the sample covariance matrix SigmaSamp of the spectral data; (2) the eigendecomposition eigenSigmaSamp of SigmaSamp using the function eigen(); (3) a tensor product smooth fitTE of SigmaSamp using gam(); and (4) a FACE smooth fitFACE of SigmaSamp: The objective here is to compare the computation of the eigenvalues and smoothed eigenvectors of the sample covariance using the functions eigen() and gam() with the same computation using fpca.face(). We will see that fpca.face() is much faster and much more accurate.

```
> library(refund) ; data(gasoline)
> wavelength <- seq(900,1700,by = 2)
> SigmaSamp <- cov(gasoline$NIR)
> eigenSigmaSamp <- eigen(SigmaSamp)
> mesh <- expand.grid(1:401,1:401) ; library(mgcv)
> fitTE <- gam(as.vector(SigmaSamp) ~ te(mesh[,1],mesh[,2],
+                k = c(25,25)))
> NIRcentered <- apply(gasoline$NIR,2,
+                    function(x){x - mean(x,na.rm = TRUE)})
> fitFACE <- fpca.face(NIRcentered,knots = 300,pve = 0.998)
> cumVariance <- cumsum(fitFACE$evalues)
```

Let $\mathscr{E}_k$ be the space spanned by the first k eigenvectors. Nine eigenvalues were returned, so, since pve was set to 0.998, at least 99.8% of the variability is in $\mathscr{E}_9$. The vector cumVariance contains the cumulative sums (the partial sums) of the first nine eigenvalues of the sample covariance matrix. Therefore, the kth element of cumVariance is the amount of variance in the subspace $\mathscr{E}_k$. For interpretability, it is useful to convert these values to percentages of variance. This is done next by multiplying by pve and dividing by the cumVariance[9]. We see that over 99% of the total variance is in $\mathscr{E}_6$.

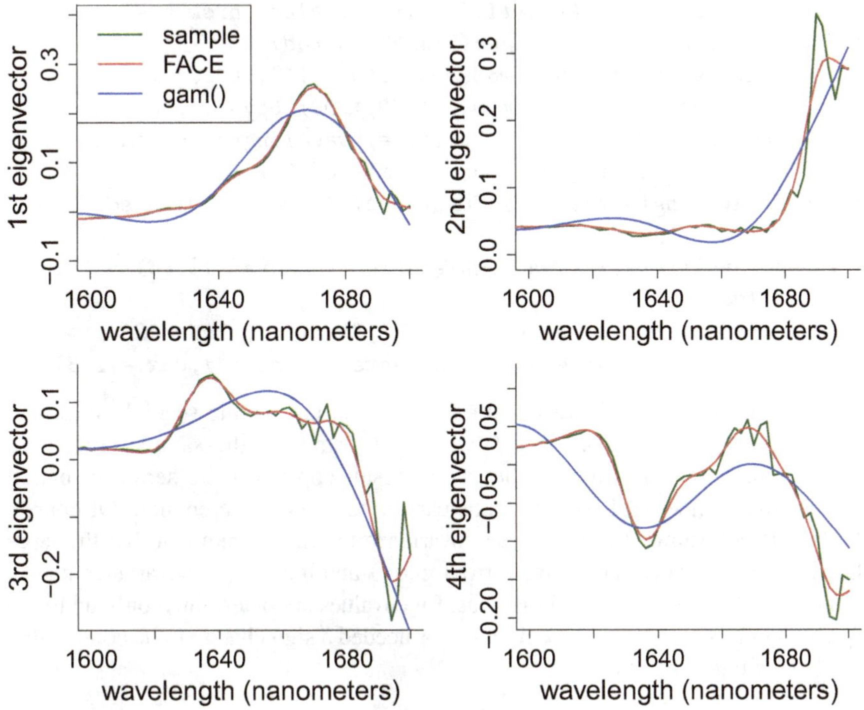

Fig. 5.20 The first four eigenvectors of the sample covariance matrix compared with the corresponding eigenvectors of sample covariance matrix smoothed by FACE and by `gam()` with a tensor product spline. The wavelength is restricted to the range 1600–1700 nanometers to show detail.

```
> print(0.998*cumVariance/cumVariance[9])
[1] 0.7522813 0.8671441 0.9359032 0.9823560 0.9891529
[6] 0.9929274 0.9952780 0.9969302 0.9980000
```

In this application, when computing the bivariate fit `fit_TE`, `gam()` ran out of memory on a laptop with 8 gigabytes of random access memory when 30 × 30 basis functions were used, but ran successfully when 25 × 25 basis functions were used. In contrast, `fpca.face()` had no trouble with 300 × 300 basis functions. In addition, while `gam()` with 25 × 25 basis functions took over 7 minutes, FACE with 300 × 300 basis functions took about half a second, showing that FACE can speed computations by several orders of magnitude.

Figure 5.20 compares the first four eigenvectors of the sample covariance matrix with the corresponding eigenvectors of the `fpca.face()` smooth and of the `gam()` smooth of this matrix. To show detail, the eigenvectors are plotted over only the limited range of wavelengths between 1600 and 1700 nanometers. The first panel of Fig. 5.20 is generated by the code:

```
> SigmaTensProd <- matrix(fitTE$fitted.values,nrow = 401)
> eigenSigmaTensProd <- eigen(SigmaTensProd)
> plot(wavelength,eigenSigmaSamp$vectors[,1], type="l",
+       col="darkgreen",xlim = c(1600,1700),bty="l",cex.lab = 2,
+       cex.axis = 1.8,lwd = 2,xlab = "wavelength (nanometers)",
+       ylab = "1st eigenvector",ylim = c(-0.1, 0.4))
> lines(wavelength,fitFACE$efunctions[,1],col="indianred3",
+        lwd = 2)
> lines(wavelength,-eigenSigmaTensProd$vectors[,1],col = "blue",
+        lwd = 2)
> legend("topleft",c("sample","FACE","gam()"),,lwd = 2,
+          col = c("darkgreen","indianred3","blue"),cex = 1.8)
```

The other panels were obtained with similar code but for the second, third, and fourth eigenvectors. We see that `fpca.face()` smooths the sample covariance matrix with little bias whereas the `gam()` result appears to be seriously biased because the number of basis functions cannot be made large enough. Of course, the bias is not known since the true covariance matrix is unknown, but the large discrepancies between the estimate from `gam()` and the sample covariance matrix (which is unbiased) suggest a large bias. Eigenvalues are determined only up to the sign, and in this case the `gam()` eigenvector needed a sign change to be comparable to the other eigenvectors.

5.7 Estimating a Covariance Function with Sparse Data

Sometimes each individual function is observed only at a small set of points that varies between functions. An example is the female spinal bone mineral density data introduced in Sect. 4.2. When data from all subjects are combined, spinal bone mineral density is observed at a dense set of ages from roughly 9 to 26 years. However, each subject is observed at only a few ages within a relatively short range.

As in Chap. 4, suppose that m denotes the number of subjects and n_i is the number of measurements on subject i. Then the data are

$$(t_{ij}, y_{ij}), \quad 1 \le i \le m, \quad 1 \le j \le n_i$$

where y_{ij} is the jth response measurement on the ith subject and t_{ij} is the time when y_{ij} was measured. Let $f(t)$ denote the mean response at time t and let $\widehat{f}(t)$ be an estimate of $f(t)$. To estimate the covariance function with sparse and irregularly sampled data, one forms all products of the form

$$C_{i;j,k} \equiv \{y_{ij} - \widehat{f}(t_{ij})\}\{y_{ik} - \widehat{f}(t_{ik})\}, \quad 1 \le i \le m, \quad 1 \le j, k \le n_i. \tag{5.3}$$

Then the point cloud

$$\{(t_{ij}, t_{ik}, C_{i;j,k}) : 1 \leq i \leq m, 1 \leq j, k \leq n_i\} \tag{5.4}$$

is smoothed as a function of the (t_{ij}, t_{ik}).

5.7.1 Example: Spinal Bone Mineral Density Data

As an illustration, we will use the subset of black subjects in the femSBMD dataset from the HRW package. The following code reads the femSBMD data, extracts the subset of black subjects, and centers the variable spnbmd using a smooth estimate of the mean obtained from gam().

```
> library(HRW) ; library(mgcv) ; data(femSBMD)
> SBMDblack <- femSBMD[femSBMD$black == 1,]
> fit <- gam(spnbmd ~ s(age),data = SBMDblack)
> SBMDblack$spnbmdCent <- residuals(fit)
```

Next, we create a dataset called covPointCloud with rows $(t_{ij}, t_{ij}, C_{i;j,k})$. In the code below, there are three loops. The outer loop is over the subject index, i, and the inner loops are over j and k. Each iteration of the third loop computes one row of covPointCloud.

```
> uniqueID <- unique(SBMDblack$idnum) ; covPointCloud <- NULL
> for (i in uniqueID)
+ {
+   currSamp <- SBMDblack[SBMDblack$idnum == i,]
+   for (j in 1:dim(currSamp)[1])
+     for (k in 1:dim(currSamp)[1])
+        covPointCloud <- rbind(covPointCloud,
+           c(currSamp$age[j],currSamp$age[k],
+             currSamp$spnbmdCent[j]*currSamp$spnbmdCent[k]))
+ }
```

Finally, we smooth these data to estimate and plot the covariance function. Because of the relatively small amount of data, smoothing can be done quickly and satisfactorily with gam().

```
> fitCov <- gam(covPointCloud[,3] ~ s(covPointCloud[,1],
+                   covPointCloud[,2]),method = "REML")
```

A plot of the fit is in Fig. 5.21. The covariance function $\boldsymbol{\Sigma}(s, t)$ cannot be estimated when s is a young age, say less than 12, and t is an older age, say greater than 20 (or vice versa), since no individual subject is observed over a wide range of ages; see Fig. 4.1. Consequently, much of Fig. 5.21 is blank. Looking at the

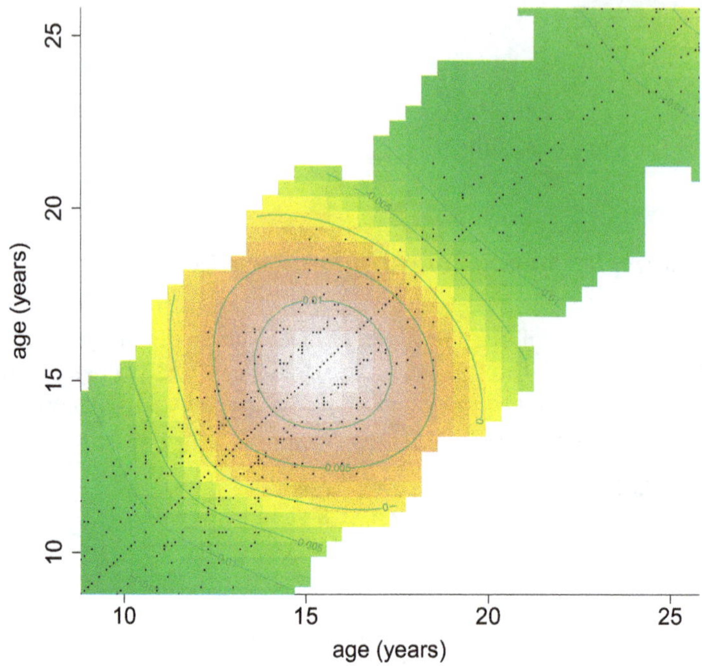

Fig. 5.21 Estimated covariance function for black subjects in the `femSBMD` dataset. The estimate is obtained by smoothing the point cloud given by (5.3) and (5.4) using bivariate penalized splines via the function `gam()` in the package `mgcv`.

diagonal in Fig. 5.21 we see that spinal bone mineral density is most variable at ages around 16. This effect is probably due to differences in the timing of the children's developments, for example, variation in the ages when growth spurts start during the teen years.

5.8 The Sandwich Smoother

Often two-dimensional data are observed on a rectangular grid. For example, a digitized two-dimensional image is a matrix giving the intensity, or intensities for color images, at each pixel. Data on a rectangular grid can be smoothed by any bivariate smoother. However, the so-called *sandwich smoother* (Xiao et al. 2013) is implemented by the `fbps()` function in the `refund` package (Goldsmith et al. 2016) and can smooth data on a rectangular grid extremely rapidly. The sandwich smoother uses a penalty that has a tensor product form similar to the tensor product form of the spline basis. This compatibility between the penalty and the spline basis allows the two-dimensional smoother to be implemented as two one-dimensional smoothers, which reduces computational time considerably.

5.8.1 Example: Brain Imaging

As an example of image smoothing using the sandwich smoother, we will use the coronal slice of a functional magnetic resonance imaging brain image. The data are from Landman et al. (2010) and are publicly available.

The code below reads the data and smooths the image with a penalized thin plate spline with 625 basis functions and a tensor product spline with 35×18 basis functions, both using `gam()`. The thin plate and tensor product splines took 84.2 and 51.9 seconds, respectively.

```
> library(mgcv) ;  library(HRW) ; data(brainImage)
> imageData <- as.matrix(brainImage)
> mesh <- expand.grid(1:80,1:37)
> mesh[,1] <- mesh[,1]/80 ; mesh[,2] <- mesh[,2]/37
> fitThinPlate <- gam(as.vector(imageData) ~
+                    s(mesh[,1],mesh[,2],k = 625))
> fitTensProd <- gam(as.vector(imageData) ~
+                    te(mesh[,1],mesh[,2],k = c(35,18)))
```

The code below computes the sandwich smoother with 50×35 basis functions using the `fbps()` function in the `refund` package (Goldsmith et al. 2016). The computation took only 2.7 seconds, an order of magnitude faster than `gam()`, which is unsurprising since `gam()` is a general purpose smoother, not specifically designed for images and other data on rectangular grids.

```
> library(refund)
> knotsSS <- list(seq(0,1,length = 50),seq(0,1,length = 35))
> fitSS <- fbps(imageData,knots = knotsSS)
```

Plots of the raw (unsmoothed) image and the three smooths are in Fig. 5.22. Here we use the `heat.colors()` palette, which leads to heat map image plots with higher values of brain activity corresponding to shades of red, then orange, then yellow, and then white. The sandwich smooth appears to provide more resolution compared to the `gam()` smooths, since the latter tend to blur the edges of features seen in the raw image.

The FACE smoother discussed in Sect. 5.6 is an implementation of the sandwich smoother designed specifically for covariance function estimation. For that purpose, FACE is much faster than the sandwich smoother, but `fpca.face()` represents special-purpose software and is not applicable to image smoothing.

5.9 Further Reading

Ruppert et al. (2003) and Wood (2006a) cover bivariate smoothing using splines. Fahrmeir and Kneib (2011) discuss spatial smoothing and geoadditive regression. Spatial smoothing is but one aspect of spatial data analysis. See Bivand et al. (2008) for an introduction to the analysis of spatial data using R. Ramsay and Silverman

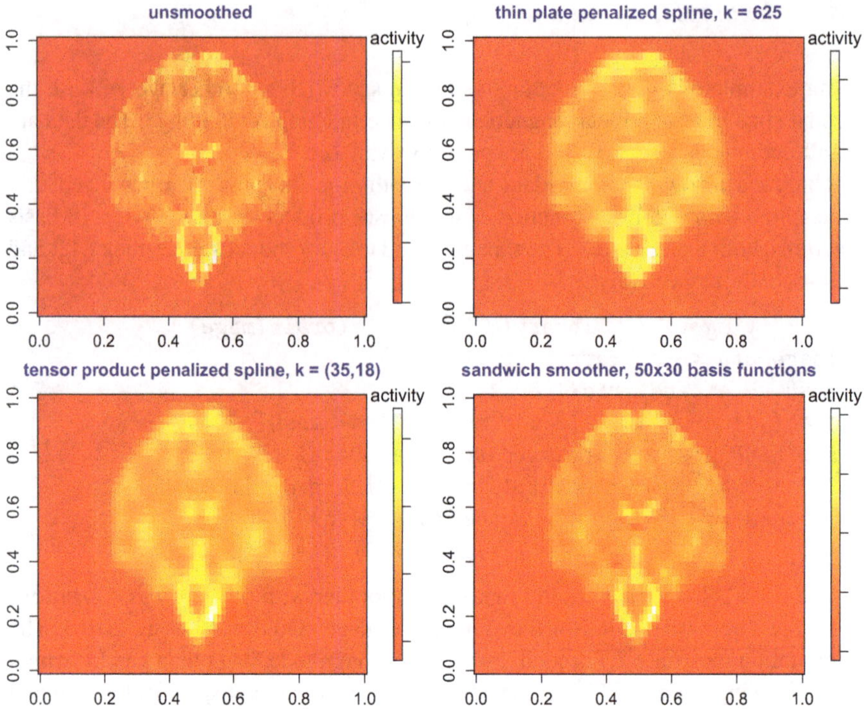

Fig. 5.22 Top right panel: The raw (unsmoothed) image of the coronal slice of a functional magnetic resonance image corresponding to a slice of the data frame. Top right panel: Image smoothed with a 625 basis function penalized thin plate spline. Bottom left panel: Image smoothed with a tensor product penalized spline with 35 by 18 basis functions. Bottom right panel: Image smoothed using the sandwich smoother with a 50 by 30 tensor product basis.

(2006) is an excellent introduction to functional data analysis and Ramsay et al. (2009) discuss functional data analysis in R and MATLAB, including the use of R's `fda` package (Ramsay et al. 2017). Horváth and Kokoszka (2012) is a more theoretically advanced treatment of functional data analysis. Recent books on geostatistics include Diggle and Ribeiro (2007), Cressie and Wikle (2011), and Cressie (2015).

5.10 Exercises

1. Visualization of bivariate penalized spline fits is of fundamental importance. As demonstrated in this exercise, there are numerous options for displaying such fits. The packages `fields` (Nychka et al. 2017) and `HRW` are required for this exercise.

a. Issue the following commands to load data on scallop abundance:
```
> library(HRW) ; data(scallop)
> x1 <- scallop$longitude ; x2 <- scallop$latitude
> y <- asinh(scallop$totalCatch)
```
b. Obtain a rudimentary visualization of the data using the following commands:
```
> library(lattice) ; cloud(y ~ x1 + x2)
```
c. Issue the following commands to obtain the `mgcv` package's `gam()` function default bivariate penalized spline fit, summarize the fit, examine the residuals, and check adequacy of the spline basis dimension:
```
> library(mgcv) ; fit <- gam(y ~ s(x1,x2))
> summary(fit) ; gam.check(fit)
```
d. Three different visualizations of the fitted surface are provided by the following commands. The first involves contour display and the second involves perspective plot display. The third and fourth involve image display with two different color schemes.
```
> par(mfrow = c(2,2)) ; plot(fit) ; plot(fit,scheme = 1)
> plot(fit,scheme = 2) ; plot(fit,scheme = 3)
```
e. The `HRW` supports customization of the boundary region in which contour lines and image pixels are shown. This is useful for applications where geographical regions containing the data have highly irregular shapes. Issue the following command and follow the instructions shown on the screen to create a suitable boundary polygon for the scallop data, corresponding to the geographical region of the observed longitude and latitude data:
```
> scallopBdry <- createBoundary(x1,x2)
```
Then issue these commands to list and then save the polygon matrix `scallopBdry` that you just obtained:
```
> print(scallopBdry)
> write.table(scallopBdry,"scallopBdry.txt",
+             row.names = FALSE,col.names = FALSE)
```
The last of these commands saves the boundary information in the file `scallopBdry.txt` for future use.

f. Obtain a contour plot restricted to the personalized boundary stored in `scallopBdry` via the commands:
```
> ngrid <- 101
> x1grid <- seq(min(x1),max(x1),length = ngrid)
> x2grid <- seq(min(x2),max(x2),length = ngrid)
> x1x2mesh <- expand.grid(x1grid,x2grid)
> names(x1x2mesh) <- c("x1","x2")
> fitmesh <- matrix(predict(fit,newdata = x1x2mesh),
+                   ngrid,ngrid)
> outInds <- (1:ngrid^2)[pointsInPoly(x1x2mesh,scallopBdry)
+                        == FALSE]
> fitmesh[outInds] <- NA
> xlimVal <- c(1.1*min(x1) - 0.1*max(x1),
```

```
+                   1.1*max(x1) - 0.1*min(x1))
> ylimVal <- c(1.1*min(x2) - 0.1*max(x2),
+                   1.1*max(x2) - 0.1*min(x2))
> contour(x1grid,x2grid,fitmesh,xlab = "degrees longitude",
+             ylab = "degrees latitude",xlim = xlimVal,
+             ylim = ylimVal,bty = "l")
> lines(scallopBdry,col = "navy")
> points(x1,x2,col = "dodgerblue",cex = 0.5)
```

g. Next issue the following commands to obtain an image display of the fitted surface with the pixels confined to the personalized boundary stored in `scallopBdry`:

```
> image(x1grid,x2grid,fitmesh,xlab = "degrees longitude",
+         ylab = "degrees latitude",xlim = xlimVal,
+         ylim = ylimVal,bty = "l")
> lines(scallopBdry,col = "navy")
> points(x1,x2,col="dodgerblue",cex=0.5)
```

h. The image plot produced in part g. uses the default heat map color scheme. However, there are many other possibilities. The following commands show the fitted surface using the `terrain.colors()` color scheme corresponding geographical maps, with green corresponding to grassy plains, shades of yellow and brown corresponding to altitudes about the tree-line and white indicating snow-capped peaks.

```
> image(x1grid,x2grid,fitmesh,col = terrain.colors(1000),
+         xlab = "degrees longitude",
+         ylab = "degrees latitude",xlim = xlimVal,
+         ylim = ylimVal,bty = "l")
> lines(scallopBdry,col = "navy")
> points(x1,x2,col="dodgerblue",cex=0.5)
```

i. The `image()` function has the disadvantage of not linking the image shades with heights of the surface being displayed. A remedy is to use the function `image.plot()` in the package `fields` (Nychka et al. 2017). Ensure that this package is available in your R environment and issue the following commands to obtain a legend bar for the image shades:

```
> library(fields)
> image.plot(x1grid,x2grid,fitmesh,
+         col = terrain.colors(1000),
+         xlab = "degrees longitude",
+         ylab = "degrees latitude",
+         legend.args = list(text = "arcsinh(total catch)"),
+         xlim = xlimVal,ylim = ylimVal,bty = "l")
> lines(scallopBdry,col = "navy")
> points(x1,x2,col="dodgerblue",cex=0.5)
```

j. The last display uses the persp() function in R for showing the surface as a perspective plot. The arguments theta and phi control the visualization angles and can be twiddled to obtain other views of the surface. Exercise 3 illustrates the used of three-dimensional spin graphics to view such a surface dynamically.

```
> persp(x1grid,x2grid,fitmesh,col = "green3",theta = 15,
+         phi = 45,xlab = "degrees longitude",
+         ylab = "degrees latitude",
+         zlab = "arcsinh(total catch)")
```

2. Make sure that the package HRW is available in your R environment and start an R session.

a. Extract data on location and price of houses in Sydney, Australia:

```
> library(HRW) ; data(SydneyRealEstate)
> x1 <- SydneyRealEstate$longitude
> x2 <- SydneyRealEstate$latitude
> y <- SydneyRealEstate$logSalePrice
```

b. Issue the command:

```
> plot(x1,x2,col = "dodgerblue",cex = 0.1)
```

This plot shows that the geographical region over which the data were observed is highly irregular, with houses located between waterways and national park boundaries around Sydney. This dataset is also very large, with recordings for 37,676 houses.

c. Next enter:

```
> data(SydneyRealEstateBdry)
> print(SydneyRealEstateBdry)
> lines(SydneyRealEstateBdry,col = "navy",lwd = 2)
```

These commands load, print, and draw the boundary polygon, stored in SydneyRealEstateBdry and consisting of 202 vertices, that encompasses the main part of the geographical data.

d. Issue the following commands to obtain the a bivariate penalized spline with 150 basis functions, using the function gam() in the mgcv package:

```
> library(mgcv) ; fit <- gam(y ~ s(x1,x2,k = 150))
> summary(fit)
```

e. Now enter these commands to obtain an image plot of the fitted surface using the terrain.colors() palette:

```
> plot(fit,scheme = 2,hcolors = terrain.colors(1000),
+        xlab = "longitude",ylab = "latitude")
```

However observe that this image plot is dominated by meaningless parts of the fitted surface that are beyond the location of the data. For example, the right region corresponds to a section of the Pacific Ocean to the east of the city of Sydney.

f. The following code uses SydneyRealEstateBdry to make another image plot of the fitted surface:

```
> ngrid <- 201
> x1grid <- seq(min(x1),max(x1),length = ngrid)
> x2grid <- seq(min(x2),max(x2),length = ngrid)
> x1x2mesh <- expand.grid(x1grid,x2grid)
> names(x1x2mesh) <- c("x1","x2")
> fitmesh <- predict(fit,newdata = x1x2mesh)
> outInds <- (1:ngrid^2)[pointsInPoly(x1x2mesh,
+                 SydneyRealEstateBdry) == FALSE]
> fitmesh[outInds] <- NA
> fitmesh  <- matrix(fitmesh,ngrid,ngrid)
> image(x1grid,x2grid,fitmesh,col = terrain.colors(1000),
+       xlab = "longitude",ylab = "latitude",bty = "l")
> lines(SydneyRealEstateBdry,col = "navy",lwd = 2)
```

This is a better plot of the fitted surface, with the image colors corresponding to the locations of Sydney's residential areas. As expected, the higher-priced parts of Sydney are near the ocean beaches and harbor on the east side.

g. This part is concerned with learning how to obtain new boundary polygons—even though a good one for the Sydney real estate data is stored as SydneyRealEstateBdry in the HRW package. Issue the following commands to obtain a new boundary file for Sydney:

```
> SydneyRealEstateBdryNew <- createBoundary(x1,x2)
> write.table(SydneyRealEstateBdryNew,
+                "SydneyRealEstateBdryNew.txt",
+                row.names = FALSE,col.names = FALSE)
```

h. Repeat the commands of part f. with SydneyRealEstateBdry replaced by SydneyRealEstateBdryNew.

3. R also supports visualization of bivariate penalized spline fits via three-dimensional spin graphics supported by the package rgl (Adler et al. 2017). Illustration for the ozone concentration example is provided in this exercise. Ensure that rgl is available in your R environment.

   a. Issue the following commands to obtain a bivariate penalized spline fit to ozone concentration as a function of longitude and latitude based on the dataset ozoneSub:

```
> library(HRW)  ; data(ozoneSub)
> x1 <- ozoneSub$longitude ; x2 <- ozoneSub$latitude
> y <- ozoneSub$ozone
> library(mgcv) ; fit <- gam(y ~ s(x1,x2,k = 60))
> ngrid <- 201
> x1grid <- seq(min(x1),max(x1),length = ngrid)
> x2grid <- seq(min(x2),max(x2),length = ngrid)
> x1x2mesh <- expand.grid(x1grid,x2grid)
> names(x1x2mesh) <- c("x1","x2")
> fitmesh <- matrix(predict(fit,newdata = x1x2mesh),
+                      ngrid,ngrid)
```

b. For a rudimentary display via rgl we transform the data and fit objects to the unit cube via the following commands:
```
> x1UC <- (x1 - min(x1))/(max(x1) - min(x1))
> x2UC <- (x2 - min(x2))/(max(x2) - min(x2))
> yUC <- (y - min(y))/(max(y) - min(y))
> x1gridUC <- (x1grid - min(x1))/(max(x1) - min(x1))
> x2gridUC <- (x2grid - min(x2))/(max(x2) - min(x2))
> fitmeshUC <- (fitmesh - min(y))/(max(y) - min(y))
```
c. Now issue these commands:
```
> library(rgl) ; rgl.bg(col = "white")
> rgl.spheres(x1UC,x2UC,yUC,col = "dodgerblue",
+             radius = 0.015)
> rgl.surface(x1gridUC,x2gridUC,fitmeshUC,
+             color = "darkgreen",alpha = 0.4)
```
This leads to a three-dimensional spin graphics display, which is best viewed with the graphics window maximized. The mouse or touchpad on your computer can be used for rotation and zooming to aid visualization of the bivariate penalized spline fit.

d. Running the script ozone3Dspin.R in the HRW package produces a more elaborate display with axes, base, and locations of cities in the region being studied. This script can be run as follows:
```
> library(HRW) ; demo(ozone3Dspin,package = "HRW")
```
To locate, and possibly copy and modify, ozone3Dspin.R note that its location is determined by:
```
> system.file("demo","ozone3Dspin.R",package = "HRW")
```

4. The dataset LakeAcidity in the R package gss (Gu 2017) has data collected by the United States Environmental Protection Agency on 112 lakes in the states of Virginia, North Carolina, South Carolina, Tennessee and Georgia in the Blue Ridge region of the USA. The main variables are ph (power of hydrogen), cal (calcium concentration), lon (longitude), and lat (latitude) respectively. Since the calcium concentration data are heavily skewed we work with their logarithms, which roughly mimics the ph variable for hydrogen concentration.

   a. Ensure that the package gss is installed in your R environment and issue the following commands:
```
> library(gss) ; data(LakeAcidity)
> x1 <- LakeAcidity$lon
> x2 <- LakeAcidity$lat
> chullInds <- chull(x1,x2)
> chullInds <- c(chullInds,chullInds[1])
> LakeAcidityConvexHullBdry <- cbind(x1,x2)[chullInds,]
> plot(x1,x2,bty="l",xlab = "longitude",ylab = "latitude",
+       col = "dodgerblue")
> lines(LakeAcidityConvexHullBdry,col = "navy")
```

The boundary matrix LakeAcidityConvexHullBdry is the convex hull of the longitude/latitude data. The convex hull is a reasonable boundary in this case since the geographic region corresponding to the LakeAcidity dataset is approximately convex.

b. Use the function gam() from the package mgcv to obtain bivariate penalized spline fits of mean log calcium concentration as a function of longitude and latitude, with both

   i. thin plate spline basis functions,
   ii. tensor product spline basis functions

   and all other spline parameters set to their defaults.

c. Obtain image plots of the thin plate spline basis and tensor product spline basis fits with the pixels restricted to the polygon defined by LakeAcidityConvexHullBdry. Which regions have the highest values of mean calcium concentration?
d. Compare the fits from the two bivariate penalized spline approaches.
e. Build an appropriate model geoadditive for predicting pH from log calcium concentration and geographical location.
f. How does calcium concentration affect mean pH?
g. Describe the effect of location on mean pH.

5. This exercise is concerned with bivariate nonparametric density estimation via penalized splines.

   a. Issue the following R commands to obtain a scatterplot of two of the variables from the plankton data frame:

```
> library(HRW) ; data(plankton)
> x1 <- plankton$redFluorBlueLight
> x2 <- plankton$greenFluorBlueLight
> plot(x1,x2,col = "dodgerblue",bty = "l",cex = 0.2,
+      xlab = "red fluorescence under blue light",
+      ylab = "green fluorescence under blue light")
```

   b. Use a bivariate extension of the approach used in Exercise 6, based on gam() function in the mgcv package, to obtain and plot a bivariate nonparametric density estimate for these data.

6. Transformations of the response variable are commonly used to ameliorate regression model assumptions. However, summaries on the original response scale often are of interest.

   a. Issue the following commands from Exercise 1 to load the scallop data, fit bivariate penalized spline to total catch as a function of location, and check the residuals:

```
> library(HRW) ; data(scallop)  ; library(mgcv)
> fitOrdinary <- gam(totalCatch ~
+    s(longitude,latitude,k = 50),data = scallop)
> gam.check(fitOrdinary)
```

b. The residual plots from part a. indicate a strong departure from the assumptions of normality and homoscedasticity of the residuals. Next issue the following commands in which the response variable is transformed:

```
> fitTransResp <- gam(asinh(totalCatch)
+            ~ s(longitude,latitude,k = 50),data = scallop)
> gam.check(fitTransResp)
```

The transformation applied to the response data, corresponding to the R function asinh(), is $\sinh^{-1}(x) \equiv \log(x+\sqrt{1+x^2})$. The residual plots have good accordance with the normality and homoscedasticity assumptions.

c. The fitted surface can be visualized via:

```
> plot(fitTransResp,scheme = 2,
+      hcolors = terrain.colors(1000))
```

Parts e.–i. of Exercise 1 provide R code for various other visualizations of this fit including those with the image plot pixels confined to the region encompassed by the data.

d. Back-transformation of the surface, via the function $\sinh(x) \equiv (e^x - e^{-x})/2$ gives a biased estimate of mean total catch. The *smearing estimate* of Duan (1983) attempts to reduce the bias. Compute the smearing estimates of the mean response surface and compare it to surface obtained via naïve back-transforming.

7. Section 5.3.1 illustrates geoadditive model analysis for the Sydney real estate data with geographic location and two additional predictors. We now consider the full set of available predictors. Ensure that the packages gam (Hastie 2017a) and HRW are available in your R environment.

a. Issue the following commands to fit a bivariate penalized spline to logarithm of sale price as a function of longitude and latitude and save the residuals of this fit:

```
> library(HRW) ; data(SydneyRealEstate)  ; library(mgcv)
> fitGeog <- gam(logSalePrice
+                   ~ s(longitude,latitude,k = 50),
+                  data = SydneyRealEstate)
> geogResids <- residuals(fitGeog)
```

b. Use the function step.Gam() in the package gam to select a suitable additive model for prediction of geogResids with each of the 38 remaining variables in the data frame SydneyRealEstate considered as possible predictors.

c. Now use the function gam() in the package mgcv, to fit the geoadditive model containing a bivariate function of longitude and latitude and non-geographic component dictated by the model selected in part b. and with GCV smoothing parameter selection for all penalized spline components. Produce numerical and graphical summaries of the fitted geoadditive model.

8. Varying-coefficient model analysis via R was demonstrated in Sect. 5.4. This exercise is concerned with a similar analysis for data from a study on exhaust emissions from ethanol fuel (Brinkman 1981) and used by Hastie and Tibshirani (1993) to illustrate varying-coefficient models.

   a. Issue the following R commands to load and visualize the `ethanol` data frame:
      ```
      > library(lattice) ; data(ethanol) ; pairs(ethanol)
      ```
      There are three variables `NOx`, `E`, and `C` which are, respectively, the concentration of NOx (nitric oxide and nitrogen dioxide) normalized to the work performed by the engine, the equivalence ratio that measures the richness of the air/ethanol mixture, and the compression ratio. The number of records is $n = 88$.
   b. The varying-coefficient model in Hastie and Tibshirani (1993) is

      $$\texttt{NOx}_i = f_0(\mathrm{E}_i) + f_1(\mathrm{E}_i)\,\mathrm{C}_i + \varepsilon_i, \quad 1 \le i \le n.$$

      where $\varepsilon_i \overset{\text{ind.}}{\sim} N(0, \sigma^2)$. Fit this model using `gam()` and obtain numerical and graphical summaries of the fit.
   c. Fit the bivariate nonparametric model

      $$\texttt{NOx}_i = f(\mathrm{E}_i, \mathrm{C}_i) + \varepsilon_i, \quad 1 \le i \le n$$

      using bivariate penalized splines. Obtain numerical and graphical summaries of the fit.
   d. Compare the two models for these data.

9. Exercise 6 was concerned with semiparametric longitudinal data analysis of the dataset `Milk` from the package `nlme`. Here we revisit these data in the context of covariance function estimation. Ensure that the package `fields` (Nychka et al. 2017) is in your R environment.

   a. Enter the following commands to load, and visualize the `Milk` data.
      ```
      > library(nlme) ; data(Milk)
      > plot(Milk,strip = FALSE,type = "l",as.table = TRUE)
      ```
      The data consist of weekly protein contents of milk from 79 cows over 19 consecutive weeks since calving. About 10% of the protein measurements are missing. There are also dietary data but these are left aside for this exercise concerning covariance function estimation.
   b. Issue these commands to obtain a matrix of protein values with the rows corresponding to cows and the columns corresponding to number of weeks since calving:
      ```
      > uniqueID <- unique(as.character(Milk$Cow))
      > proteinMatrix <- matrix(NA,79,19)
      > for (i in 1:79)
      ```

```
+     (proteinMatrix[i,Milk$Time[Milk$Cow == uniqueID[i]]]
+                  <- Milk$protein[Milk$Cow == uniqueID[i]])
```

c. Enter the following commands to obtain and visualize the sample covariance function:

```
> sampCovMat <- var(proteinMatrix,na.rm = TRUE)
> library(fields)
> image.plot(1:19,1:19,sampCovMat,
+            col = terrain.colors(1000),
+            xlab = "time since calving (weeks)",
+            ylab = "time since calving (weeks)",
+            legend.args = list(text="sample cov. func."))
```

d. Estimate the sample covariance function using bivariate penalized spline smoothing of the sample covariance matrix.

10. The variable cca in the dataset DTI2 in the refund package (Goldsmith et al. 2016) has diffusion tensor imaging profiles of 340 subjects. Each tract has 93 observations, some with missing values. Make sure that the packages refund and fields (Nychka et al. 2017) are in your R environment.

a. Issue the following commands to store the data in the standard longitudinal format and visualize it using lattice graphics:

```
> library(refund) ; data(DTI2) ; ccaLongitDF <- NULL
> for (i in 1:340)
+     ccaLongitDF <- rbind(ccaLongitDF,cbind(rep(i,93),
+                           1:93,DTI2$cca[i,]))
> ccaLongitDF <- as.data.frame(ccaLongitDF)
> names(ccaLongitDF) <- c("idnum","pixel", "diffusion")
> library(lattice)
> ccaLongitVis <- xyplot(diffusion ~ pixel|idnum,
+                 group = idnum,data = ccaLongitDF,
+                 strip  = FALSE,layout = c(20,17),
+                 as.table = TRUE,
+                 panel = function(x,y,subscripts,groups)
+                 {
+                    panel.grid()
+                    panel.superpose(x,y,subscripts,groups,
+                    col = "darkgreen",type = "l")
+                 })
> print(ccaLongitVis)
```

The panels in this plot correspond to the diffusion tensor imaging profiles of the 340 subjects.

b. Enter these commands to obtain and display the sample covariance matrix:

```
> sampCovMat <- var(DTI2$cca,na.rm = TRUE)
> library(fields)
> image.plot(1:93,1:93,sampCovMat,
+        col = terrain.colors(1000),
```

```
+         xlab = "pixel number",ylab = "pixel number",
+         legend.args = list(text="sample cov. func."))
```

c. Obtain estimates of the covariance matrix using:

   i. thin plate penalized splines,
   ii. tensor product penalized splines,
   iii. the sandwich smoother.

Chapter 6
Selection of Additional Topics

6.1 Introduction

Chapters 2–5 deal with the most fundamental semiparametric regression topics and implementation in R. There are numerous other topics but, of course, not all of them can be covered in a single book. Instead we cover a selection of additional topics in this final chapter that we feel are worthy of some mention. These concern robust and quantile regression, functional data, kernel machines and classification, missing data, and measurement error. Other themes with some semiparametric regression overlap, such as survival analysis, time series analysis, and wavelet regression are not covered, but we give some pointers regarding R implementation in the further readings at the end of the chapter.

Towards the end of this chapter we describe the *graphical models* approach to semiparametric regression. This paradigm allows arbitrarily complicated Bayesian semiparametric regression analyses to be entertained, using so-called Bayesian inference engines. R now provides good access to some of the most versatile Bayesian inference engines, such as BUGS (Lunn et al. 2013) and Stan (Carpenter et al. 2017). Use of Stan for semiparametric regression was introduced in Sect. 2.10.

6.2 Robust and Quantile Semiparametric Regression

It is well known that regression methods based on least squares are prone to erratic behavior when outliers are present in the data. Methods that down-weight the effect of outliers are usually called *robust*, whereas those that ignore gross outliers are also labeled as *resistant*. An introduction to parametric robust regression is provided by Maronna et al. (2006).

Quantile regression is another alternative to least-squares regression and is summarized by Koenker (2005). Rather than targeting mean functions, quantile

J. Harezlak et al., *Semiparametric Regression with R*, Use R!,
https://doi.org/10.1007/978-1-4939-8853-2_6

regression targets *quantile* functions. In certain applications such functions are more appealing and interpretable. However, quantile regression methods could also be classified as robust or resistant due to their imperviousness to outliers in the response data.

The principles behind robust and quantile regression extend to semiparametric regression and many such methods are supported in R. We give a flavor of robust and quantile regression in R in this section.

6.2.1 Robust and Resistant Scatterplot Smoothing

We now introduce the notions of robust and quantile semiparametric regression in the simple *scatterplot smoothing* setting. We use this informal term, rather than nonparametric regression, since the target function may be a mean function but, instead, could be a median function.

As a running example we will use data from the 1987 cross-section of the Michigan Panel Study of Income Dynamics, which is in the data frame `Workinghours` in the R package `Ecdat`. The data consist of several measurements on 3382 households. Further details are given in Sect. 3 of Lee (1995). The following code extracts and plots data with the predictor data (stored in `x`) set to wife's age (years) and the response data (stored in `y`) set to household income that is not from the wife's work. Note that `Workinghours$income` is in hundreds of dollars. We divide this by 10 so that the income data are in thousands of U.S. dollars.

```
> library(Ecdat) ; data(Workinghours)
> x <- Workinghours$age ; y <- Workinghours$income/10
> plot(x,y,col = "dodgerblue",cex = 0.2,bty = "l",
+ xlab = "wife's age (years)",
+ ylab = "other household income (thousands of U.S. dollars)")
```

The R function `loess()` is one of the oldest functions for smoothing a scatterplot, and based on local polynomial ideas. An ordinary fit to data (x_i, y_i), $1 \le i \le n$, stored in x and y can be achieved via:

```
> fitOrdinary <- loess(y ~ x,span = 0.25)
> xg <- seq(min(x),max(x),length = 201)
> fitOrdinaryg <- predict(fitOrdinary,xg)
> lines(xg,fitOrdinaryg,col = "darkgreen",lwd = 2)
```

The designation `span = 0.25` means that one quarter of the data are used in each local polynomial fit. A robust alternative is obtained and plotted via:

```
> fitRobust <- loess(y ~ x,family = "symmetric",span = 0.25)
> fitRobustg <- predict(fitRobust,xg)
> lines(xg,fitRobustg,col = "indianred3",lwd = 2)
```

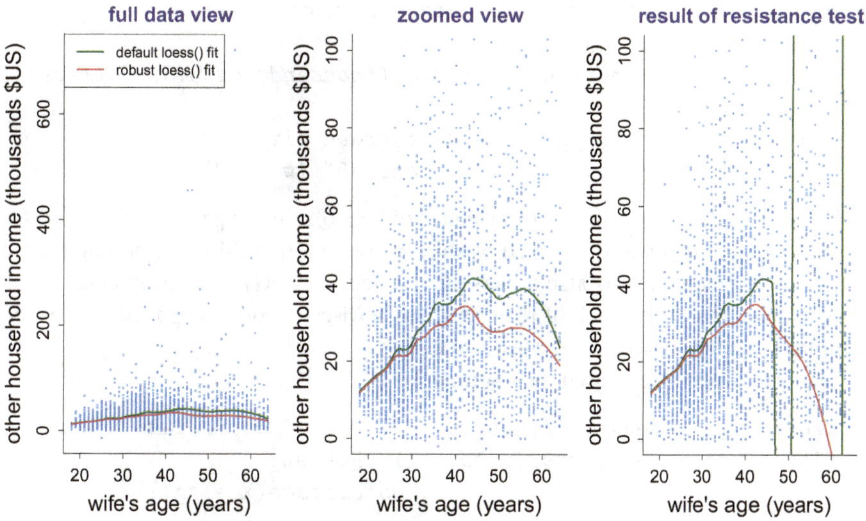

Fig. 6.1 Left panel: default and robust `loess()` fits to data from the 1987 cross-section of the Michigan Panel Study of Income Dynamics. Middle panel: zoomed view of the left panel fits. Right panel: fits when the largest response observation is replaced by a value 1000 times larger, as a test of resistance.

and replaces the least-squares criterion by one based on M-estimation with Tukey's biweight function (e.g. Huber 1981). The plot produced by this code is in the first panel of Fig. 6.1.

The following code produces a zoomed view of the same fits so that the differences are more easily seen:

```
> plot(x,y,col = "dodgerblue",cex = 0.2,bty = "l",
+   xlab = "wife's age (years)",
+   ylab = "other household income (thousands of U.S. dollars)",
+   ylim = c(0,100))
> lines(xg,fitOrdinaryg,col = "darkgreen",lwd = 2)
> lines(xg,fitRobustg,col = "indianred3",lwd = 2)
```

The robust fit is considerably lower for wife's ages in the range of about 40–55 years due its robustness against the high income outliers. This zoomed view is in the middle panel of Fig. 6.1.

Finally, we carry out a resistance test. This involves conversion of the largest outlier, another household income value of 722,000 U.S. dollars, into a *gross* outlier by multiplying it by 1000. The following code does this test:

```
> y[which.max(y)] <- 1000*y[which.max(y)]
> fitOrdinary <- loess(y ~ x,span = 0.25)
> fitOrdinaryg <- predict(fitOrdinary,xg)
> fitRobust <- loess(y ~ x,family = "symmetric",span = 0.25)
> fitRobustg <- predict(fitRobust,xg)
```

```
> plot(x,y,col = "dodgerblue",cex = 0.2,bty = "l",
+   xlab = "wife's age (years)",
+   ylab = "other household income (thousands of U.S. dollars)",
+   ylim = c(0,100))
> lines(xg,fitOrdinaryg,col = "darkgreen",lwd = 2)
> lines(xg,fitRobustg,col = "indianred3",lwd = 2)
```

and the result is shown in the right panel of Fig. 6.1. As expected, the `loess()` least-squares fit is severely affected by this one observation. More surprising is that the Tukey's biweight fit is also adversely affected. However, the documentation of `loess()` admits to this lack of resistance since M-estimation is applied iteratively with initialization based on the least-squares fit.

In Fig. 6.2 we instead used the commands:

```
> library(quantreg) ; fitRobust <- rqss(y ~ qss(x,lambda = 5))
> fitRobustg <- as.vector(predict(fitRobust,
+                                 data.frame(x = xg)))
```

corresponding to the R functions `qss()` and `rqss()` from the package quantreg (Koenker 2017). This is a *quantile smoothing spline* (Koenker et al. 1994) fit to the data, with smoothing parameter $\lambda = 5$. The default quantile is the median, corresponding to `tau = 0.5` inside `qss()`. Hence curves in Fig. 6.2 correspond to an estimate of the

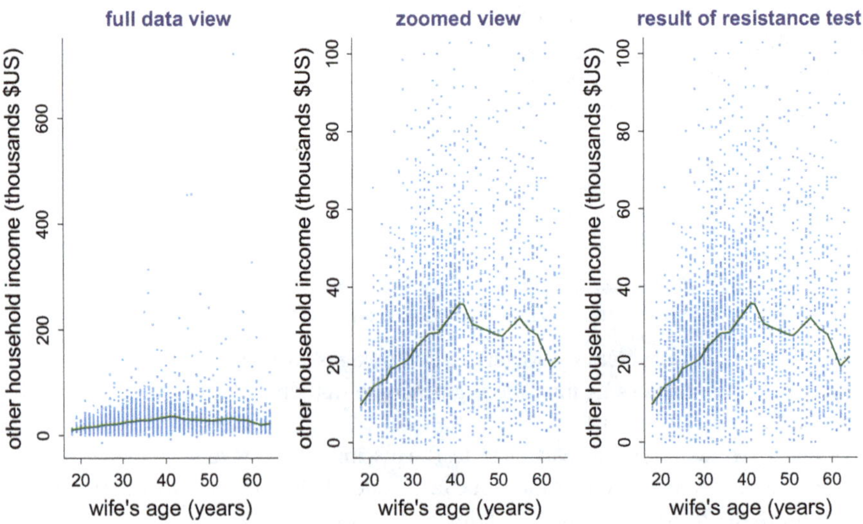

Fig. 6.2 Left panel: `rqss()` conditional median fit to data from the 1987 cross-section Middle panel: zoomed view of the left panel fit. Right panel: fit when the largest response observation is replaced by a value 1000 times larger, as a test of resistance.

median other household income, given the wife's age.

Note that `rqss()` passes the resistance test with flying colors. This is due to medians being unaffected by gross outliers.

6.2.2 Robust Semiparametric Regression

Robust semiparametric regression is an extension of robust and resistant scatterplot smoothing, as described in Sect. 6.2.1, in which

- the fits are robust and/or resistant to outliers,
- models for handling multiple predictors (e.g., additive models) are included,
- and, ideally, automatic selection of smoothing parameters is included in the methodology.

R packages such as `rgam` (Salibian-Barrera et al. 2014) and `robustgam` (Wong et al. 2013) provide some support for robust semiparametric regression models. However, their current releases are restricted to binary response and Poisson response models, so cannot handle data from the Michigan Panel Study of Income Dynamics with a continuous response such as other household income.

Another approach to robust semiparametric regression in R involves modeling the response variable to have a heavy-tailed distribution. The `VGAM` (Yee 2017) supports responses having a (Student's) t distribution, which has arbitrarily heavy tails depending on the degrees of freedom parameter. As explained in Lange et al. (1989), the likelihood corresponding to the t distribution induces a type of robustness on the regression fitting. Note that the t distribution with location μ, scale $\sigma > 0$, and degrees of freedom $\nu > 0$ has density function

$$p(x; \mu, \sigma, \nu) = \frac{\Gamma\left(\frac{\nu+1}{2}\right)}{\sigma\sqrt{\pi\nu}\,\Gamma(\nu/2)[1 + \{(x-\mu)/\sigma\}^2/\nu]^{\frac{\nu+1}{2}}}. \tag{6.1}$$

We write $x \sim t(\mu, \sigma, \nu)$ for a random variable x having this density function.

We now work through an example t distribution response fit, via the `vgam()` function in `VGAM`. First load in the data frame `Workinghours` via:

```
> library(Ecdat) ; data(Workinghours)
```

Next create the data frame `MichInc` via use of the `transform()` function:

```
> MichInc <- data.frame(husbandManager=as.numeric(
+              as.character(Workinghours$occupation) == "mp"))
```

```
> MichInc <- transform(MichInc,
+                 otherIncome = Workinghours$income/10,
+                 nonWhite = Workinghours$nonwhite,
+                 homeOwned = Workinghours$owned,
+                 unemployRate = Workinghours$unemp,
+                 wifeAge = Workinghours$age,
+                 wifeEducationYears = Workinghours$education,
+                 numChildren = with(Workinghours,
+                               child5 + child13 + child17))
```

The fit is then obtained using:

```
> library(VGAM)
> fit <- vgam(otherIncome ~ s(wifeAge,df=10)
+                        + s(unemployRate,df = 4)
+                        + s(wifeEducationYears, df = 4)
+                        + s(numChildren,df = 4) + nonWhite
+                        + homeOwned + husbandManager,
+                        family = studentt3,data = MichInc)
```

Here `family = studentt3` specifies that the response is a t distribution with location parameter corresponding to the additive model on the right-hand side of the $\sim$. The default in `studentt3()` of `zero = -(2:3)` specifies that the scale and degrees of freedom parameters are constants. The data are created from the data frame `Workinghours` in the `Ecdat`, corresponding to the 1987 cross-section of the Michigan Panel Study of Income Dynamics. Details of the variable creation are in the R script `MichInctAddMod.R` in the `HRW` package. The definitions are:

`otherIncome` = household income from sources other than the wife in thousands of U.S. dollars,

`wifeAge` = age of wife in years,

`unemployRate` = local unemployment rate as a percentage,

`wifeEducationYears` = wife's number of years of education,

`numChildren` = number of children in the household,

`nonWhite` = indicator that the wife is not white,

`homeOwned` = indicator that the household home is owned

and `husbandManager` = indicator that the husband is a manager.

The estimate of the degrees of freedom parameter is

$$\widehat{\nu} = 2.56,$$

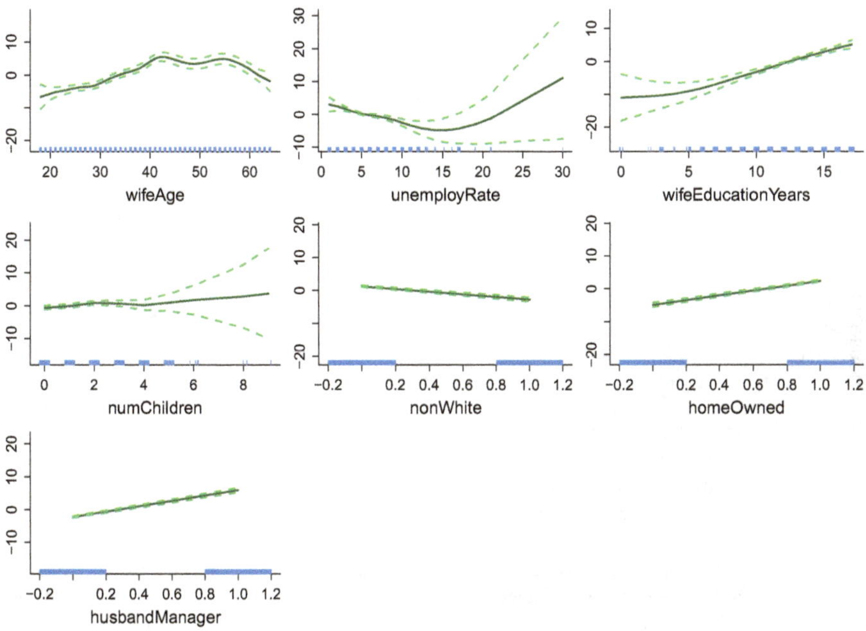

Fig. 6.3 Estimated additive components from a robust additive model fit to data from the 1987 cross-section of the Michigan Panel Study of Income Dynamics. The fit was obtained via the package VGAM with `family = studentt3`, which models the response variable to be a t distribution. The dashed curves correspond to ± 2 standard errors.

corresponding to a t distribution with tails decaying at the rate $|y|^{-3.56}$. Such tails are quite heavy, in keeping with the large positive outliers in the response (e.g., Fig. 6.1, left panel). Figure 6.3 shows the estimated additive model components with ± 2 standard errors. The R code for this plot is

```
> plotvgam(fit,se = TRUE,noxmean = TRUE,bty = "l",
+              lcol = c("darkgreen"),scol = "green3",
+              rcol = "dodgerblue",llwd = 2,slwd = 2,
+              ylab = "",cex.axis = 1.5,cex.lab = 1.8)
```

and the arguments of `plotvgam()` are explained via `help(plotvgam)`. The effect of wife's age is seen to be similar to that of the median regression fit in Fig. 6.2. The other effects are as one might expect. For example, the husband being a manager has a positive effect on other household income.

A drawback of t distribution response semiparametric regression via `vgam()` is the requirement to specify `vgam()` the effective degrees of freedom for each smooth function in the additive model. One way to have smoothing parameter selection automated is to adopt a Bayesian approach with mixed model-based penalized splines, and then use one of R's Markov chain Monte Carlo (MCMC) packages for fitting. Automatic smoothing parameter selection is encompassed in

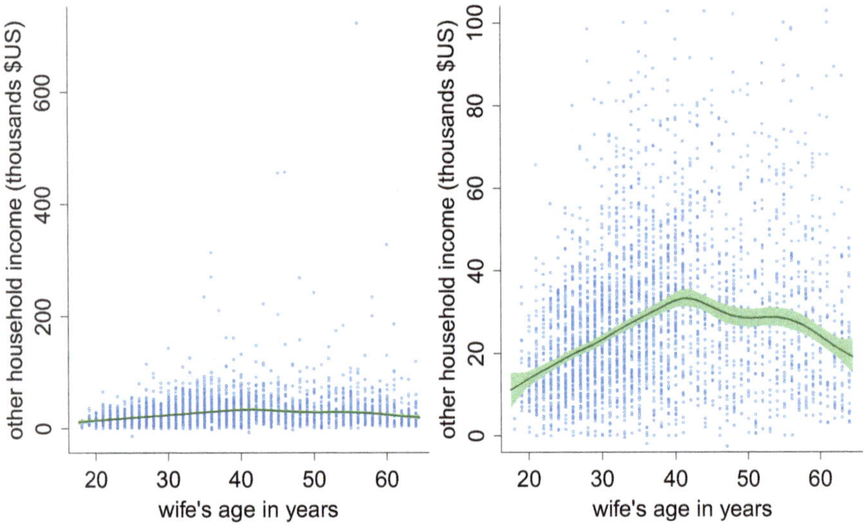

Fig. 6.4 Left panel: Bayesian t distribution response nonparametric regression fit to data from the 1987 cross-section of the Michigan Panel Study of Income Dynamics. Right panel: zoomed view of the fit shown in the left panel.

the Bayesian fitting and inference. In this vein, a class of Bayesian t distribution response semiparametric regression models is:

$$
\begin{gathered}
y_i \mid \boldsymbol{\beta}, \boldsymbol{u}, \sigma_\varepsilon, \nu \overset{\text{ind.}}{\sim} t\big((\boldsymbol{X\beta} + \boldsymbol{Zu})_i, \sigma_\varepsilon, \nu\big), \\
\boldsymbol{\beta} \sim N(0, \sigma_\beta^2 \boldsymbol{I}), \quad \boldsymbol{u}_1, \ldots, \boldsymbol{u}_d \mid \sigma_{u1}, \ldots, \sigma_{ud} \overset{\text{ind.}}{\sim} N(0, \underset{1\le j\le d}{\text{blockdiag}}(\sigma_{uj}^2 \boldsymbol{I})), \\
\sigma_\varepsilon \sim \text{Half-Cauchy}(A_\varepsilon), \quad \sigma_{uj} \overset{\text{ind.}}{\sim} \text{Half-Cauchy}(A_{uj}), \quad 1 \le j \le d, \\
\nu \sim \text{Uniform}(\nu_{\text{low}}, \nu_{\text{upp}}).
\end{gathered}
\tag{6.2}
$$

The standard deviation parameters $\sigma_{u1}, \ldots, \sigma_{ud}$ control the amount of smoothing for each of the d functions. The hyperparameters σ_β, A_ε and A_{uj} are set, nominally, to large values to impose noninformativity. A reasonable choice for the prior on ν is Uniform($\frac{1}{100}$, 100) since it encompasses a wide range of heavy-tailed and light-tailed distributions. An alternative family of degrees of freedom priors is described by Verdinelli and Wasserman (1991).

Figure 6.4 shows the results of fitting a $d = 1$ version (6.2) to the Michigan panel study of Income Dynamics data with wife's age as the only predictor. The `stan()` function in the `rstan` is used for MCMC-based fitting and inference. The Stan code for model specification is:

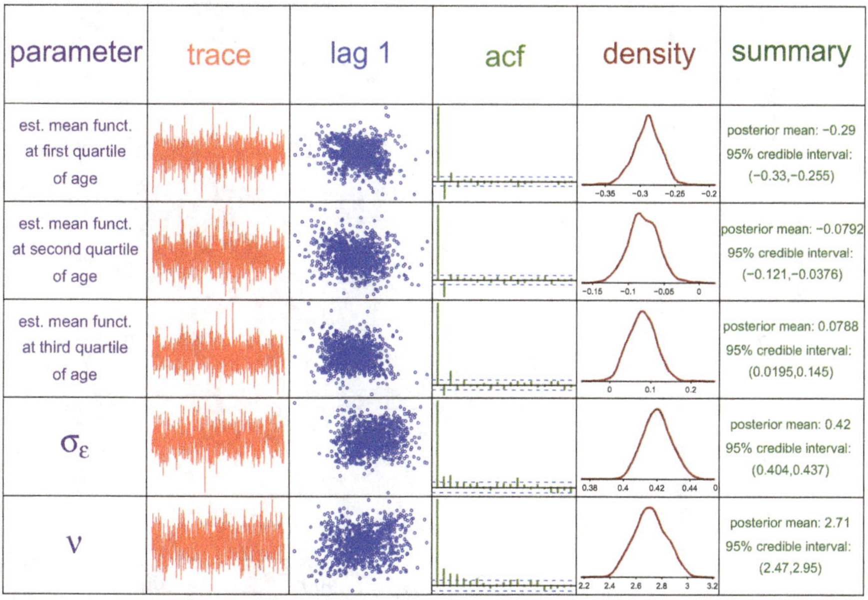

Fig. 6.5 Plots and summaries of MCMC samples for parameters of interest from the *t* distribution response nonparametric regression example: trace plot, lag-1 plot, estimated autocorrelation function, estimated posterior density function, and numerical summary.

```
model
{
   y ~ student_t(nu,X*beta+Z*u,sigmaEps) ;
   u ~ normal(0,sigmaU);  beta ~ normal(0,sigmaBeta);
   sigmaEps ~ cauchy(0,Aeps); sigmaU ~ cauchy(0,Au);
   nu ~ uniform(nuLow,nuUpp);
}
```

Note that Stan uses the ordering (ν, μ, σ) in its t distribution designation, rather than the ordering (μ, σ, ν) used in our t distribution notation introduced in (6.1).

Figure 6.5 summarizes the MCMC samples corresponding to the fit shown in Fig. 6.4. Note that the Bayes estimate of ν is 2.71, with a corresponding 95% credible interval of $(2.47, 2.95)$. The R package HRW contains a script named MichIncMCMCt.R that carries out the analysis corresponding to Figs. 6.4 and 6.5. To run this script and obtain the relevant file issue:

```
> library(HRW) ; demo(MichIncMCMCt,package = "HRW")
```

Its location can be obtained from the commands:

```
> system.file("demo","MichIncMCMCt.R",package = "HRW")
```

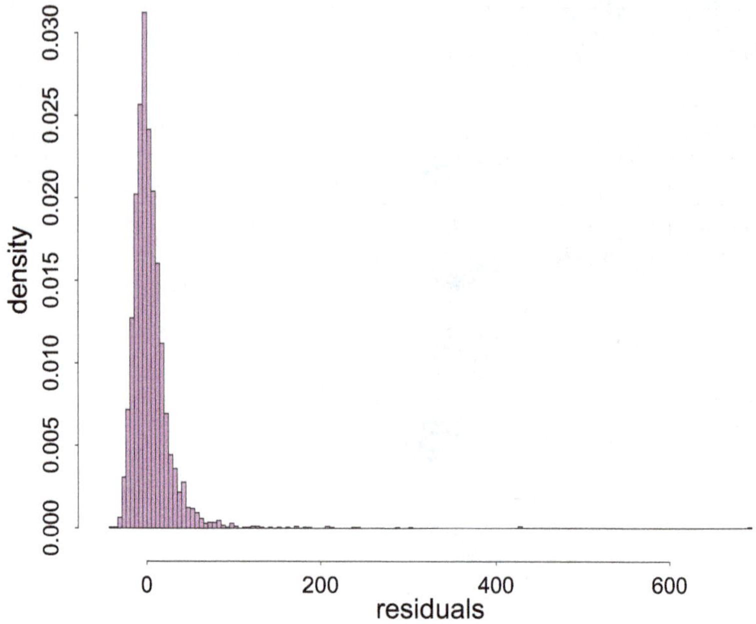

Fig. 6.6 Histogram of the residuals from the quantile smoothing spline fit to 1987 cross-section of the Michigan Panel Study of Income Dynamics shown in the left and middle panels of Fig. 6.2.

Figure 6.6 is a histogram of the residuals from the quantile smoothing spline fit shown in the left and middle panels of Fig. 6.2. It shows that the outliers tend to be positive-valued, which suggests that an asymmetric error distribution is appropriate as opposed to the symmetric $t(0, \sigma_\varepsilon, \nu)$ distribution used in (6.2).

The skewed t density function with ν_{left} and ν_{right} degrees of freedom, devised by Jones and Faddy (2003), is

$$\begin{aligned} p_{\text{JF}}(x;\, \nu_{\text{left}}, \nu_{\text{right}}) = & \frac{\Gamma(\nu_{\text{left}}+\nu_{\text{right}})}{\Gamma(\nu_{\text{left}})\Gamma(\nu_{\text{right}})2^{\nu_{\text{left}}+\nu_{\text{right}}-1}\sqrt{\nu_{\text{left}}+\nu_{\text{right}}}} \\ & \times \left(1 - \frac{x+M_{\nu_{\text{left}},\nu_{\text{right}}}}{\sqrt{\nu_{\text{left}}+\nu_{\text{right}}+(x+M_{\nu_{\text{left}},\nu_{\text{right}}})^2}}\right)^{\nu_{\text{left}}+\frac{1}{2}} \\ & \times \left(1 + \frac{x+M_{\nu_{\text{left}},\nu_{\text{right}}}}{\sqrt{\nu_{\text{left}}+\nu_{\text{right}}+(x+M_{\nu_{\text{left}},\nu_{\text{right}}})^2}}\right)^{\nu_{\text{right}}+\frac{1}{2}} \end{aligned} \tag{6.3}$$

where

$$M_{\nu_{\text{left}},\nu_{\text{right}}} \equiv \frac{(\nu_{\text{right}} - \nu_{\text{left}})\sqrt{\nu_{\text{right}} + \nu_{\text{left}}}}{\sqrt{(2\nu_{\text{right}} + 1)(2\nu_{\text{left}} + 1)}}.$$

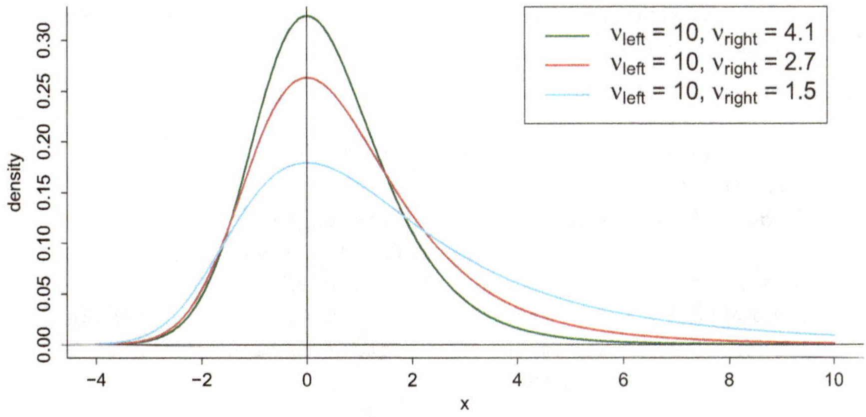

Fig. 6.7 Jones–Faddy skewed t density functions with degrees of freedom pairs $(\nu_{\text{left}}, \nu_{\text{right}}) = (10, 4.1), (10, 2.7), (10, 1.5)$.

Note that (6.3) is a centered version of the density function given in Jones and Faddy (2003) so that $p_{\text{JF}}(x; \nu_{\text{left}}, \nu_{\text{right}})$ has its mode at $x = 0$. The left and right tail decay rates are, respectively,

$$|x|^{-(2\nu_{\text{left}}+1)} \quad \text{and} \quad |x|^{-(2\nu_{\text{right}}+1)} \quad \text{if } \nu_{\text{left}} \geq \nu_{\text{right}}.$$

The tail decay rates are reversed if $\nu_{\text{left}} < \nu_{\text{right}}$. If $\nu_{\text{left}} = \nu_{\text{right}} = \omega$, then (6.1) reduces to the ordinary t distribution with 2ω degrees of freedom. Figure 6.7 shows three different versions of $p_{\text{JF}}(\cdot; \nu_{\text{left}}, \nu_{\text{right}})$, with ν_{left} fixed at 10, but $\nu_{\text{right}} \in \{4.1, 2.7, 1.5\}$. The differing tail behavior on the left and right sides of the densities when ν_{left} and ν_{right} differ is apparent.

Since Fig. 6.6 is strongly suggestive of differing tail decay rates, we investigate R fitting of the model:

$$y_i|\boldsymbol{\beta}, \boldsymbol{u}, \sigma_\varepsilon, \nu_{\text{left}}, \nu_{\text{right}} \stackrel{\text{ind.}}{\sim} \frac{1}{\sigma_\varepsilon} p_{\text{JF}}\left(\frac{y_i - (\boldsymbol{X\beta} + \boldsymbol{Zu})_i}{\sigma_\varepsilon}; \nu_{\text{left}}, \nu_{\text{right}}\right). \tag{6.4}$$

The full Bayesian model that we consider is analogous to (6.2), although we put the following uniform priors on the degrees of freedom parameters:

$$\nu_{\text{left}}, \nu_{\text{right}} \stackrel{\text{ind.}}{\sim} \text{Uniform}\left(\tfrac{1}{100}, 100\right).$$

However, for MCMC fitting in `rstan`, we need to work around the fact that the family of density functions described by (6.3) is not one of the families supported by Stan. We overcome this by using the `target +=` facility, in which the log-likelihood is incremented via the logarithm of (6.3), up to additive constants. The relevant Stan code is:

```
transformed parameters
{
   real JFmode;              real logNormFac;
   vector[n] arg;
   JFmode = (nuLft-nuRgt)*sqrt((nuLft+nuRgt)/
                              ((2*nuLft+1)*(2*nuRgt+1)));
   arg = JFmode + (y - X*beta - Z*u)/sigmaEps;
   logNormFac = (nuLft+nuRgt-1)*log(2) + 0.5*log(nuLft+nuRgt)
                - lgamma(nuLft+nuRgt) + lgamma(nuLft)
                + lgamma(nuRgt);
}
model
{
   for (i in 1:n)
      target += (nuLft+0.5)*log1p(arg[i]/
                     sqrt(nuLft+nuRgt+square(arg[i])))
                      + (nuRgt+0.5)*log1m(arg[i]/
                         sqrt(nuLft+nuRgt+square(arg[i])))
                      - log(sigmaEps) - logNormFac;
   u ~ normal(0,sigmaU); beta ~ normal(0,sigmaBeta);
   sigmaU ~ cauchy(0,Au); sigmaEps ~ cauchy(0,Aeps);
   nuLft ~ uniform(0.01,100); nuRgt ~ uniform(0.01,100);
}
```

Note that, in Stan, `log1p` and `log1m` are, respectively, the functions $\log(x+1)$ and $\log(x-1)$ in x.

Figure 6.8 shows the estimated mode function according to the Jones–Faddy skewed t response nonparametric regression model (6.4) to the same data. The curve is lower than that for the t distribution model, shown in Fig. 6.4, since the mode is below the mean for right-skewed distributions.

Figure 6.9 shows the summaries of the MCMC samples for the (6.4) fit. The Bayes estimates of ν_{left} and ν_{right} are 7.4 and 1.7, respectively. This corresponds to tail decay rates of $|y|^{-15.8}$ on the left and $|y|^{-4.4}$ on the right. This difference reflects the propensity for income data to have positive outliers that are much more extreme than outliers in the negative direction.

The script `MichIncMCMCskewt.R` in the `HRW` package carries out the analysis corresponding to Figs. 6.8 and 6.9. To run `MichIncMCMCskewt.R` and obtain the relevant file issue:

```
> library(HRW) ; demo(MichIncMCMCskewt,package = "HRW")
> system.file("demo","MichIncMCMCskewt.R",package = "HRW")
```

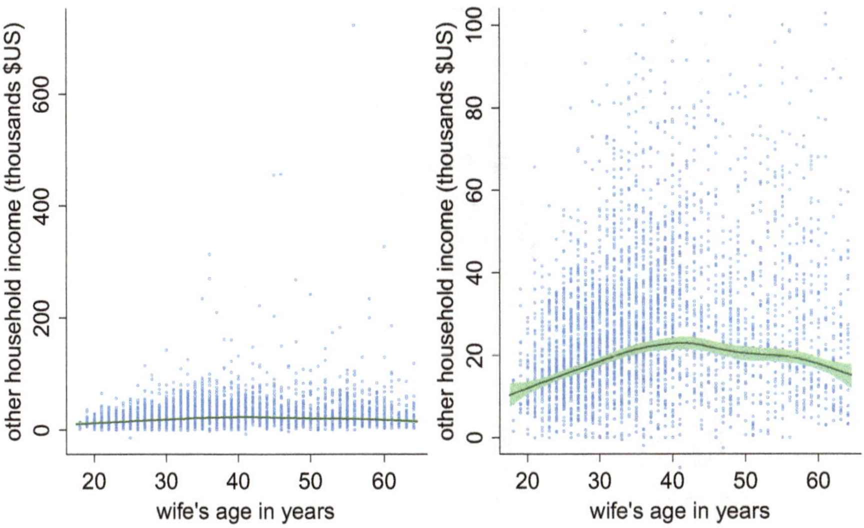

Fig. 6.8 Left panel: Bayesian Jones–Faddy skewed t response nonparametric regression fit to data from the 1987 cross-section of the Michigan Panel Study of Income Dynamics. Right panel: zoomed view of the fit shown in the left panel.

6.2.3 Quantile Semiparametric Regression

Versions of quantile semiparametric regression are supported by R packages such as `quantreg` (Koenker 2017) and `VGAM` (Yee 2017). We now describe some of their capability via additional examples from the 1987 cross-section of the Michigan Panel Study of Income Dynamics.

First we address the fitting of multiple quantile functions to a single predictor. There are two general approaches to multiple quantile regression:

- applying single quantile regression methodology several times with the quantile value set to various values.
- fitting functions of the predictor corresponding to transformation parameters, such as those for the Box–Cox family of transformations. The transformed responses then have a fixed distribution such as the standard Normal. The back-transformations of the quantiles of this fixed distribution are then estimates of the conditional quantiles of the original response.

The second approach, which goes back to Cole and Green (1992), is more elaborate and has the advantage that the fitted quantile curves cannot cross each other. These authors dubbed it the *LMS method* after the initials of the Box–Cox transformation parameters: λ, μ, and σ. The generic form of the $100\tau\%$ quantile curve, for $0 < \tau < 1$, is

$$\widehat{\mu}(x)\{1+\widehat{\lambda}(x)\widehat{\sigma}(x)\Phi^{-1}(\tau)\}^{1/\widehat{\lambda}(x)}, \quad (6.5)$$

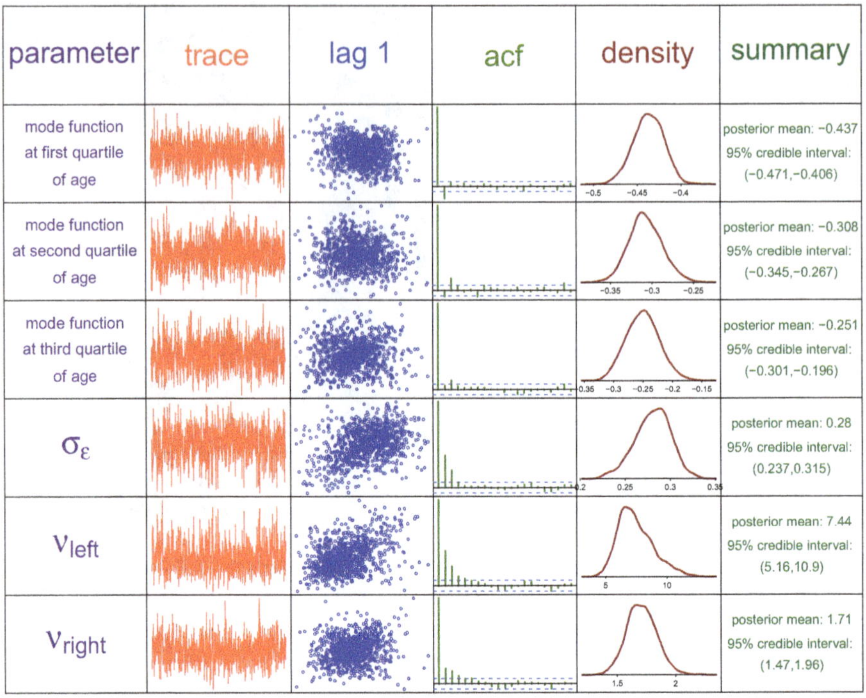

Fig. 6.9 Plots and summaries of MCMC samples for parameters of interest from the Jones–Faddy skewed t response nonparametric regression example: trace plot, lag-1 plot, estimated autocorrelation function, estimated posterior density function, and numerical summary.

where $\widehat{\lambda}(x)$, $\widehat{\mu}(x)$ and $\widehat{\sigma}(x)$ are the estimated Box–Cox transformation parameters at x. Further details are given in Yee (2004).

Both approaches are supported in R. The first approach can, of course, be accomplished by calling `rqss()` from Sect. 6.2.1 and Fig. 6.2, with its `tau` parameter set to each of the desired quantile values. The upper panels of Fig. 6.10 show the result corresponding to the 1, 5, 25, 50, 75, 95, and 99% quantiles. The relevant code is given in the `MichIncMultQSS.R`, and involves looping through `tau` set to entries of the vector

$$(0.01, 0.05, 0.25, 0.5, 0.75, 0.95, 0.99).$$

To run `MichIncMultQSS.R` issue:

```
> library(HRW) ; demo(MichIncMultQSS,package = "HRW")
```

Its location for possible copying and modifying is determined by:

```
> system.file("demo","MichIncMultQSS.R",package = "HRW")
```

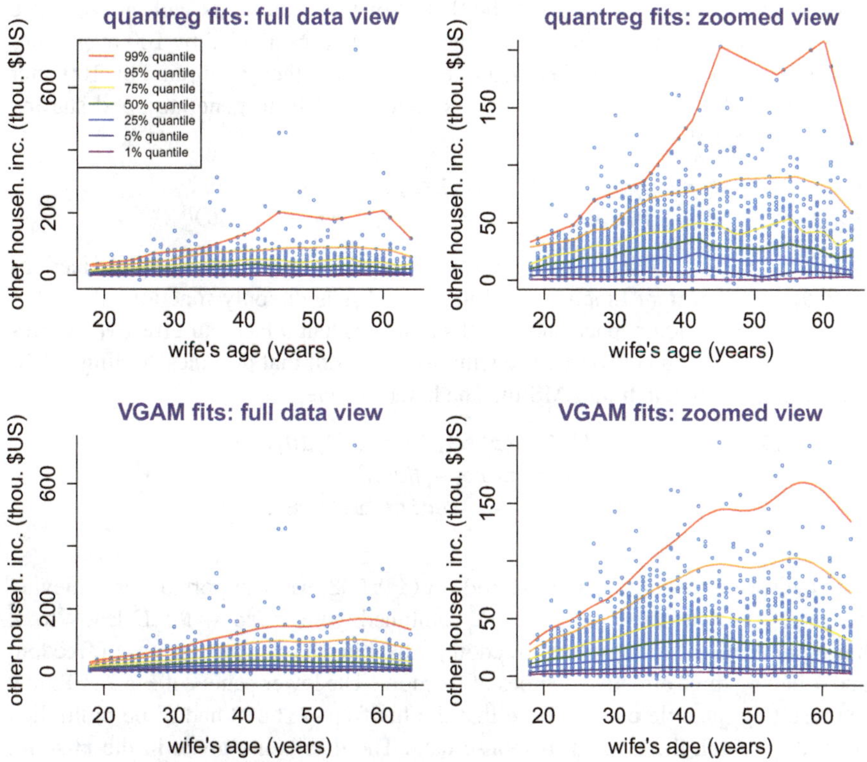

Fig. 6.10 Upper panels: full data and zoomed views of quantile smoothing spline fits to data on other household income versus wife's age from the 1987 cross-section of the Michigan Panel Study of Income Dynamics. The fits were obtained using the function `rqss()` from the package as described in the text. Lower panels: full data and zoomed views of quantile fits according to the LMS method for the same data. The fits were obtained using the functions `vgam()` and `lms.bcn()` from the package `VGAM` as described in the text.

The lower panels of Fig. 6.10 are obtained from the implementation of the LMS method within the function `vgam()` of the package `vgam()` `VGAM` (Yee 2017). The relevant sub-function of `vgam()` is named `lms.bcn()`. However, the LMS fits shown in Fig. 6.10 took some effort, as we now explain. First, suppose that the scatterplot data are stored in R arrays x and y as follows:

```
> library(Ecdat)          ; data(Workinghours)
> x <- Workinghours$age ; y <- Workinghours$income/10
```

The data used in the call to `lms.bcn()` within `vgam()` involves the arrays xLMS and LMS obtained via:

```
> npi <- which(y<=0)  ; yLMS <- y[-npi]/100 ; xLMS <- x[-npi]
```

The [-npi] leads to omission of the 0.56% of the households with non-positive other household income values. Working with responses divided by 100 overcomes numerical instability issues. The quantile curves were then multiplied by 100 after fitting. It is also necessary to obtain a simple initial fit to generate good starting values. This involves:

```
> fitLMSinit <- vgam(yLMS ~ s(xLMS,df = 4),
+                    lms.bcn(zero = c(1,3)),maxit = 50)
```

The zero = c(1,3) specification inside lms.bcn() forces the $\widehat{\lambda}$ and $\widehat{\sigma}$ functions in (6.5) to be constant functions. It follows that $\widehat{\mu}$ is the only function allowed to vary over the predictor space and the df=4 dictates that it has four effective degrees of freedom. Thus fitLMSinit is a simplistic LMS fit that provides starting values for the more elaborate final LMS fit. The latter involves:

```
> fitLMS  <- vgam(yLMS ~ s(xLMS,df = c(3,10,3)),
+                 lms.bcn(zero = NULL),
+                 etastart = predict(fitLMSinit),
+                 maxit = 500)
```

with the specification etastart=predict(fitLMSinit) important for achieving convergence. The df = c(3,10,3) combined with zero = NULL leads to $\widehat{\lambda}$ having three effective degrees of freedom, $\widehat{\mu}$ having ten effective degrees of freedom and $\widehat{\sigma}$ having three effective degrees of freedom. The lower panels of Fig. 6.10 show the resulting quantile curves. Note that the fits from fitLMS had to be multiplied by 100 to match the original response data. The details are given in the R script MichIncLMS.R in the HRW package. To run and locate the script on the computer on which HRW resides issue:

```
> library(HRW) ; demo(MichIncLMS,package = "HRW")
> system.file("demo","MichIncLMS.R",package = "HRW")
```

Our final quantile regression example involves the additive model extension of quantile smoothing splines for multiple predictors, via rqss() in the package quantreg. We use the same predictors as those used for the vgam(family = studentt3) example given in Sect. 6.2.2. An example call to rqss() for an additive model 5% quantile fit is:

```
> library(quantreg) ; library(Ecdat) ; data(Workinghours)
> rqss(otherIncome ~ qss(wifeAge,lambda = 3.5)
+                  + qss(unemployRate,lambda = 3.5)
+                  + qss(wifeEducationYears,lambda = 3.5)
+                  + qss(numChildren,lambda = 3.5) + nonWhite
+                  + homeOwned + husbandManager,tau = 0.05,
+                    data = Workinghours)
```

Figure 6.11 shows the fitted 5, 50, and 95% quantile curves. In each panel, the quantile curves are the slices of the quantile surfaces for the panel predictor, with all other predictors set to their averages. The quantile curves are consistent with

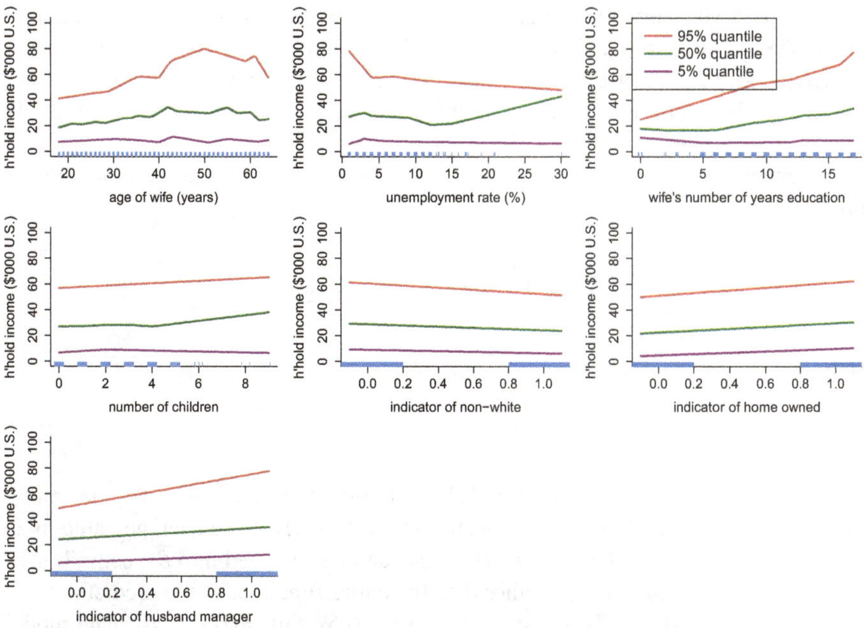

Fig. 6.11 Estimated additive components from 5, 50, and 95% quantile smoothing spline-based additive model fits to data from the 1987 cross-section of the Michigan Panel Study of Income Dynamics. The fit is via the function within the package. In each panel, the quantile curves are the slices of the quantile surfaces for the panel predictor, with all other predictors set to their averages.

intuition, such as median income increasing with the wife's number of education years. The vertical axis range is fixed across all panels so that visual comparisons can be made regarding the relative effects of the predictors.

The VGAM package also has LMS quantile regression functions, `lms.yjn()` and `lms.yjn2()`, using the Yeo and Johnson (2000) family of transformations instead of the Box–Cox family. Yee (2004) contains relevant details. Exercise 9 is concerned with Yeo–Johnson LMS quantile regression.

6.3 Scalar-on-Function Linear Regression

It has become increasingly common for either the response or the predictor variable in a regression model to be a function. Consider first the case where the response y is scalar and the predictor is a function $x(t)$ defined on an interval $[a, b]$. The *scalar-on-function linear model* is

$$y_i = \beta_0 + \int_a^b x_i(t)\beta_1(t)\,dt + \varepsilon_i, \quad 1 \le i \le n, \tag{6.6}$$

where y_i and $x_i(t)$ are the response and predictor for the ith case, β_0 is an intercept, and $\beta_1(t)$ is a smooth coefficient function. The integral in (6.6) is the inner product of $x_i(t)$ and $\beta_1(t)$ in a space of functions and is the analog of the dot or inner product of the vector of predictor variables and the coefficient vector in a multiple linear regression model.

To understand model (6.6), consider first a simple linear regression model with predictors

$$\overline{x}_i \equiv \int_a^b x_i(t)\,dt, \quad 1 \le i \le n,$$

so that

$$y_i = \beta_0 + \overline{\beta}_1 \overline{x}_i + \varepsilon_i, \tag{6.7}$$

where β_0 and $\overline{\beta}_1$ are the intercept and slope. Integration of a function is the analog of summation of the elements of a vector, so $\overline{x}_i/(b-a)$ is the average value of x_i over the interval (a, b). The factor $1/(b-a)$ can be subsumed into $\overline{\beta}_1$, so (6.7) uses only the average value of $x_i()$, rather than the entire function, as the predictor.

Notice that model (6.7) is a special case of (6.6) with $\beta_1(t) \equiv \overline{\beta}_1$. The model implies that the effect of $x_i(t)$ on y_i is constant, that is, it does not depend on t. Suppose one suspects that the effect of $x(t)$ depends on t in a smooth way, so that it does not change much over any short interval. Then one might use J predictors where the ith value of the jth predictor $\overline{x}_{ji}$ is either

$$\overline{x}_{ji} \equiv \int_{I_j} x_i(t)\,dt, \quad \text{where} \quad I_j \equiv \left[a + \frac{(j-1)(b-a)}{J}, a + \frac{j(b-a)}{J}\right], \quad 1 \le j \le J,$$

or, as an approximation to this integral,

$$\overline{x}_{ji} \equiv x_i(M_j)\left(\frac{b-a}{J}\right) \quad \text{where} \quad M_j \equiv \text{mid-point of } I_j = a + \frac{(j-\frac{1}{2})(b-a)}{J}.$$

Then the model is

$$y_i = \beta_0 + \sum_{j=1}^{J} \overline{\beta}_{1,j}\,\overline{x}_{ji} + \varepsilon_i \tag{6.8}$$

with slopes $\overline{\beta}_{1,1}, \ldots, \overline{\beta}_{1,J}$. Model (6.8) is the special case of (6.6) with

$$\beta_1(t) \equiv \overline{\beta}_{1,j} \text{ on } I_j.$$

Model (6.8) is sensitive to the choice of J and is prone to overfitting if J is large. Also, it is not realistic to assume that the effect of $x(t)$ is constant on intervals and jumps between intervals. A better approach is to use model (6.6) with a roughness penalty to keep $\beta_1(t)$ smooth and prevent overfitting.

In practice, functional variables are not observed continuously but instead only at a finite set of points. We will only consider the so-called "dense data" case where all functions are observed on the same equally spaced, fine grid. Regression with sparsely observed functional predictors requires that the unobserved values of the function on a dense grid be predicted from the sparsely observed values; doing this is beyond the scope of this book.

6.3.1 Example: Diffusion Tensor Imaging Data

As an example, we will use the `DTI` data frame in the `refund` package (Goldsmith et al. 2016). The scalar response is the *Paced Auditory Serial Addition Task (PASAT)* score, which is a test of cognitive ability. Subjects are given a number every 3 seconds and are asked to add the number just heard to the previous number. The score is the number of correct answers out of 60.

The predictor function is `cca` which is fractional anisotropy along the corpus callosum tract of the brains of multiple sclerosis patients and t is location along the tract, with $t = 0$ and $t = 1$ indicating the two ends of the tract. Diffusion tensor imaging is a magnetic resonance imaging technique used to locate and study white matter tracts in the brain. A white matter tract consists of axons that connect different parts of the brain and are insulated by myelin, a white, fatty substance. The corpus callosum is the largest white matter tract and connects the left and right hemispheres of the brain. Diffusion tensor imaging measures diffusion of water along a white matter tract. The result is a tensor that can be expressed as a 3×3 symmetric, positive-definite matrix. If diffusion was isotropic, then this matrix would be a multiple of the identity matrix with all three eigenvalues equal. Fractional isotropy measures the variation in the eigenvalues (relative to their sizes) and is

$$\left[\frac{3\left\{(\lambda_1-\overline{\lambda})^2+(\lambda_2-\overline{\lambda})^2+(\lambda_3-\overline{\lambda})^2\right\}}{2(\lambda_1^2+\lambda_2^2+\lambda_3^2)}\right]^{1/2} \tag{6.9}$$

where $\lambda_1 \geq \lambda_2 \geq \lambda_3$ are the eigenvalues of this matrix and $\overline{\lambda} = (\lambda_1 + \lambda_2 + \lambda_3)/3$. The numerator of (6.9) measures the variation between the three eigenvalues and the denominator serves to make fractional anisotropy scale free.

In a healthy patient, the myelin should insulate the axons so that diffusion is mostly parallel to them and fractional anisotropy is high. Multiple sclerosis is a disease of the myelin that degrades the ability of myelin to insulate and, consequently, of white matter tracts to conduct signals. Since the corpus callosum is involved with cognition, one might hypothesize that higher values of fractional anisotropy along the corpus callosum tract would indicate healthier patients who would have higher PASAT scores. The DTI data frame also includes controls, but PASAT was not measured on them. We will study the relationship between PASAT scores and fractional anisotropy along the corpus callosum in multiple sclerosis patients.

The scalar-on-function linear model can be fit by several R functions. In this example, we will use the function pfr() in the refund package (Goldsmith et al. 2016). *Penalized functional regression* (Goldsmith et al. 2011) is a simple yet efficient scalar-on-function regression method that utilizes the principal components analysis eigendecomposition of $\boldsymbol{\Sigma}$ to provide the low-dimensional representation (5.2) of $x_i(t)$ in (6.6), noting that $\widehat{y}_i(t)$ in (5.2) corresponds to $x_i(t)$ in (6.6), and models $\beta_1(t)$ with a spline basis. Penalized functional regression can be implemented by the pfr() function in the refund package (Goldsmith et al. 2016). The pfr() function fits both scalar-on-function linear regression models and their generalized additive model extensions, which will be introduced in Sect. 6.4 and include the scalar-on-function linear model as a special case.

The following code fits a scalar-on-function linear model using the subjects where PASAT is not missing, i.e., multiple sclerosis patients. Here lf() in y ~ lf(X) specifies a scalar-on-function linear model fit of y to X.

```
> library(refund) ; data(DTI) ; indsNonMiss <- !is.na(DTI$pasat)
> y <- DTI$pasat[indsNonMiss] ; X <- DTI$cca[indsNonMiss,]
> fitlf <- pfr(y ~ lf(X),method = "REML")
```

A plot of $\widehat{\beta}_1(t)$ is obtained using:

```
> plot(fitlf,xlab = "tract location",
+    ylab=expression(paste(widehat(beta)[1],"(tract location)")),
+    shade= TRUE,col = "darkgreen",bty = "l",
+    cex.lab = 1.5,shade.col = "palegreen",
+    cex.axis = 1.5)
> abline(h = 0,col = "slateblue")
```

and is shown in Fig. 6.12. We see that this estimated coefficient function is positive, which is consistent with our hypothesis that higher values of fractional anisotropy should indicate healthier myelin and better performance on PASAT. Given the width of the variability band, it is reasonable to hypothesize that $\beta_1(t)$ is constant so that model (6.7) holds. It is easy to fit model (6.7):

```
> library(mgcv) ; meanx <- rowMeans(X)
> fitMeanx <- gam(y~meanx) ; print(summary(fitMeanx))

Family: gaussian
Link function: identity

Formula:
y ~ meanx
```

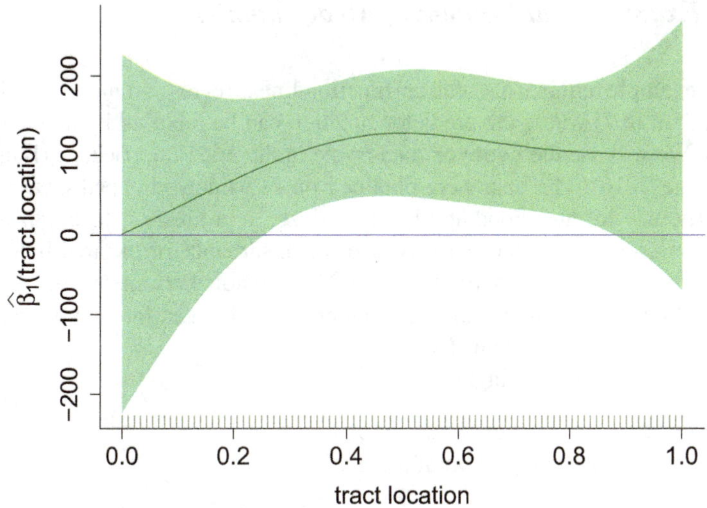

Fig. 6.12 Estimate of $\beta_1(t)$ when Paced Auditory Serial Addition Task (PASAT) score is regressed on fractional anisotropy in the corpus callosum tract.

```
Parametric coefficients:
            Estimate Std. Error t value Pr(>|t|)
(Intercept)  -0.2577     6.0520  -0.043    0.966
meanx        95.6330    12.1214   7.890 4.43e-14


R-sq.(adj) =  0.155   Deviance explained = 15.8%
GCV = 131.96  Scale est. = 131.17    n = 334
```

Next, we compare models (6.6) and (6.7) using AIC.

```
> fitlf$aic
[1] 2582.062
> fitMeanx$aic
[1] 2580.585
```

From the results, we see that the simpler model (6.7) has a smaller (better) AIC. Thus, in this example, one really does not need to use scalar-on-function linear regression, but of course this would not be known unless a scalar-on-function linear model fit was tried.

6.3.2 *Example: Fat Content of Meat Samples*

The next example requires a scalar-on-function linear regression model that is more complex than (6.7) where the predictor function can be replaced by its integral (or average). We will use the tecator data frame in the fda.usc package (Bande and de la Fuente 2016). The data were obtained on 215 finely chopped meat samples using a Tecator Infratec Food and Feed Analyzer, a product of the company *Foss* (www.fossna.com). The analyzer recorded the absorbance of infrared light at 100 equally spaced wavelengths from 850 to 1050 nanometers in the near infrared spectrum. In addition, three responses, percentages of water, fat, and protein, were determined by analytical chemistry.

The data are obtained using:

```
> library(fda.usc) ; data(tecator)
> X <- tecator$absorp.fdata$data
```

Next, we set the grid of wavelength values corresponding to the absorbance function ordinates, which were just stored as rows in X:

```
> wavelength <- seq(850,1050,length = 100)
```

We will use percentage of fat as the response and the absorbance spectrum, or one of its first two derivatives, as the predictor. As shown below, derivatives can be computed using the functions:

create.bspline.basis(), smooth.basisPar() and eval.fd()

in the fda package (Ramsay et al. 2017). The first of these three functions creates a B-spline basis and smooth.basisPar() fits these splines to the data with a small amount of smoothing; the argument "2" specifies a penalty on the second derivative and "1e-09" is the value of the smoothing parameter. Then eval.fd() differentiates the spline fits. The third argument of eval.fd() specifies the order of the derivative. Details on usage of the fda package are given in Ramsay et al. (2009). The matrices Xderiv and Xderiv2 containing the first and second derivatives, respectively, of the functions in X are then obtained via:

```
> library(fda)
> bbt <- create.bspline.basis(rangeval = range(wavelength),
+                              nbasis = 20)
> Xfd <- smooth.basisPar(wavelength,t(X),bbt,2,1e-9)
> Xderiv <- t(eval.fd(wavelength,Xfd$fd,1))
> Xderiv2 <- t(eval.fd(wavelength,Xfd$fd,2))
```

Figure 6.13 displays the first 10 of these absorbance functions and their derivatives. We see that the first and, especially, the second derivatives have sharp peaks and valleys that might have useful information for the prediction of fat concentration. The script absorbFuncsDrvs.R contains the code that produced Fig. 6.13. It can be run and located using:

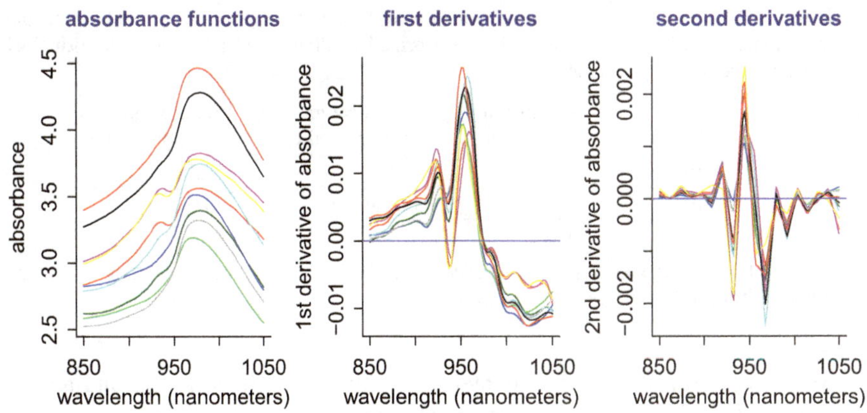

Fig. 6.13 Left panel: Plot of the first ten absorbance functions. Middle panel: Plot of their first derivatives. Right panel: Plot of their second derivatives.

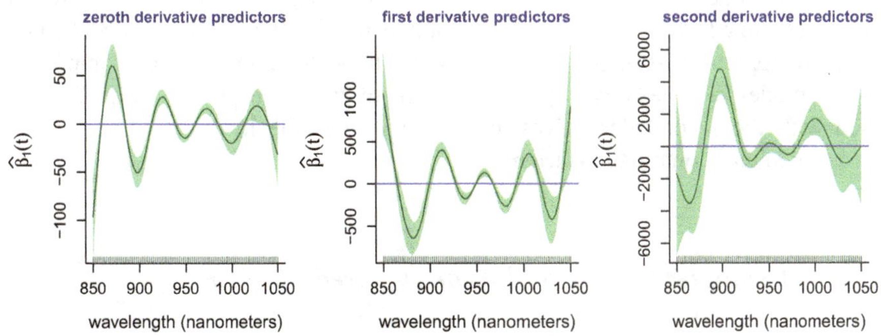

Fig. 6.14 Left panel: $\widehat{\beta}_1(t)$ using the absorbance functions as predictors. Middle panel: $\widehat{\beta}_1(t)$ using their first derivatives of the absorbance functions as predictors. Right panel: $\widehat{\beta}_1(t)$ using their second derivatives of the absorbance functions as predictors.

```
> library(HRW) ; demo(absorbFuncsDrvs,package = "HRW")
> system.file("demo","absorbFuncsDrvs.R",package = "HRW")
```

We fit scalar-on-function linear models with the zero-order, first and second derivatives as predictors using the `pfr()` function as in Sect. 6.3.1. The estimates of $\beta_1(t)$ for the three predictors are shown in Fig. 6.14. The code that produced the estimates and plots is in the script `absorbScalOnFuncAna.R` in the `HRW` package. To run the script issue the commands:

```
> library(HRW) ; demo(absorbScalOnFuncAna,package = "HRW")
```

Its location is determined by:

```
> system.file("demo","absorbScalOnFuncAna.R",package = "HRW")
```

Table 6.1 AIC values for four scalar-on-function regression estimators: linear, additive in the original predictor variable, additive in the transformed predictor, and additive in the principal component scores of the predictor.

Derivative order	Linear	Additive in orig. predic.	Additive in transf. predic.	Additive in princ. compon. scores
0	1046	1034	1292	822
1	1035	712	854	816
2	1056	655	865	588

Derivatives of the absorbance functions of orders 0, 1, and 2 were used as predictors of percentage of fat in the Tecator dataset

An object returned by `pfr()` contains the AIC of the fitted model and allows us to compare different models. The column labeled "linear" in Table 6.1 contains the AIC for linear scalar-on-function regression using zero-order, first and second derivatives. We see that the first derivatives provide the best fit when the model is linear. The models corresponding to the other columns of Table 6.1 are discussed in Sects. 6.4 and 6.5. They produce considerably better fits than the linear model when the predictors are first or second derivatives. The additive in principal component scores model has the lowest AIC with each of the three predictors and, when used with the second derivatives as predictors, has the lowest AIC among all model/predictor combinations tried.

6.3.3 *Example: Octane and Near Infrared Spectra*

The `gasoline` dataset contains octane and near infrared spectra measured on 60 samples of gasoline. Octane is expensive to measure directly, so a method for estimating octane using near infrared spectra could save money. We will use `pfr()` to estimate the octane of gasoline from near infrared spectra.

```
> library(refund) ; data(gasoline)
> wavelength <- seq(900,1700,by = 2)
> fitPFR <- pfr(octane ~ lf(NIR,argvals = wavelength,
+                 presmooth = "fpca.face",
+                 presmooth.opts = list(knots=50)),
+                 data = gasoline)
```

Because the argument `presmooth` in `lf()` was set to `fpca.face`, fast covariance estimation (FACE) (see Sect. 5.6) was used for principal components analysis and was, in fact, extremely fast, taking less than 1 second.

A plot of the estimated coefficient function is obtained using:

```
> wavelength <- seq(900,1700,by = 2)
> plot(fitPFR,xlab="wavelength (nanometers)",
+      ylab = expression(paste(beta[1](t))),
```

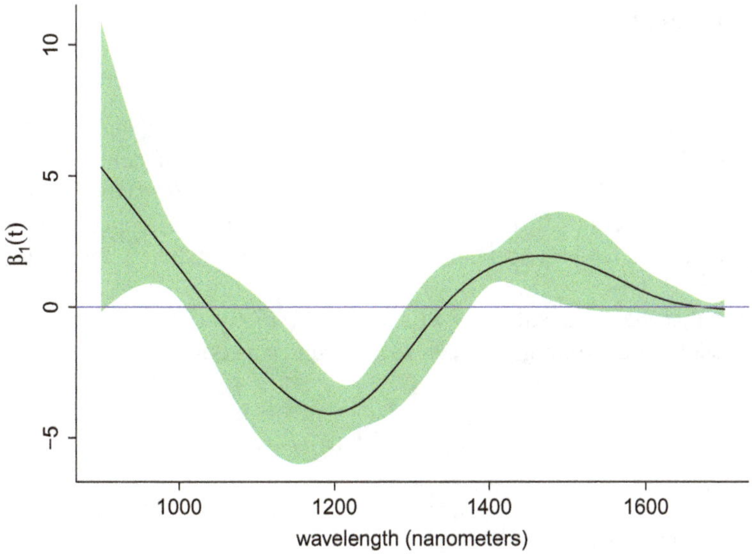

Fig. 6.15 Estimate of the coefficient function for predicting `octane` from `NIR` and 95% pointwise confidence bounds.

```
+        shade = TRUE,shade.col = "palegreen",rug = FALSE,
+        cex.axis = 1.5,cex.lab = 1.5,bty = "l")
> abline(h = 0,col = "slateblue")
```

and is shown in Fig. 6.15.

We see in Fig. 6.15 that the coefficient function is estimated with considerable uncertainty as the variability band contains zero except around 1000, 1200, and 1400 nanometers. Lower values of the variable `NIR` around 1200 nanometers and, to a lesser extent, higher values around 1000 and 1400 nanometers predict higher octane values.

Dr. Luo Xiao (personal communication) conducted a cross-validation study of the prediction of `octane` by `NIR`. In that study, penalized functional regression using FACE had a sum of squared prediction errors equal to 280. In contrast, penalized functional regression using a penalized thin plate spline to smooth the covariance function had a sum of squared prediction errors equal to 410. Penalized functional regression with FACE predicted more accurately than penalized functional regression with a penalized thin plate spline, because FACE could accommodate a much higher dimensional basis for smoothing the covariance matrix and estimating the eigenvectors used to represent the `NIR` functions. The lower dimensional basis used by the `gam()`-default, which is a penalized thin plate spline, causes substantial bias in this example.

6.4 Scalar-on-Function Additive Models

So far we have only considered scalar-on-function linear models. Just as linear models with scalar predictors can be extended to additive models, scalar-on-function linear models can be extended to scalar-on-function additive models. There are two distinct types of additive functional regression models. The first type is additive in the predictive functions themselves and is introduced in this section. The second type is additive in the principal component scores of the predictors and is discussed in Sect. 6.5.

Recall that the linear model (6.8) was used to motivate the scalar-on-function linear model. Model (6.8) generalizes to the additive model

$$y_i = \beta_0 + \sum_{j=1}^{d} f_j(\overline{x}_{ji}) + \varepsilon_i. \tag{6.10}$$

The functional analog of (6.10) is

$$y_i = \beta_0 + \int_a^b f\big(x_i(t), t\big)\, dt + \varepsilon_i, \tag{6.11}$$

where $f(x, t)$ is smooth in both x and t. Summation over the index j in (6.10) is replaced by integration over the variable t in (6.11).

To accommodate Binomial, Poisson, and other non-Gaussian responses, one can use a scalar-on-function generalized additive model which assumes that the conditional distribution of y_i given x_i is in an exponential family and

$$g\{E(y_i|x_i)\} = \beta_0 + \int_a^b f\big(x_i(t), t\big)\, dt$$

for some monotonic link function g. Model (6.11) is the special case of the scalar-on-function generalized additive model with an identity link function and, at least as a working assumption, a Gaussian response.

If model (6.11) holds for the functions $x(t)$, then it also holds with $x(t)$ replaced by $G_t\{x(t)\}$ for any smooth (in t and x) function $G_t(x)$ that is strictly monotonically increasing in x for each t. The reason this is true is that $f(x, t)$ can be replaced by $f(G_t^{-1}(x), t)$ for any $G_t(x)$ that is smooth in (t, x) since $f(x, t)$ is nonparametric. We have found it often useful to transform $x(t)$ by its empirical cumulative distribution function, since then the range of $G_t\{x(t)\}$ is the interval $[0, 1]$ and so the range does not depend on t. The empirical cumulative distribution function is not smooth in x, but we have not found that to be a problem. Moreover, the cumulative distribution function could be smoothed if that were desired.

A scalar-on-function generalized additive model can be fit using pfr() by replacing lf(), which is used for a linear fit, by af(), which specifies an additive fit. If the argument Qtransform is FALSE, then the functions are not transformed by the empirical cumulative distribution functions.

6.4.1 Example: Fat Content of Meat Samples

Next we fit additive models to the Tecator data without and with transformation of the second derivatives of the absorbances. The code is:

```
> fitPFRnoTrans <-pfr(y ~ af(Xderiv2,argvals=wavelength,
+     bs = "ps",k = c(7,7),m = list(c(2,2),c(2,2)),
+     Qtransform = FALSE), method = "REML")
> fitPFRtrans <- pfr(y ~ af(Xderiv2,argvals=wavelength,
+     bs = "ps",k = c(7,7),m = list(c(2,2),c(2,2)),
+     Qtransform = TRUE),method = "REML")
```

The default, which is used here, is a tensor product spline model for $f(x,t)$, k specifies the dimension of each basis, and m specifies the orders of the bases and penalties. Here each basis is a cubic (order = 2) B-spline of size (number of basis functions) 7 and a second-order difference penalty is imposed.

Perspective and heat map plots of the fitted functions are in Fig. 6.16. Setting the argument scheme of plot.gam() to one or two produces a perspective or a heat map plot, respectively, for two-dimensional smooths. The code that produced Fig. 6.16 is in the script absorbBivarFigs.R in the HRW package. This script can be run and located using:

```
> library(HRW) ; demo(absorbBivarFigs,package = "HRW")
> system.file("demo","absorbBivarFigs.R",package = "HRW")
```

Recall that in Table 6.1, the linear model fit was compared via AIC with additive model fits. It can be seen in that table that the additive model using untransformed second derivatives of any of the three predictors produced a noticeable smaller AIC than the corresponding linear model. However, the additive model to be introduced next has an even smaller AIC.

6.5 Additive Models Using Principal Component Scores

Another approach to scalar-on-function additive model analysis is to transform the functions to a small number of principal component scores and then fit a linear or additive model to the scores.

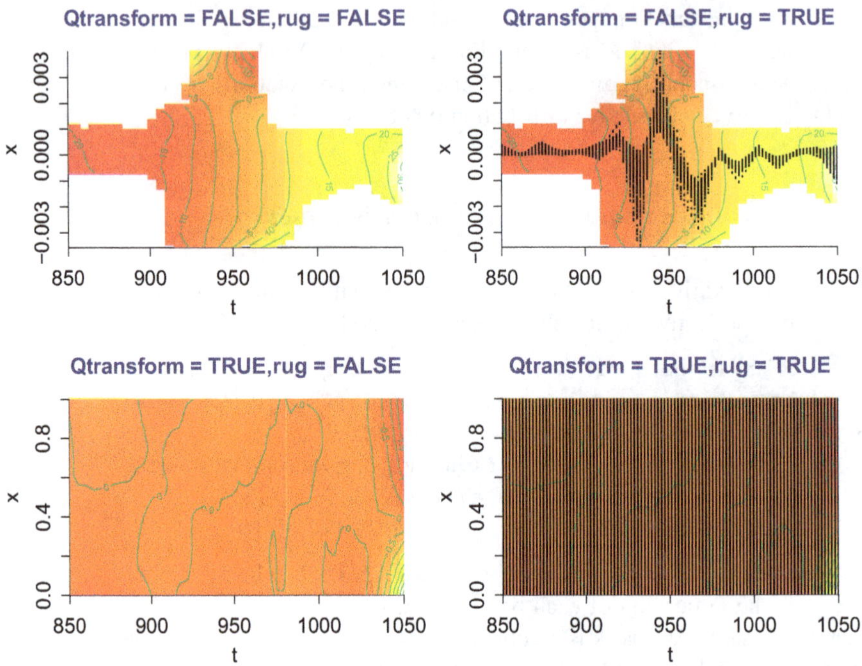

Fig. 6.16 Top left panel: heat map of the estimate of $f(x, t)$ with `rug = FALSE`. Top right panel: heat map of the estimate of $f(x, t)$ with `rug = TRUE`. Having `rug = TRUE` results in the locations of the data being shown. Bottom panels: Similar to the top panels but with `Qtransform = TRUE`. In the case where x is transformed, the rug obscures the heat map but serves to show that the transformed data are nearly uniform on a rectangular region.

A model that is additive in the original functions is not additive in the principal component scores. Therefore, fitting an additive model to the scores is potentially much different than fitting an additive model to the original functions. Which model fits best will depend on the data, so it is worthwhile trying both.

6.5.1 *Example: Fat Content of Meat Samples*

We now illustrate fitting an additive model to the principal component scores with the Tecator data. We will use the second derivatives of the absorbances. First we do a principal components analysis and create a *scree plot* which is in Fig. 6.17. A scree plot is a bar graph of the eigenvalues from a principal components analysis. Recall that the kth eigenvalue is the percentage of the total variance attributed to the direction of the kth eigenvector.

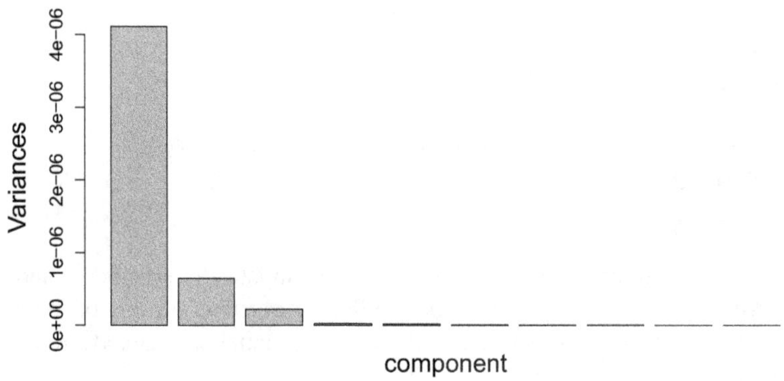

Fig. 6.17 Scree plot second derivatives of the absorbances in the Tecator data.

```
> pc <- prcomp(Xderiv2)
> plot(pc,cex.lab = 1.5,cex.main = 1.8,col.main = "navy",
+       main = "",xlab = "component")
```

We see from the scree plot that much of the variation is in the first three principal component directions and nearly all of it is in the first six or seven directions.

Next we fit an additive model using the first six scores:

```
> PCscores <- pc$x[,1:6] ; PCscore1 <- PCscores[,1]
> PCscore2 <- PCscores[,2] ; PCscore3 <- PCscores[,3]
> PCscore4 <- PCscores[,4] ; PCscore5 <- PCscores[,5]
> PCscore6 <- PCscores[,6]
> fitPCaddMod <- gam(y ~ s(PCscore1) + s(PCscore2)
+     + s(PCscore3) + s(PCscore4) + s(PCscore5) + s(PCscore6))
> summary(fitPCaddMod) ; print(fitPCaddMod$aic)

Family: gaussian
Link function: identity

Formula:
y ~ s(PCscore1) + s(PCscore2) + s(PCscore3) +
    s(PCscore4) + s(PCscore5) + s(PCscore6)

Parametric coefficients:
            Estimate Std. Error t value Pr(>|t|)
(Intercept) 18.14233    0.06035   300.6   <2e-16

Approximate significance of smooth terms:
              edf Ref.df        F  p-value
s(PCscore1) 6.618  7.722 3861.091  < 2e-16
s(PCscore2) 6.919  7.994  216.237  < 2e-16
s(PCscore3) 7.627  8.476   23.705  < 2e-16
```

```
s(PCscore4) 2.363  2.963   15.426 4.14e-09
s(PCscore5) 3.411  4.255    6.229 7.57e-05
s(PCscore6) 3.352  4.177    7.035 2.19e-05

R-sq.(adj) =  0.995   Deviance explained = 99.6%
GCV = 0.91628  Scale est. = 0.78293    n = 215
```

```
[1] 588.2958
```

We see from the summary that all six scores are statistically significant and R^2 is 0.995. Also, AIC for this model is about 588.3, lower than for any of the models previously examined. (The value of AIC for this model was already reported in Table 6.1.)

The following commands lead to a plot of the fitted functions, as displayed in Fig. 6.18:

```
> par(mfrow=c(2,3),mai = c(0.55,0.55,0.05,0.05));
> plot(fitPCaddMod, bty="l",cex.lab = 1.5,lwd = 2,
+       cex.axis = 1.5,col = "darkgreen",shade = TRUE,
+       shade.col="palegreen")
```

We see from Fig. 6.18 that the first three scores have large effects, especially the first two. Also, the effect of the first score is clearly nonlinear, which shows the need for an additive model.

Finally, to further illustrate this excellent fit, we plot the fitted values versus the responses. The plot is in Fig. 6.19 and is produced using the commands:

```
> xylimVal <- range(c(fitPCaddMod$fitted,tecator$y[,1]))
> plot(fitPCaddMod$fitted,tecator$y[,1],xlim = xylimVal,
+       ylim = xylimVal,xlab = "fitted values",
+       ylab = "responses",bty = "l",cex.lab = 1.5,
+       cex.axis = 1.5,col = "dodgerblue")
> abline(0,1,col = "darkgreen")
```

The responses range from 0 to 50% and the fitted values follow the responses very closely. We have not looked at out-of-sample prediction errors; this is left as an exercise (Exercise 4.)

6.6 Function-on-Function Linear Regression

Suppose now that both the predictor and the response are functions with domains S and T, respectively. The predictor function is $x(s)$, $s \in S$, and the response function is $y(t)$, $t \in T$. How can we predict $y(t)$ from $x(s)$? To gain insight, let us first fix t to obtain a scalar response $y(t)$. Then we can use the scalar-on-function regression model

$$y_i(t) = \beta_0 + \int_S \beta_1(s) x_i(s)\, ds + \varepsilon_i, \quad t \in T. \tag{6.12}$$

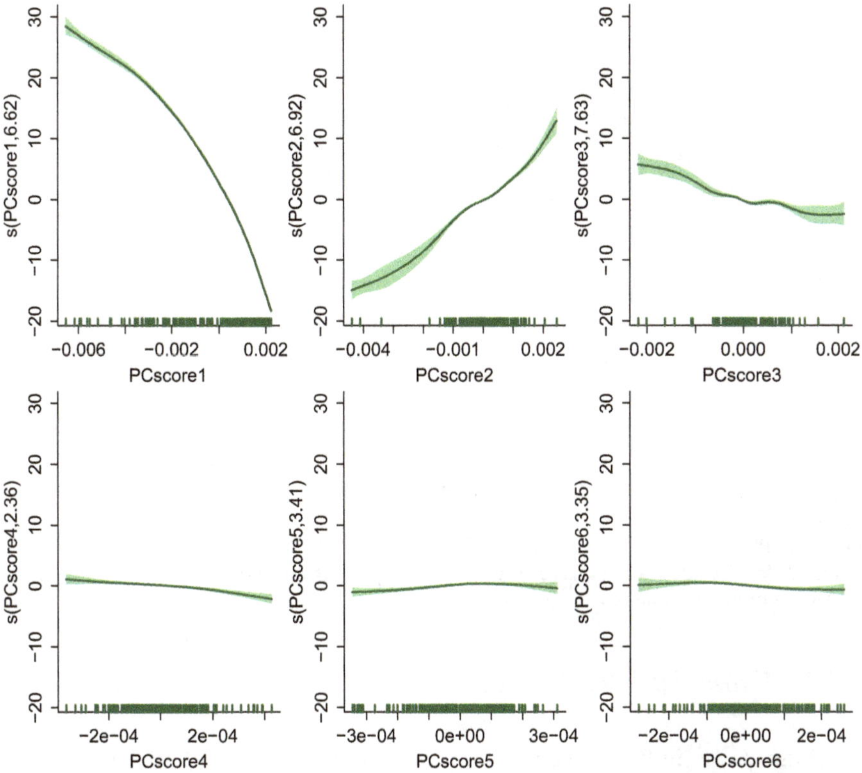

Fig. 6.18 Component functions of an additive model fit to the first six principal component scores of the second derivatives of the absorbances in the Tecator data.

Of course, β_0 and $\beta_1(s)$ will change as we change t, so it is better to rewrite (6.12) as

$$y_i(t) = \beta_0(t) + \int_S \beta_1(s,t)x_i(s)\,ds + \varepsilon_i, \quad t \in T. \tag{6.13}$$

One could estimate $\beta_0(t)$ and $\beta_1(s,t)$ by scalar-on-function regression on a grid of values of t and then interpolate to all values of t. However, $\beta_0(t)$ should be a smooth function, and $\beta_1(s,t)$ should be smooth in both t and s. Using repeated scalar-on-function regression in this way imposes smoothness on $\beta_1(s,t)$ only as a function of s and no smoothness on $\beta_0(t)$. Function-on-function regression is a methodology that imposes smoothness in both s and t and can be implemented in R with several functions. We will use the function `linmod()` within the `fda` package (Ramsay et al. 2017) and the function `pffr()` within the `refund` package in the upcoming example.

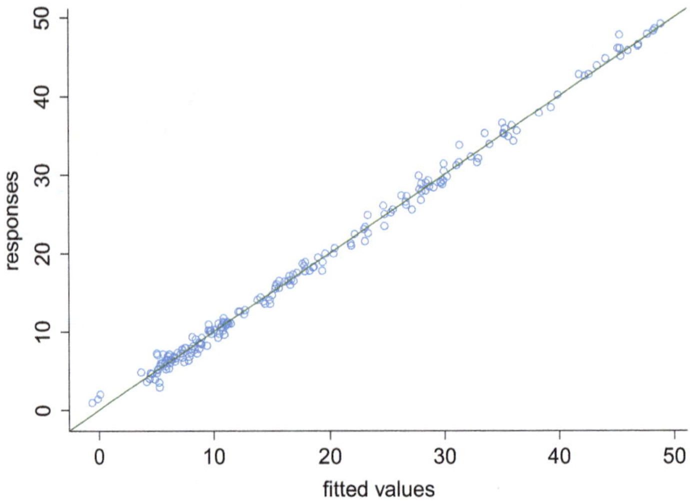

Fig. 6.19 Plot of fitted values from a model additive in the principal component scores and responses for the Tecator data. The fit uses the first six principal component scores of second derivatives of the absorbances.

6.6.1 *Example: Yield Curves*

As an illustration of function-on-function regression, we will predict changes in the U.S. daily interest-rate yield curves using changes in the European yield curves on the same day as the predictor functions. The yield curve at maturity t is the average continuously compounded interest rate that will be earned on a bond that matures at time t. The yield curve depends on the type of bond and changes each day as bond prices change; the market determines bond prices and the yield curve is computed from prices.

6.6.1.1 Implementation Using the `fda` Package

Daily observations of yield curves are in the dataset `yields` in the `HRW` package. This file has 91 columns. The first column is the date, columns 2–31 are European yields at maturities from 1 to 30 years, columns 32–61 are Japanese yields at these maturities, and columns 62–91 are U.S. yields at the same maturities.

The following code reads the data file, deletes the first column of dates, differences the yields, and creates vectors `EUdiffs` and `USdiffs` of changes in European and U.S. yields, respectively. The vector `t` contains the maturities.

```
> library(HRW) ; data(yields)
> diffs <- apply(as.matrix(yields[,-1]),2,diff)
> diffs <- na.omit(diffs)
```

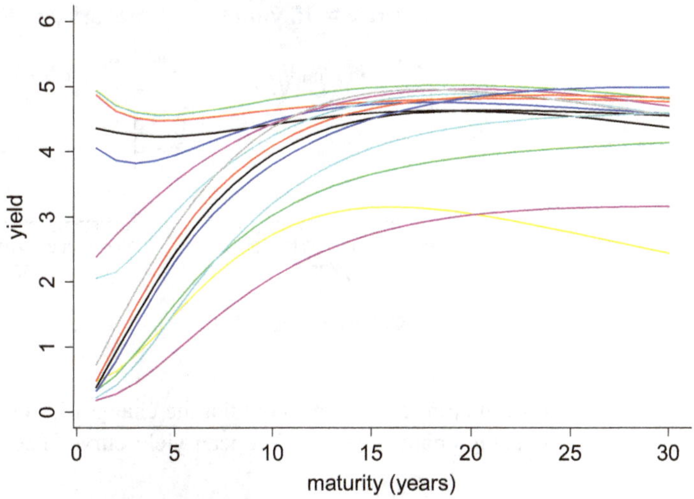

Fig. 6.20 14 U.S. yield curves spaced 100 days apart.

```
> EUdiffs <- as.matrix(diffs[,1:30])
> USdiffs <- as.matrix(diffs[,61:90]) ; t <- 1:30
```

The shapes of the yield curves change with time. This can be seen in Fig. 6.20 which contains 14 U.S. yield curves spaced 100 days apart and was produced using:

```
> yieldsCleaned <- na.omit(yields)[,-1]
> plot(t,yieldsCleaned[1,61:90], type="l",ylim = c(0,6),
+       lwd = 2,xlab = "maturity (years)",ylab = "yield",
+       bty = "l",cex.lab = 1.5,cex.axis = 1.5)
> for (i in 2:14) lines(t,yieldsCleaned[100*i+1,61:90],
+                         col  = i,lwd  = 2)
```

The yield curves in this dataset were smoothed using the six-parameter Svensson (1994) model. Raw data or yield curves that were spline-smoothed would be preferable, but such data are difficult to find in standard repositories.

Figure 6.21 contains time series plots of U.S. yields with 1-, 15-, and 30-year maturities. The code for the first panel of Fig. 6.21, for 1-year maturities, is:

```
> year <- 2006 + (1:1406)*6/1406
> plot(year,yieldsCleaned[,61],type = "l",col = "darkgreen",
+       xlab = "year",ylab = "yield",ylim = c(0,5.5),
+       main = "maturity = 1 year",bty = "l",col.main = "navy")
```

The other panels are similar but with ordinates `yieldsCleaned[,75]` for 15-year maturities and `yieldsCleaned[,90]` for 30-year maturities.

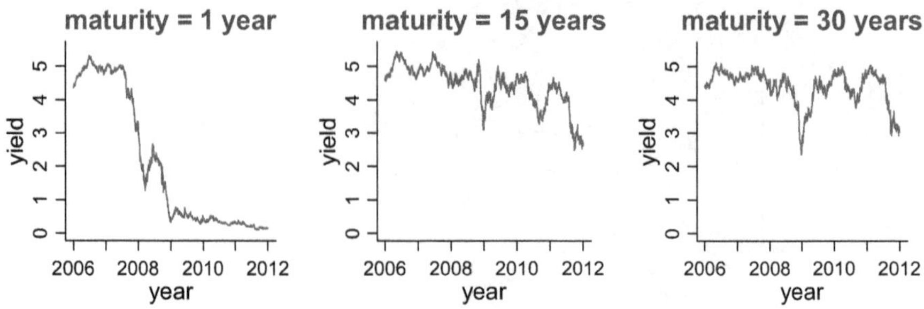

Fig. 6.21 Time series plots of U.S. yields with maturities of 1, 15, and 30 years.

First we use scalar-on-function regression to predict the changes in the 1-, 15-, and 30-year U.S. yields using changes in the European yield curve. The relevant code is:

```
> library(refund) ; yscalar1 <- USdiffs[,1]
> fitSF1 <- pfr(yscalar1 ~ lf(EUdiffs,argvals = 1:30),
+                    method = "REML")
> yscalar15 <- USdiffs[,15]
> fitSF15 <- pfr(yscalar15 ~ lf(EUdiffs,argvals = 1:30),
+                     method = "REML")
> yscalar30 <- USdiffs[,30]
> fitSF30 <- pfr(yscalar30 ~ lf(EUdiffs,argvals = 1:30),
+                     method = "REML")
```

The coefficient functions are plotted in Fig. 6.22. The code for the first panel is:

```
> plot(fitSF1, shade = TRUE, shade.col = "palegreen",
+        xlab = "maturity",ylab = expression(widehat(beta)[1](s)),
+        main = "response is y(1)",bty = "l",ylim = c(-0.05,0.15),
+        rug = TRUE,col.main = "navy")
```

The other two panels are similar, with `fitSF1` replaced by `fitSF15` and then `fitSF30`.

Notice that changes in 1-year U.S. yields are affected positively by changes in short-term European yields but are not affected significantly by changes in medium-to long-term European yields. Changes in 15-year and 30-year U.S. yields are positively affected by changes in European yields of all maturities, although the effect is larger for changes in short-term European yields.

Next, we use the `fda` package (Ramsay et al. 2017) to perform function-on-function regression for model (6.13) and obtain an estimate of the coefficient function $\beta_1(s,t)$ that is smooth in both European and U.S. maturities (s and t, respectively). The intercept function $\beta_0(t)$ will also be estimated. First, we use the

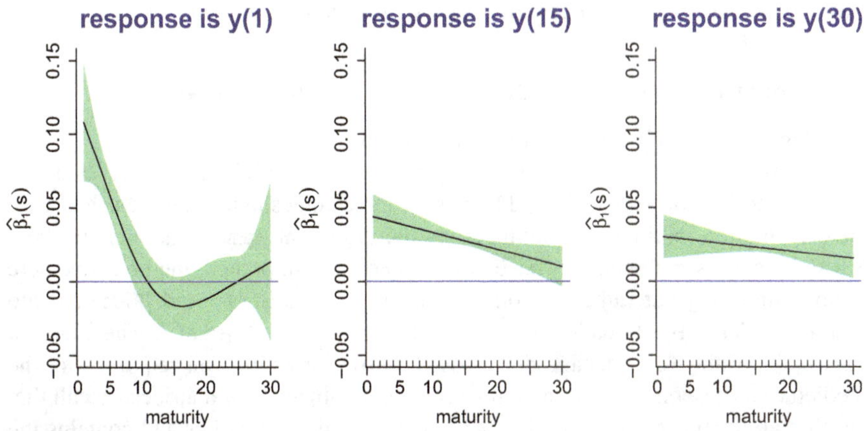

Fig. 6.22 Estimated coefficient functions $\widehat{\beta_1}(s)$ when predicting changes in 1-, 15-, and 30-year U.S. yields from changes in the European yield curve.

function `create.bspline.basis()` in the `fda` package to set up a B-spline basis called `yieldBetaBasis` with 20 basis functions over the interval [0, 30]. This basis is used for $\beta_0(t)$, and $\beta_1(s, t)$ is modeled as a tensor product of this basis with itself. We begin with:

```
> library(fda)
> yieldBetaBasis <- create.bspline.basis(c(0,30),20)
```

The first line of next code chunk uses the function `fdPar()` to create `yieldBeta0Par`, a *functional parameter object*, to use the language of the `fda` package. As explained in the help files for that package, functional parameter objects are used as arguments to functions that estimate functional parameters. After estimation, `yieldBeta0Par` will contain the spline coefficients of the functional parameter $\beta_0(t)$ in model (6.13). The arguments "2" and "0.00001" specify a penalty on the second derivative and a smoothing parameter equal to 0.00001, which implies minimal smoothing. The second line uses the function `bifd()` to create a *bivariate functional data object* `yieldBeta1fd`, that will contain the 20×20 coefficients of the tensor product of `yieldBetaBasis` with itself. The third line uses `bifdPar()` to create a "bivariate functional parameter object" with second derivative penalties on both of its bases and both penalties equal to 500. After estimation, `yieldBeta1fd` will contain spline coefficients of the estimate of $\beta_1(s, t)$ in (6.13). The code is:

```
> yieldBeta0Par <- fdPar(yieldBetaBasis,2,0.00001)
> yieldBeta1fd  <- bifd(matrix(0,20,20),yieldBetaBasis,
+                            yieldBetaBasis)
> yieldBeta1Par <- bifdPar(yieldBeta1fd,2,2,500,500)
```

Next we create a list that, after estimation, will contain the coefficients of estimates of $\beta_0(t)$ and $\beta_1(s,t)$:

```
> yieldBetaList <- list(yieldBeta0Par,yieldBeta1Par)
```

The elements in this list have just been discussed.

With the following code, a function parameter object D2fdPar is created that will be used for both EUdiffs and USdiffs, the changes in the European and U.S. yields. There is minimal smoothing since the argument lambda is only 0.00001. Since the curves in the dataset have already been pre-smoothed, the objective here is not smoothing but rather to convert the matrices EUdiffs and USdiffs into functional data objects called x2 and y2 using smooth.basis(). The function linmod() in the fda package that is used below for estimation requires that the predictor and response function be functional data objects, not matrices. Recall that the first argument, t, of smooth.basis() was computed earlier and contains the maturities of the yields, which are the arguments of these functions.

```
> D2fdPar <- fdPar(yieldBetaBasis,lambda = 0.00001)
> x2 <- smooth.basis(t,t(EUdiffs),D2fdPar)$fd
> y2 <- smooth.basis(t,t(USdiffs),D2fdPar)$fd
```

Finally, we use the function linmod() to fit model (6.13) and the function eval.bifd() to extract the estimate of $\beta_1(s,t)$ as a 30×30 matrix:

```
> linmodSmooth <- linmod(x2,y2,yieldBetaList)
> yieldbeta1mat <- t(eval.bifd(t,t,linmodSmooth$beta1estbifd))
```

We plot the estimated coefficient function $\widehat{\beta}_1(s,t)$ via:

```
> library(fields)
> xlab1 <- "s (argument of predictor function"
> xlab2 <- "= change in European yields)"
> ylab1 <- "t (argument of response function"
> ylab2 <- "= change in U.S. yields)"
> image.plot(t,t,yieldbeta1mat,col = terrain.colors(1000),
+            cex.main = 1.8, col.main = "navy",
+            xlab = paste(xlab1,xlab2),
+            ylab = paste(ylab1,ylab2),
+            legend.args = list(text =
+            expression(widehat(beta)[1](s,t)),adj = 0.8))
```

Figure 6.23 shows the result. As in Chap. 5, we use terrain colors. White indicates the largest values of $\widehat{\beta}_1(s,t)$ and green corresponds to smallest values of this function. The bottom of this plot corresponds to $t = 1$ (change in U.S. yield at 1-year maturity). We see there that $\widehat{\beta}_1(s,1)$ assumes large positive values for small s and decreases as s increases, in agreement with the left panel of Fig. 6.22. There is also reasonably good agreement of Fig. 6.23 with the middle and right panels of Fig. 6.22.

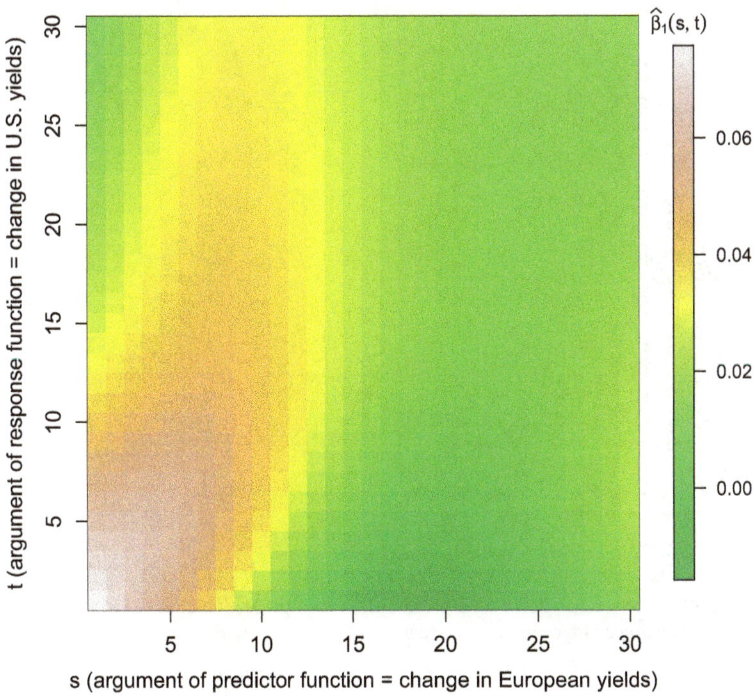

Fig. 6.23 Plot of estimated coefficient function $\widehat{\beta}_1(s,t)$ when predicting changes in U.S. yield curve (a function of t) using changes in the European yield curve (a function of s).

We see in Fig. 6.23 that changes in U.S. yields at any maturity depend little on changes in European yields at long maturities, e.g., greater than 15 years. Changes in short-term (under 5 years) U.S. yields depend most highly on changes in short-term European yields. Changes in long-term (over 20 years) U.S. yields depend most highly on medium-term (5–10 years) European yields.

The estimated intercept $\widehat{\beta}_0(t)$ is plotted in Fig. 6.24 using:

```
> par(mai = c(1,1.2,0.2,0.1))
> plot(linmodSmooth$beta0estfd,xlab = "s",
+       ylab = expression(widehat(beta)[0](s)),bty = "l",
+       lwd = 2,col = "darkgreen",cex.lab = 1.5,
+       cex.axis = 1.5,ask = FALSE)
```

To see how well function-on-function regression can predict changes in U.S. yield curves, we divided the dataset into training and test data, so, for example `EUdiffsTrain` and `EUdiffsTest` are the changes in the European yield curves in the training and test data, respectively. The training data consisted of the first 1000 days and the test data the next 100 days.

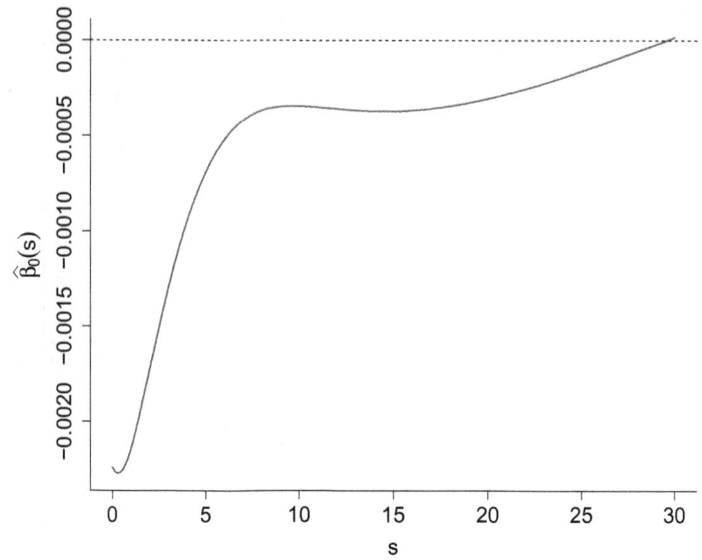

Fig. 6.24 Plot of estimated intercept function $\widehat{\beta}_0(t)$ when predicting changes in U.S. yield curve (a function of t) using changes in the European yield curve (a function of s).

The function-on-function regression model just fit to the entire dataset was refit using only the training data. The new fitted model is in the object `linmodSmooth`.

The following code predicts the changes in yields in the test data using the fit to the training data. Thus, these are *out-of-sample* predictions, and, unlike in-sample predictions could have a negative R^2. The object `RsqFDA` contains the out-of-sample R^2 using the changes in the European yield curves as predictors. We see that `RsqFDA` is 0.128, which is small but, at least, positive indicating some predictive power.

First we set up a new basis for prediction:

```
> xtextEU <- smooth.basis(t,t(as.matrix(EUdiffsTest)),
+                         D2fdPar)$fd
```

Then we set up linear predictors bivariate functional data object coefficients:

```
> C <- linmodSmooth$beta1estbifd$coef
```

The basis inner product with `xtextEU` is:

```
> B <- inprod(yieldBetaBasis,xtextEU)
```

`C%*%B` are the coefficients for the linear part of the functional prediction. We need to replicate the `beta0` coefficients over the columns of `C%*%B` and add the matrices:

```
> predCoefs <- C%*%B + matrix(linmodSmooth$beta0estfd$coef,20,
+                              nTest,byrow = FALSE)
```

Now we create the functional data object:

```
> predFDA <- fd(predCoefs,yieldBetaBasis)
```

The object `fittedFDA` contains the prediction of the U.S. yields test data:

```
> fittedFDA <-  t(eval.fd(1:30,fdobj = predFDA))
```

The prediction errors are

```
> errors <-  fittedFDA - USdiffsTest
```

and an out-of-sample R^2 is

```
> RsqFDA <- 1 - sum(errors^2)/sum((USdiffsTest
+                          - colMeans(USdiffsTrain))^2)
> print(round(RsqFDA,3))
```

```
[1] 0.128
```

Figure 6.25 shows every ninth change of the U.S. yield curve in the test dataset and the prediction of that change using the European yield curve change.

6.6.1.2 Implementation Using the `refund` Package

For function-on-function regression, the `linmod()` function of the `fda` package is limited to a single function predictor. In contrast, the `pffr()` function in the `refund` package also performs function-on-function regression and is built upon functions in the `mgcv` package and so, like the `gam()` function in that package, can fit additive models. The predictors in the additive model can be functional or scalar. As with `gam()`, the scalar predictors can enter linearly, as smooth functions of a single predictor, or as smooth functions of multiple predictors. Another advantage of `pffr()` is that it is easy to use, especially if one is already familiar with `gam()`.

The function `pffr()` can use either `ff()` or `ffpc()` to construct a function-on-function term. These differ in that `ff()` represents the coefficient function $\beta_1(s,t)$ as a tensor product of cubic B-splines whereas `ffpc()` represents both the functional predictor and $\beta_1(s,t)$ using the principal components basis, i.e., the eigenfunctions of the covariance operator of the functional predictors. As an example, suppose y is a functional response, x_1 and x_2 are functional predictors, and z_1, z_2, z_3, and z_4 are scalar predictors. Then the code `y <- pffr(y ~ ff(x1) + ffpc(x2) + s(z1) + z2 + s(z3,z4))` fits an additive model with the effect of x_1 modeled by `ff()`, the effect of x_2 modeled by `ffpc()`, z_1 having a smooth effect, z_2 entering linearly, and z_3 and z_4 entering as a bivariate effect. The function `pffr()` implements penalized function-on-function regression (Ivanescu et al. 2015).

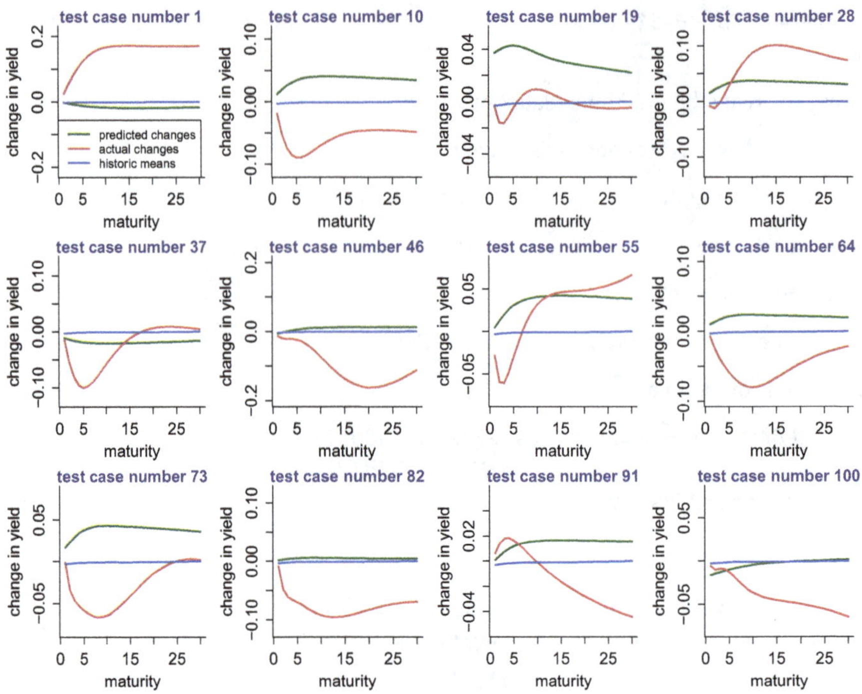

Fig. 6.25 Prediction of changes in every ninth U.S. yield curve in the test dataset. The color-coding of the curves is according to the predicted changes, the actual changes, and the historic means, i.e., mean changes in the training data. The predictions are somewhat closer to the true curves compared to the historic means, because the out-of-sample R^2 is positive.

In this section, we illustrate `pffr()` with a single functional predictor using again changes in the European and U.S. yield curves. The following code fits the function-on-function mode to the training data using `ff()`, predicts the yield changes in the test data, and computes an out-of-sample R^2 called `RsqFonF`. We see that the R^2 is 0.150, somewhat better than R^2 using `linmod()`, which, as we just saw, is 0.128.

A function-on-function regression is fit via:

```
> fitFonF <- pffr(USdiffs ~ ff(EUdiffs), data = dataTrain)
> predsFonF <- predict(fitFonF,newdata = dataTest)
> RsqFonF <- 1 - mean((dataTest$USdiffs - predsFonF)^2) /
+     mean((t(dataTest$USdiffs) - colMeans(dataTrain$USdiffs))^2)
```

with the R^2 given by:

```
> print(round(RsqFonF,3))
```

```
[1] 0.15
```

When estimating the function-on-function regression model using `ff()`, a warning was issued that the predictors have a very low effective rank and that `ffpc()` might be a better choice. However, when we reran the code above with `ff()` replaced by `ffpc()`, the out-of-sample R^2 did not improve and, in fact, dropped slightly to 0.145:

```
> fitFFPC <- pffr(USdiffs ~ ffpc(EUdiffs),data = dataTrain)
> predsFFPC <- predict(fitFFPC,newdata = dataTest)
> RsqFFPC <- 1 - (mean((dataTest$USdiffs - predsFFPC)^2)/
+   mean((t(dataTest$USdiffs) - colMeans(dataTrain$USdiffs))^2))
> print(round(RsqFFPC,3))

[1] 0.145
```

6.7 Kernel Machines

Kernel machines form a very general class of function estimation procedures that includes many of the estimators treated in the earlier chapters as special cases. One of the main uses of kernel machines is classification and the subclass known as *support vector machines* comprises the most common approach. Thus far we have not discussed the important problem of classification, so we will focus on this particular use of kernel machines in the examples. Since their inception in the 1990s, the kernel machine literature has grown enormously. Books on the topic include Cristianini and Shawe-Taylor (2000) and Schölkopf and Smola (2002). Gaussian processes (e.g. Rasmussen and Williams 2006) are also closely related to kernel machines.

There are various ways by which kernel machines can be derived. Perhaps the most elegant approach is that based on the functional analytic structure known as a *reproducing kernel Hilbert space*, and details can be found in the references mentioned in the previous paragraph. In addition, Pearce and Wand (2006, 2009) explain how many of the semiparametric regression estimators considered in this book are kernel machines within the reproducing kernel Hilbert space framework. From now on we focus on the concrete aspects of kernel machines and their implementation in R.

Suppose we observe the pairs $(\boldsymbol{x}_i, y_i)$, $1 \leq i \leq n$, where $\boldsymbol{x}_i \in \mathbb{R}^d$ and $y_i \in \mathbb{R}$. A kernel machine fit to such data is a real-valued function $\widehat{f}(\boldsymbol{x})$, $\boldsymbol{x} \in \mathbb{R}^d$, that has the following ingredients: a loss function $\mathscr{L}$, a set of linearly independent functions $h_0(\boldsymbol{x}), \ldots, h_p(\boldsymbol{x})$ on $\mathbb{R}^d$ and kernels $\mathscr{K}_1, \ldots, \mathscr{K}_q$ with corresponding penalties $\lambda_1, \ldots, \lambda_q > 0$. A kernel is a symmetric, positive definite, function that maps $\mathbb{R}^d \times \mathbb{R}^d$ into $\mathbb{R}$. There are numerous options for their choice. Some examples of kernels are

$$\mathscr{K}(\boldsymbol{x},\boldsymbol{x}') = (1+\boldsymbol{x}^T\boldsymbol{x}')^7,$$

$$\mathscr{K}(\boldsymbol{x},\boldsymbol{x}') = \exp\{-20\|\boldsymbol{x}-\boldsymbol{x}'\|\}$$

$$\text{and}\quad \mathscr{K}(\boldsymbol{x},\boldsymbol{x}') = \frac{1}{1+\|\boldsymbol{x}-\boldsymbol{x}'\|^{13}}.$$

Given the aforementioned ingredients, the corresponding kernel machine is chosen from the class of functions

$$f(\boldsymbol{x}) = \sum_{j=0}^{p} \beta_j h_j(\boldsymbol{x}) + \sum_{i=1}^{n} \{\alpha_{1i}\mathscr{K}_1(\boldsymbol{x}_i,\boldsymbol{x}) + \ldots + \alpha_{qi}\mathscr{K}_q(\boldsymbol{x}_i,\boldsymbol{x})\} \tag{6.14}$$

with the coefficients $\boldsymbol{\beta} \equiv (\beta_0,\ldots,\beta_p)$ and $\boldsymbol{\alpha}_\ell \equiv (\alpha_{\ell 1},\ldots,\alpha_{\ell,n})$, $1 \le \ell \le q$, obtained by solving the optimization problem

$$\begin{aligned}(\widehat{\boldsymbol{\beta}},\widehat{\boldsymbol{\alpha}}_1,\ldots,\widehat{\boldsymbol{\alpha}}_q) = \operatorname{argmin}_{\boldsymbol{\beta},\boldsymbol{\alpha}_1,\ldots,\boldsymbol{\alpha}_q}\Big[&\textstyle\sum_{i=1}^n \mathscr{L}(y_i, f(\boldsymbol{x}_i))\\ &+\lambda_1\boldsymbol{\alpha}_1^T\boldsymbol{\mathcal{K}}_1\boldsymbol{\alpha}_1 + \ldots + \lambda_q\boldsymbol{\alpha}_q^T\boldsymbol{\mathcal{K}}_q\boldsymbol{\alpha}_q\Big]\end{aligned} \tag{6.15}$$

where $\boldsymbol{\mathcal{K}}_\ell = [\mathscr{K}_\ell(\boldsymbol{x}_i,\boldsymbol{x}_j)]_{1\le i,j\le n}$ are known as *Gram matrices*. The kernel machine $\widehat{f}$ takes the form of (6.14) with the coefficients replaced by their estimated values (6.15). In many kernel machine applications $q = 1$, in which case there is a single kernel $\mathscr{K}_1$ and a single penalty λ_1. However, many useful kernel machines, such as those corresponding to generalized additive models (Chap. 3) and longitudinal data analysis (Chap. 4), require the more general infrastructure described here.

The penalized spline scatterplot smoother, introduced in Chap. 2, for univariate predictor data is a $q = 1$ kernel machine. Let (x_i, y_i), $1 \le i \le n$, denote a $d = 1$ regression dataset. Set the loss function to be

$$\mathscr{L}(a,b) = (a-b)^2,\quad a,b\in\mathbb{R},$$

known as *squared-error loss*, and put

$$h_0(x) \equiv 1,\quad h_1(x) \equiv x \quad\text{and}\quad \mathscr{K}_1(x,x') \equiv \sum_{k=1}^{K} z_k(x)z_k(x') \tag{6.16}$$

where $\{z_k : 1 \le k \le K\}$ is an appropriate spline basis such as the linearly transformed cubic O'Sullivan splines described in Sect. 2.2. Then the resulting *least-squares* kernel machine is

$$\widehat{f}(x) = \widehat{\beta}_0 + \widehat{\beta}_1 x + \sum_{i=1}^{n} \widehat{\alpha}_{1i}\,\mathscr{K}_1(x_i,x) \tag{6.17}$$

where, from (6.15) and some algebra,

$$\widehat{\boldsymbol{\alpha}}_1 = \{(\boldsymbol{I} - \boldsymbol{H})\mathcal{K}_1 + \lambda_1 \boldsymbol{I}\}^{-1}(\boldsymbol{I} - \boldsymbol{H})\,\boldsymbol{y}$$
$$\text{and} \quad \begin{bmatrix} \widehat{\beta}_0 \\ \widehat{\beta}_1 \end{bmatrix} = (\boldsymbol{X}^T\boldsymbol{X})^{-1}\boldsymbol{X}^T\,(\boldsymbol{y} - \mathcal{K}_1\widehat{\boldsymbol{\alpha}}_1)\,. \tag{6.18}$$

Here $\mathcal{K}_1 \equiv \boldsymbol{Z}\boldsymbol{Z}^T$ is the Gram matrix, $\boldsymbol{H} \equiv \boldsymbol{X}(\boldsymbol{X}^T\boldsymbol{X})^{-1}\boldsymbol{X}^T$ is the linear component 'hat' matrix, with

$$\boldsymbol{X} \equiv [1\; x_i]_{1\le i\le n} \quad \text{and} \quad \boldsymbol{Z} \equiv [z_k(x_i)]_{\substack{1\le i\le n \\ 1\le k\le K}} \tag{6.19}$$

denoting the linear component and spline basis design matrices. Note that (6.17) can be shown to coincide with the standard penalized spline estimator given in Chap. 2.

If we replace (6.16) by

$$h_0(x) \equiv 1 \quad \text{and} \quad \mathscr{K}_1(x, x') \equiv \exp\{-\gamma(x - x')^2\}, \quad \text{for some } \gamma > 0, \tag{6.20}$$

then we arrive at a different kernel machine for fitting the same data. The $\mathscr{K}_1$ in (6.20) is known as the *radial basis function kernel*. The fitted kernel machine is given by (6.17) and (6.18) but with

$$\mathcal{K}_1 \equiv [\mathscr{K}_1(x_i, x_j)]_{1\le i,j,\le n} = [\,\exp\{-\gamma(x_i - x_j)^2\}]_{1\le i,j,\le n}, \quad \boldsymbol{X} \equiv \boldsymbol{1}$$

where $\boldsymbol{1}$ is the $n \times 1$ vector of ones. In this case, $\boldsymbol{H} = \boldsymbol{1}(\boldsymbol{1}^T\boldsymbol{1})^{-1}\boldsymbol{1}^T = \frac{1}{n}\boldsymbol{1}\boldsymbol{1}^T$.

Figure 6.26 shows kernel machine fits to the data on area per million złoty versus construction date for the Warsaw apartment data described in Chap. 2, for the penalized spline and radial basis function kernel machines, with ingredients (6.16) and (6.20). The z_k correspond to O'Sullivan cubic spline basis functions, as described in Sect. 2.2, and $K = 25$. The radial basis function scale parameter is $\gamma = 0.1$. The penalty parameters are such that there are 16 effective degrees of freedom for each fit. The two kernel machine fits are similar, but with some noticeable differences—especially where the predictor data are sparse.

Figure 6.26 illustrates the fact that kernel machines provide an alternative approach to the semiparametric regression models treated in the earlier chapters with the loss function $\mathscr{L}$ corresponding to the log-likelihood of the response model. However, kernel machines also provide an elegant framework for the construction of classification methods, via a different choice for $\mathscr{L}$. Such is the focus of the remainder of this section.

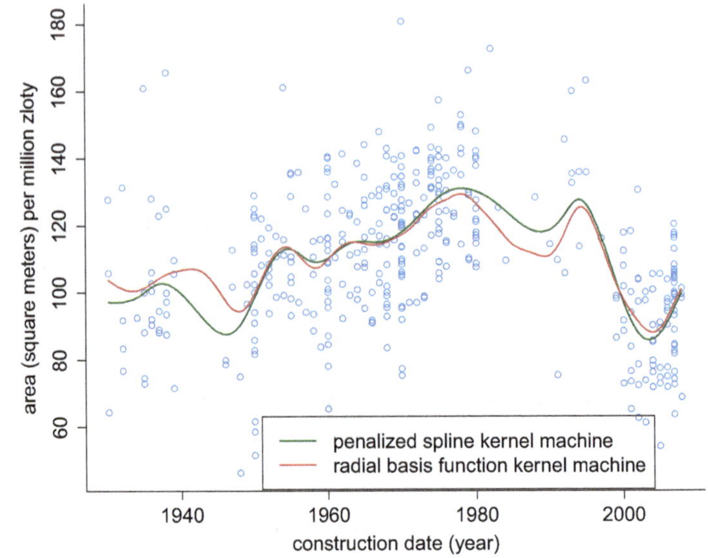

Fig. 6.26 Two different least-squares kernel machine fits to data on price per square meter versus construction date for the Warsaw apartment data described in Sect. 2. The effective degrees of freedom of each fit is 16.

6.7.1 Support Vector Machine Classification

A support vector machine is a kernel machine with the loss function set to

$$\mathscr{L}(a, b) = (1 - ab)_+, \quad a, b \in \mathbb{R},$$

commonly known as *hinge loss*. This particular loss is tailored towards *classification* rather than standard regression analysis, with details provided in, e.g., Moguerza and Muñoz (2006). Since the late 1990s, support vector machines have become one of the most popular "off-the-shelf" classifiers and have proven to be very successful in a variety of applications. As before, the data are $(\boldsymbol{x}_i, y_i)$, $1 \le i \le n$, where $\boldsymbol{x}_i \in \mathbb{R}^d$, but now the y_i are coded according to $\boldsymbol{x}_i$ belonging to one of two classes, $\mathscr{C}_-$ and $\mathscr{C}_+$, via:

$$y_i = \begin{cases} -1 \text{ if } \boldsymbol{x}_i \text{ belongs to } \mathscr{C}_-, \\ 1 \text{ if } \boldsymbol{x}_i \text{ belongs to } \mathscr{C}_+. \end{cases} \tag{6.21}$$

In this classification context, the $(\boldsymbol{x}_i, y_i)$ are usually referred to as the *training data*. The kernel machine fit to these data, $\widehat{f}$, classifies a new $\boldsymbol{x}_{\text{new}} \in \mathbb{R}^d$ according to

$$\boldsymbol{x}_{\text{new}} \text{ classified as belonging to } \mathscr{C}_- \text{ if } \widehat{f}(\boldsymbol{x}_{\text{new}}) < 0$$
$$\boldsymbol{x}_{\text{new}} \text{ classified as belonging to } \mathscr{C}_+ \text{ if } \widehat{f}(\boldsymbol{x}_{\text{new}}) \geq 0.$$

Figure 6.27 shows a dataset for which classification is of interest. It is a color-coded scatterplot matrix of data on several variables pertaining to cars sold at auction. The scatterplot point colors denote whether a car bought at auction by an automobile dealership is considered a good or bad buy. The origin of the data is a classification competition titled "Don't Get Kicked!" that ran on the kaggle platform (www.kaggle.com) during 2011–2012. The data are stored in the data frame `carAuction` within the `HRW` package. For display purposes, only 300 cars and 15 predictor variables are shown in Fig. 6.27 and jittering has been applied to

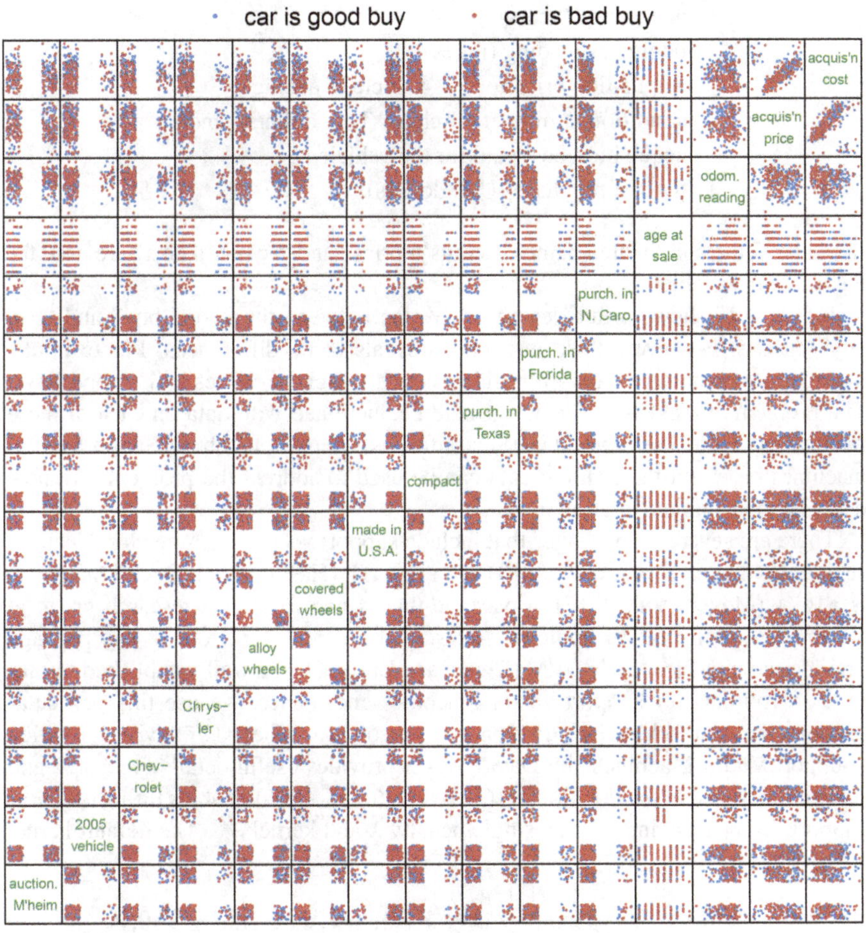

Fig. 6.27 Color-coded scatterplot matrix of 15 predictors and 300 cars from the kaggle platform's "Don't Get Kicked!" competition.

the 0/1 indicator variables (e.g., `compact` = 1 if the car is compact and `compact` = 0 otherwise). There are 11 indicator variables in total, corresponding to whether or not the car:

was auctioned by the company Manheim,	is a 2005 vehicle,	
is a Chevrolet,	is a Chrysler,	
has alloy wheels,	has covered wheels,	(6.22)
was made in the USA,	is compact,	
was purchased in Texas,	was purchased in Florida	
and was purchased in North Carolina.		

There are also four continuous variables:

age at sale (years),
odometer reading (miles),
acquisition price for this vehicle in average condition at time of purchase (U.S. dollars), and
acquisition cost paid for the vehicle at time of purchase (U.S. dollars). (6.23)

In Fig. 6.27 the abbreviated names "acquis'n price" and "acquis'n cost" are used for the last two variables.

Figure 6.27 shows a fair degree of overlap between the "good buy" and "bad buy" classes. However, close inspection reveals some differences. For example, there is a tendency for cars with higher cost at purchase values to be good buys. The problem at hand is: if a new car is to be auctioned with data on each of these features available, then should it be classified as a good or bad buy? Support vector machines based on the training data can be used to address the problem. We now describe their implementation in R.

There are several R packages that include support vector machine classification, such as `probsvm` (Zhang et al. 2013) and `svmpath` (Hastie 2016). One of the oldest is `e1071` (Meyer et al. 2017), a package that is named after a research group at Vienna University of Technology, Austria. The `svm()` function of this package provides an interface to LIBSVM (Chang and Lin 2011)—a well-established library of C++ routines for support vector machines and, hence, we use this particular support vector machine R function for illustration. The `e1071` vignette titled "Support Vector Machines" by David Meyer provides useful details on `svm()` and related functions. In the notation of (6.14) and (6.15), `svm()` enables fitting of $q = 1$ support vector machines with a single penalty λ and kernel $\mathscr{K}$. The default kernel of `svm()` is

$$\mathscr{K}(\boldsymbol{x}, \boldsymbol{x}'; \gamma) = \exp\{-\gamma \|\boldsymbol{x} - \boldsymbol{x}'\|^2\}, \quad \boldsymbol{x}, \boldsymbol{x}' \in \mathbb{R}, \quad \gamma > 0,$$

the family of d-dimensional *radial basis function* kernels. Some other kernel families are also supported by `svm()`. Lagrange multiplier theory can be used to show that solving (6.15) with hinge loss reduces to the quadratic programming problem

$$\min_{\boldsymbol{\alpha}} \left[\tfrac{1}{2}\boldsymbol{\alpha}^T\{(\boldsymbol{y}\boldsymbol{y}^T) \odot \mathcal{K}_{(\gamma)}\}\boldsymbol{\alpha} - \mathbf{1}^T\boldsymbol{\alpha} \right]$$
$$\text{subject to } 0 \le \alpha_i \le C, \quad 1 \le i \le n \text{ and } \boldsymbol{y}^T\boldsymbol{\alpha} = 0$$

where $\mathcal{K}_{(\gamma)} \equiv [\mathscr{K}(\boldsymbol{x}_i, \boldsymbol{x}'_j; \gamma)]$ is the Gram matrix and $C > 0$, known as the *cost parameter*, is a monotone transformation of the penalty λ. Here, and throughout this section, $\boldsymbol{A} \odot \boldsymbol{B}$ denotes the element-wise product of matrices $\boldsymbol{A}$ and $\boldsymbol{B}$ having the same dimensions.

Our specific illustration of the `svm()` involves taking all 8976 "bad buy" cars in the original kaggle dataset and adding the same number of randomly chosen "good buy" cars. We then hold back 1000 randomly selected cars for misclassification rate estimation, resulting in a training data sample size of $n = 16{,}952$. Working with such a "balanced" training dataset aids exposition, but also circumvents the problem that `svm()` does not scale well to very large n. See Exercise 8 for one possible remedy.

In R, set `yTrain` to be the array containing the 16,952 indicators of whether a car is a good buy (coded as -1) or a bad buy (coded as 1). Then construct `predMatTrain` to be the $16{,}952 \times 15$ R data frame containing the training sample values of each of the $d = 15$ predictors (6.22) and (6.23), corresponding to Fig. 6.27. If `gammaVal` and `costVal` are set to positive numbers, then a support vector machine fit is achieved via the commands:

```
> trainDF <- data.frame(yTrain,predMatTrain)
> fitSVM <- svm(factor(yTrain) ~ .,data = trainDF,
+                   gamma = gammaVal,cost = costVal)
```

If `predMatTest` is the 1000×15 data frame defined analogously to the data frame `predMatTrain`, but containing the held back predictor values, then the vector of classifications is obtained from the command:

```
> yHat <- 2*as.numeric(predict(fitSVM,predMatTest)) - 3
```

Note that the linear transformation used in this command maps the $\{1, 2\}$ values, produced from `as.numeric(predict(fitSVM,predMatTest))`, to values in the set $\{-1, 1\}$ corresponding to the coding convention of (6.21).

As pointed out by, e.g., Hastie and Zhu (2006), the Gaussian kernel machine parameter vector $(\gamma, C) \in \mathbb{R}^2_+$ can have a pronounced effect on the classification performance of a radial basis function support vector machine. The `svm()` function is accompanied by function named `tune.svm()` that aids good selection of (γ, C), defaulted to be ten-fold cross-validation. An example call to `tune.svm()` is:

```
> estParamVec <- tune.svm(factor(yTrain) ~ ., data = trainDF,
+                              gamma = 10^(-5:-1),
```

```
+                            cost = 10^seq(1,2,length=5)
+                            )$best.parameters
```

which conducts a search over the grid

$$(\gamma, C) \in \{10^{-5}, 10^{-4}, \ldots, 10^{-1}\} \times \{10^{1}, 10^{5/4}, 10^{6/4}, 10^{7/4}, 10^{2}\}.$$

For the $n = 16{,}952$ car auction data with 15 predictors `tune.svm()` was found to be very slow, so a random subsample of size 1000 is used instead. This resulted in the choice

$$(\widehat{\gamma}, \widehat{C}) = (10^{-2}, 10^{5/4}) \approx (0.01, 17.7828).$$

These values were used for `gammaVal` and `costVal` in the above call to `svm()`. The resulting *confusion matrix* is

	Classified good buy	Classified bad buy
Actual good buy	385	121
Actual bad buy	181	313

and the estimated misclassification rate is

$$\frac{181 + 121}{1000} = 30.2\%.$$

Even though about 30% of cars are misclassified according to the support vector machine, it is much better than 50% from a random guess. We have carried out more detailed analyses involving other variables from the original kaggle "Don't Get Kicked!" dataset but found that it is very difficult to improve, significantly, beyond a 30% misclassification rate for such balanced training data.

The R scripts `carAucRadialSVMtune.R` and `carAucRadialSVM.R` in the `HRW` package carry out the full radial basis function support vector machine analysis presented here. These scripts can be run by issuing the following R commands:

```
> library(HRW) ;demo(carAucRadialSVMtune,package="HRW")
> demo(carAucRadialSVM,package = "HRW")
```

The locations of these script files can be determined using:

```
>system.file("demo","carAucRadialSVMtune",package="HRW")
>system.file("demo","carAucRadialSVM",package = "HRW")
```

The "Support Vector Machines" vignette in `e1071` by David Meyer concludes by pointing out that `svm()` "scales rather badly with the data size," and this had been our experience when trying to fit support vector machines to all 72,983 cars in the full car auction dataset. One remedy to this problem is to use a *low-rank* support vector machine, such as the penalized spline versions that we discuss next.

6.7.1.1 Penalized Spline Support Vector Machines

The radial basis kernel is the most commonly used kernel in support vector machine classification and typically produces a high quality classifier when properly tuned. However, such classifiers are "black boxes," in that they provide little or no insight into the effects of the predictors and their relative importance. *Penalized spline support vector machines* (Pearce and Wand 2006) combine the ideas of additive models, as described in Sect. 3.3, and support vector machine classification to produce interpretable classifiers. As alluded to earlier, their low-rank aspect can overcome problems with more common support vector machine methodology. Ormerod et al. (2008) provide a detailed description and computational details of penalized spline support vector machines. The corresponding R package, `LowRankQP` (Ormerod and Wand 2018), allows straightforward implementation in R.

We now describe fitting a penalized spline support vector machine to the $n = 16{,}952$ car auction data in R. The first steps are to set up linear and spline basis design matrices $\boldsymbol{X}$ and $\boldsymbol{Z}$, in exactly the same way as the mixed model approach to nonparametric regression in Sect. 2.7. The $\boldsymbol{X}$ matrix is the $16{,}952 \times 16$ matrix with a column of ones followed by columns containing the values of the indicator predictors (6.22) and continuous predictors (6.23). The $\boldsymbol{Z}$ matrix is

$$\boldsymbol{Z} = [\boldsymbol{Z}_{\text{age}} \mid \boldsymbol{Z}_{\text{odom.}} \mid \boldsymbol{Z}_{\text{ave. acq.}} \mid \boldsymbol{Z}_{\text{cost}}]$$

where

$$\boldsymbol{Z}_{\text{age}} \equiv [\; z_k(\text{age}_i) \;]_{\substack{1 \le i \le 16{,}952 \\ 1 \le k \le K_{\text{age}}}}$$

contains the spline basis function $\{z_k : 1 \le k \le K_{\text{age}}\}$ evaluated at each of the age at sale observations. The matrices $\boldsymbol{Z}_{\text{odom.}}$, $\boldsymbol{Z}_{\text{price}}$ and $\boldsymbol{Z}_{\text{cost}}$ are defined analogously for the other continuous predictors given at (6.23). Each of the continuous predictors require smoothing parameters

$$\lambda_{\text{age}},\ \lambda_{\text{odom.}},\ \lambda_{\text{price}},\ \lambda_{\text{cost}} > 0.$$

Once these have been set, then forms the vector

$$\boldsymbol{\lambda} = [\lambda_{\text{age}} \mathbf{1}_{K_{\text{age}}}^T, \lambda_{\text{odom.}} \mathbf{1}_{K_{\text{odom.}}}^T, \lambda_{\text{price}} \mathbf{1}_{K_{\text{price}}}^T, \lambda_{\text{cost}} \mathbf{1}_{K_{\text{cost}}}^T]^T \tag{6.24}$$

where $K_{\text{odom.}}$, K_{price} and K_{cost} are the numbers of spline basis functions for each of the other continuous predictors and $\mathbf{1}_d$ denotes the $d \times 1$ vector of ones. Then construct

$$\widetilde{\boldsymbol{Z}} \equiv \boldsymbol{Z}\,\{2\,\text{diag}(\boldsymbol{\lambda})\}^{-1/2}.$$

In our analysis the smoothing parameters are set to be equal to $\lambda = 44.7$ after transforming each of the continuous predictors to the unit interval. The R script carAucPenSplSVMtune.R in the HRW package uses two-fold cross-validation to justify the choice of λ. This script can be run and located using:

```
> library(HRW);demo(carAucPenSplSVMtune,package = "HRW")
> system.file("demo","carAucPenSplSVMtune.R",package = "HRW")
```

Having equal smoothing parameters means that $\text{diag}(\boldsymbol{\lambda})$ is multiple of the identity matrix. The quadratic programming problem is then

$$\begin{aligned}&\min_{\boldsymbol{\alpha}}\left[\tfrac{1}{2}\boldsymbol{\alpha}^T\{(\boldsymbol{y}\boldsymbol{y}^T)\odot(\widetilde{\boldsymbol{Z}}\widetilde{\boldsymbol{Z}}^T)\}\boldsymbol{\alpha}-\mathbf{1}^T\boldsymbol{\alpha}\right]\\&\text{subject to } 0\le\alpha_i\le 1,\quad 1\le i\le n \text{ and } \boldsymbol{X}^T(\boldsymbol{y}\odot\boldsymbol{\alpha})=0\end{aligned}\tag{6.25}$$

with $n = 16{,}952$ in the current example. Since the Gram matrix $\widetilde{\boldsymbol{Z}}\widetilde{\boldsymbol{Z}}^T$ has rank equal to the number of columns of $\widetilde{\boldsymbol{Z}}$, (6.25) is a low-rank quadratic programming problem. Section 3 of Ormerod et al. (2008) describes a streamlined solution strategy. We used $K_{\text{odom.}} = 5$ and $K_{\text{odom.}} = K_{\text{ave. acq.}} = K_{\text{cost}} = 15$ which corresponds to a rank of 50. The classifier is then

$$\begin{aligned}&\boldsymbol{x}_{\text{new}} \text{ classified as being a good buy if } \boldsymbol{X}_{\text{new}}\widehat{\boldsymbol{\beta}}+\boldsymbol{Z}_{\text{new}}\widehat{\boldsymbol{u}}<0\\&\boldsymbol{x}_{\text{new}} \text{ classified as being a bad buy if } \boldsymbol{X}_{\text{new}}\widehat{\boldsymbol{\beta}}+\boldsymbol{Z}_{\text{new}}\widehat{\boldsymbol{u}}\ge 0.\end{aligned}$$

where $\boldsymbol{X}_{\text{new}}$ and $\boldsymbol{Z}_{\text{new}}$ are the design matrices formed from $\boldsymbol{x}_{\text{new}}$ and $\widehat{\boldsymbol{u}} = \widetilde{\boldsymbol{Z}}^T(\widehat{\boldsymbol{\alpha}}\odot\boldsymbol{y})$. The $\widehat{\boldsymbol{\beta}}$ vector is obtained from Karush–Kuhn–Tucker conditions of the quadratic programming problem, as described in Sect. 6 of Pearce and Wand (2006).

Suppose that the array lambda is set up in R according to (6.24). Then the main R commands for fitting the penalized spline support vector machine are:

```
> Ztilde <- Z*(1/sqrt(2*lambda))
> yX <- as.matrix(y*X) ; yZtilde <- as.matrix(y*Ztilde)
> library(LowRankQP)
> LowRankQPobj <- LowRankQP(yZtilde,rep(-1,n),t(yX),
+                           rep(0,ncX),rep(1,n),niter=500)
> betaHat <- as.vector(LowRankQPobj$beta)
> uHat <-  as.vector(crossprod(yZtilde,LowRankQPobj$alpha))
```

The classifications for the held-back data are found via:

```
> yHat <- sign(Xnew%*%betaHat + Znew%*%uHat)
```

where Xnew and Znew are the R matrices containing $\boldsymbol{X}_{\text{new}}$ and $\boldsymbol{Z}_{\text{new}}$, respectively.

The R script carAucPenSplSVM.R in the HRW package contains the full set of commands for obtaining the penalized support vector machine classification vector yHat for the same training and held-back data used in Sect. 6.7.1. The ensuing confusion matrix is:

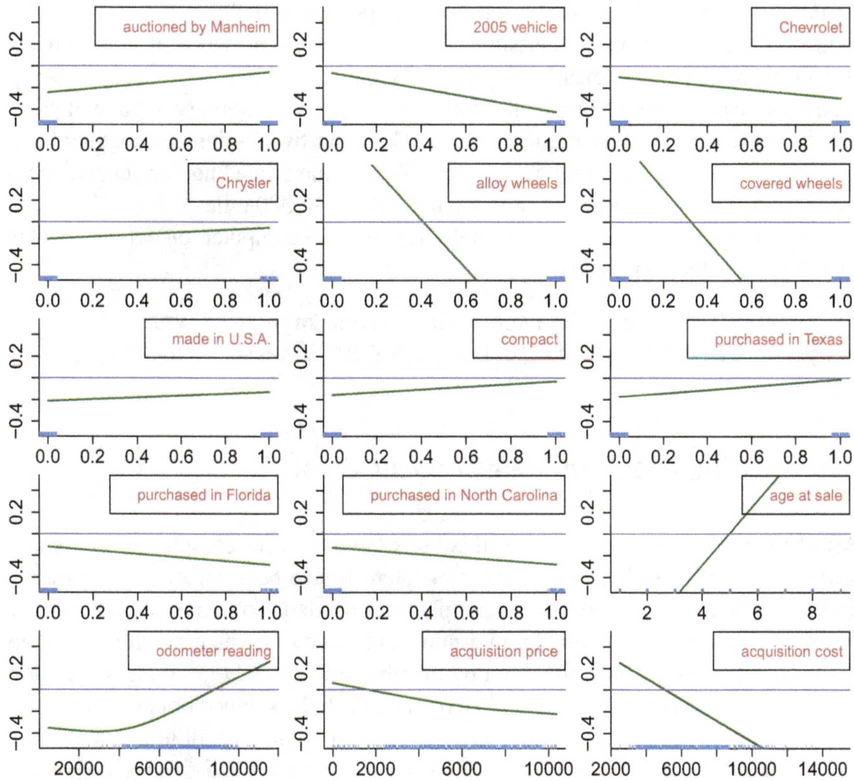

Fig. 6.28 Plot of penalized spline support vector classifier fit to the car auction training data of size $n = 16{,}952$ described in the text. In each panel, the classifier is evaluated over a grid corresponding to that panel's predictor and all other predictors are set at their averages. In the rug plots at the base of each panel, jittering has been added to the binary predictor data to aid visualization.

	Classified good buy	Classified bad buy
Actual good buy	404	102
Actual bad buy	205	289

and the estimated misclassification rate is

$$\frac{205 + 102}{1000} = 30.7\%,$$

which is very close to that obtained from the radial basis function support vector machine. However, we can visualize, and possibly interpret, this penalized spline support vector machine fit by plotting the additive components. This is done in Fig. 6.28. In each panel, a grid corresponding to that panel's predictor is created and all other predictors are set at their averages. The vertical axis limits are fixed at $(-0.5, 0.5)$.

Perusal of Fig. 6.28 shows the relative importance of the predictors and the nature of their effect on the classifier. For example, the presence of alloy wheels is seen to be strongly indicative of the car being a good buy. The effect of being a compact car is weaker. Out of the four continuous predictors, two ("age at sale" and "acquisition cost") have linear effects. The other two ("odometer reading" and "acquisition price") have nonlinear effects. The odometer reading does not ramp up for classification of a bad buy at auction until about 40,000 miles.

To run and locate `carAucPenSplSVM.R` on the computer on which `HRW` is installed issue the commands:

```
> library(HRW) ; demo(carAucPenSplSVM,package = "HRW")
> system.file("demo","carAucPenSplSVM.R",package = "HRW")
```

6.8 Missing Data and Measurement Error

All of the examples given so far in this book have been for complete datasets and with the assumption that the variables of interest have been measured accurately. Unfortunately, many applications are plagued by data with missingness and/or measurement error. For example, in a nutritional study in which much of the data comes from questionnaires filled out by members of the study group, it is common for questions involving smoking and alcohol intake to be skipped or under-reported. A potentially important biomarker, such as Vitamin D intake, will not be measured by the questionnaire process but a surrogate could be formed from some of the questionnaire responses. The problems of how to best account for such impurities in the data have spawned major areas of statistical research. Books devoted to handling missing data and/or measurement error include Fuller (1987), Little and Rubin (2002), Carroll et al. (2006), and Daniels and Hogan (2008).

While the missing data and measurement error literatures are large and varied, many of its tenets, models, and methodologies apply to semiparametric regression. In this section we will give a flavor for ways by which missing data and measurement error can be accommodated, in a principled fashion, in R-based analyses. This involves couching semiparametric regression in a *graphical models* framework and using Bayesian inference engines for fitting and inference.

6.8.1 Graphical Models Approach to Bayesian Semiparametric Regression

Complicated semiparametric regression models and analyses benefit from probabilistic graphical representations of Bayesian semiparametric models. Before returning to the handling of missing data and measurement error, we discuss the

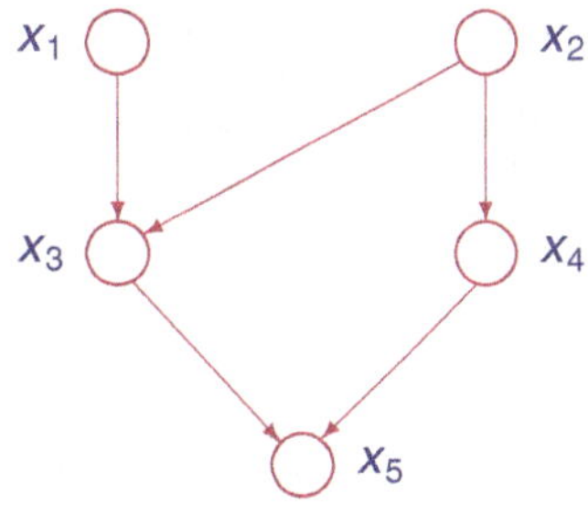

Fig. 6.29 An elementary probabilistic directed acyclic graph.

graphical models approach to Bayesian semiparametric regression in general terms. A more detailed exposition is given in Wand (2009).

Figure 6.29 shows a *directed graph*. The circles in the graph are known as *nodes*. The arrows, known as *directed edges*, impose "parent"–"child" relationships between the nodes that they join. For example, the node labeled x_3 is a parent of x_5. Conversely, x_5 is a child of x_3. The directed graph is also *acyclic* since there are no *cycles*—that is, one cannot follow the arrows away from a given node and eventually return to that node.

Now suppose that $x_1, \ldots, x_5$ are continuous random variables with joint density function satisfying

$$\begin{aligned} p(x_1, x_2, x_3, x_4, x_5) &= \textstyle\prod_{i=1}^{5} p(x_i | \text{parents of } x_i) \\ &= p(x_1)\, p(x_2)\, p(x_3|x_1, x_2)\, p(x_4|x_2)\, p(x_5|x_3, x_4). \end{aligned} \tag{6.26}$$

Then the graph in Fig. 6.29 is a *probabilistic directed acyclic graph* with respect to $p(x_1, x_2, x_3, x_4, x_5)$. This definition applies to general directed acyclic graphs having nodes corresponding to random vectors $\boldsymbol{x}_1, \ldots, \boldsymbol{x}_k$, with (6.26) replaced by

$$p(\boldsymbol{x}_1, \ldots, \boldsymbol{x}_k) = \prod_{i=1}^{k} p(\boldsymbol{x}_i | \text{parents of } \boldsymbol{x}_i).$$

For the remainder of this book, all directed acyclic graphs are assumed to be probabilistic and this adjective is omitted.

Next, consider the Bayesian nonparametric regression model

$$\begin{aligned} &\boldsymbol{y} | \boldsymbol{x}, \boldsymbol{\beta}, \boldsymbol{u}, \sigma_\varepsilon \sim N(\boldsymbol{X\beta} + \boldsymbol{Zu}, \sigma_\varepsilon^2 \boldsymbol{I}), \quad x_i | \mu_x, \sigma_x^2 \stackrel{\text{ind.}}{\sim} N(\mu_x, \sigma_x^2) \\ &\boldsymbol{u} \sim N\left(\boldsymbol{0}, \sigma_u^2 \boldsymbol{I}\right), \quad \boldsymbol{\beta} \sim N\left(\begin{bmatrix} 0 \\ 0 \end{bmatrix}, \begin{bmatrix} 100 & 10 \\ 10 & 100 \end{bmatrix}\right), \\ &\mu_x \sim N(0, 100), \quad \sigma_\varepsilon, \sigma_x, \sigma_u \stackrel{\text{ind.}}{\sim} \text{Half-Cauchy}(25), \end{aligned} \tag{6.27}$$

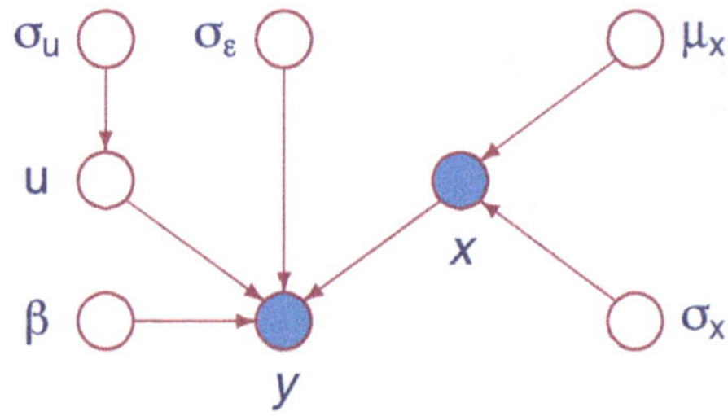

Fig. 6.30 Directed acyclic graph corresponding to model (6.27). The shaded nodes correspond to the observed data.

with prior mutual independence assumed among each of $\boldsymbol{\beta}$, $\boldsymbol{u}$, μ_x, σ_ε, σ_x, and σ_u. The vectors in (6.27) are

$$\boldsymbol{x} \equiv \begin{bmatrix} x_1 \\ \vdots \\ x_n \end{bmatrix}, \quad \boldsymbol{y} \equiv \begin{bmatrix} y_1 \\ \vdots \\ y_n \end{bmatrix}, \quad \boldsymbol{\beta} \equiv \begin{bmatrix} \beta_0 \\ \beta_1 \end{bmatrix} \quad \text{and} \quad \boldsymbol{u} \equiv \begin{bmatrix} u_1 \\ \vdots \\ u_K \end{bmatrix}.$$

The design matrices $\boldsymbol{X}$ and $\boldsymbol{Z}$ are given by (6.19). The joint density function of the random variables in (6.27) is

$$\begin{aligned} p(\boldsymbol{y}, \boldsymbol{x}, \boldsymbol{\beta}, \boldsymbol{u}, \sigma_u, \sigma_\varepsilon, \mu_x, \sigma_x) = p(\boldsymbol{y}|\boldsymbol{x}, \boldsymbol{\beta}, \boldsymbol{u}, \sigma_\varepsilon)\, p(\boldsymbol{x}|\mu_x, \sigma_x) p(\boldsymbol{\beta}) \\ \times\, p(\boldsymbol{u}|\sigma_u)\, p(\sigma_u)\, p(\mu_x)\, p(\sigma_\varepsilon)\, p(\sigma_x) \end{aligned} \tag{6.28}$$

where, for example,

$$p(\boldsymbol{y}|\boldsymbol{x}, \boldsymbol{\beta}, \boldsymbol{u}, \sigma_\varepsilon) = (2\pi\sigma_\varepsilon^2)^{-n/2} \exp\left\{-\tfrac{1}{2}\|\boldsymbol{y} - \boldsymbol{X\beta} - \boldsymbol{Zu}\|^2/\sigma_\varepsilon^2\right\}$$

and

$$p(\boldsymbol{x}|\mu_x, \sigma_x) = (2\pi\sigma_x^2)^{-n/2} \exp\left\{-\tfrac{1}{2}\|\boldsymbol{x} - \boldsymbol{1}\mu_x\|^2/\sigma_x^2\right\}.$$

Here $\boldsymbol{1}$ is the $n \times 1$ vector of ones. Then Fig. 6.30 is a directed acyclic graph representation of (6.28). The shading of the $\boldsymbol{x}$ and $\boldsymbol{y}$ nodes signifies that they correspond to observed data.

There are several advantages to such graphical representations of semiparametric regression models. First, graphs such as the one in Fig. 6.30 provide a visualization of the hierarchical structure of its corresponding Bayesian model—in this case model (6.27). Another advantage is that graph theoretic results can be used to determine parts of the model that can be separated from other parts. For example, the result

$$\{\boldsymbol{\beta}, \boldsymbol{u}, \sigma_u, \sigma_\varepsilon\} \text{ is independent of } \{\mu_x, \sigma_x\} \,\Big|\, \{\boldsymbol{x}, \boldsymbol{y}\} \tag{6.29}$$

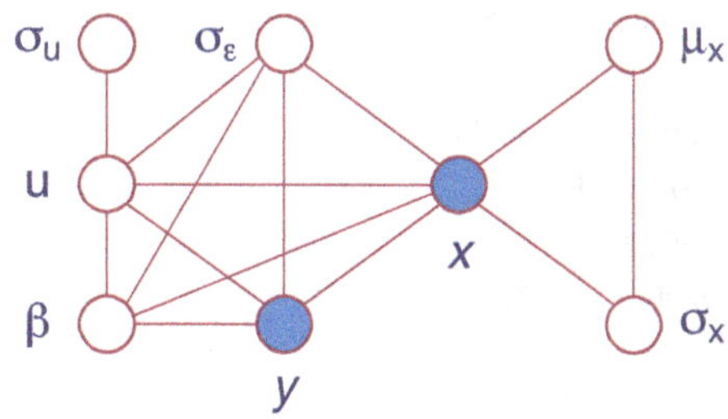

Fig. 6.31 Moral graph of the directed acyclic graph in Fig. 6.30.

asserts that, given the data, the regression parameters are independent of the predictor parameters. This implies that

$$p(\boldsymbol{\beta}, \boldsymbol{u}, \sigma_u, \sigma_\varepsilon | \boldsymbol{x}, \boldsymbol{y}, \boldsymbol{\mu}_x, \sigma_x) = p(\boldsymbol{\beta}, \boldsymbol{u}, \sigma_u, \sigma_\varepsilon, | \boldsymbol{x}, \boldsymbol{y}),$$

meaning that the values of the predictor distribution parameters have no effect on Bayesian inference for the regression parameters via the posterior density $p(\boldsymbol{\beta}, \boldsymbol{u}, \sigma_u, \sigma_\varepsilon | \boldsymbol{x}, \boldsymbol{y})$. Proposition (6.29) can be proven using Corollary 3.23 of Lauritzen (1996). The essence of the proof is that the $\boldsymbol{x}$ node blocks all paths between $\{\boldsymbol{\beta}, \sigma_\varepsilon\}$ and $\{\mu_x, \sigma_x\}$ in the *moral graph* (e.g. Lauritzen 1996) of Fig. 6.30. The moral graph is shown in Fig. 6.31 and is formed from the directed acyclic graph in Fig. 6.30 by drawing an edge between all parent nodes that have a child node in common and then removing all arrows. (For propositions similar to (6.29) but involving a subset of the nodes then the appropriate moral graph applies to a particular subgraph of the directed acyclic graph; Corollary 3.23, Lauritzen 1996).

Bayesian inference in the graphical model given by (6.27) and depicted in Fig. 6.30 involves determination of posterior density functions such as

$$p(\sigma_\varepsilon | \boldsymbol{x}, \boldsymbol{y}), \quad p(\mu_x | \boldsymbol{x}, \boldsymbol{y}), \quad p(\sigma_x | \boldsymbol{x}, \boldsymbol{y}) \quad \text{and} \quad p(\boldsymbol{a}^T \boldsymbol{\beta} + \boldsymbol{b}^T \boldsymbol{u} | \boldsymbol{x}, \boldsymbol{y}),$$

for arbitrary constant vectors $\boldsymbol{a}$ and $\boldsymbol{b}$. However, these density functions are difficult to obtain and approximation methods, based on either Monte Carlo sampling or deterministic calculations for simplified versions of the moral graph, are typically employed in practice. These approximation methods also benefit greatly from theoretic results such as Markov blanket simplification of full conditional distributions (e.g. Pearl 1988). In recent years, *Bayesian inference engines* have emerged for facilitating such approximate inference for general classes of graphical models. Examples include BUGS (Lunn et al. 2013) and Stan (Carpenter et al. 2017) based on Monte Carlo sampling, and Infer.NET (Minka et al. 2014) based on *mean field variational Bayes* and *expectation propagation* (e.g. Bishop 2006). Currently, the Monte Carlo packages have limitations regarding computational speed and the deterministic packages have versatility limitations. These drawbacks are likely to be alleviated in the future by the emergence of new Bayesian inference engines and improved versions of existing ones.

The BUGS engine can be accessed from R using the wrapper packages `BRugs` (Ligges et al. 2017) and `R2WinBUGS` (Gelman et al. 2015). Similarly, Stan is well supported by the R package `rstan` (Guo et al. 2017). Crainiceanu et al. (2005b) and Marley and Wand (2010) illustrate the use of the BUGS wrapper packages for a selection of semiparametric regression models. The remaining examples of this book are in a similar vein.

6.8.2 Nonparametric Regression with a Partially Observed Gaussian Predictor

We now introduce a useful class of models for conveying the main ideas of semiparametric regression with missing data and/or measurement error, via the graphical models approach and with implementation in R. It is called *nonparametric regression with a partially observed Gaussian predictor.*

First consider Bayesian nonparametric regression with Gaussian predictors:

$$\boldsymbol{y}|\,\boldsymbol{x},\boldsymbol{\beta},\boldsymbol{u},\sigma_\varepsilon \sim N(\boldsymbol{X\beta}+\boldsymbol{Zu},\sigma_\varepsilon^2),\qquad x_i\mid\mu_x,\sigma_x^2 \overset{\text{ind.}}{\sim} N(\mu_x,\sigma_x^2),$$

$$\boldsymbol{u}|\sigma_u \sim N(\boldsymbol{0},\sigma_u^2\,\boldsymbol{I})$$

where $\boldsymbol{X}$ and $\boldsymbol{Z}$ are given by (6.19) and appropriate priors are placed on μ_x, σ_x, $\boldsymbol{\beta}$, σ_ε and σ_u. In model (6.27), corresponding to directed acyclic graph (Fig. 6.30), the x_is are completely observed. Now, instead, suppose that we only observe $\boldsymbol{x}_{\text{obs}}$ where

$$\boldsymbol{x}=\begin{bmatrix}\boldsymbol{x}_{\text{obs}}\\ \boldsymbol{x}_{\text{unobs}}\end{bmatrix}$$

is a partition of the $n\times 1$ predictor vector into components of dimensions $n_{\text{obs}}\times 1$ and $n_{\text{unobs}}\times 1$. The $\boldsymbol{x}_{\text{unobs}}$ either corresponds to predictor data that are missing or measured with error. In the missing data case we define the vector

$$\boldsymbol{r}\equiv\begin{bmatrix}r_1\\ \vdots\\ r_n\end{bmatrix}\quad\text{where}\quad r_i=\begin{cases}1 \text{ if } x_i \text{ is observed,}\\ 0 \text{ if } x_i \text{ is missing.}\end{cases}$$

In the measurement error case, we observe a surrogate

$$\boldsymbol{w}=\begin{bmatrix}\boldsymbol{w}_{x,\text{obs}}\\ \boldsymbol{w}_{x,\text{unobs}}\end{bmatrix}$$

where $\boldsymbol{w}_{x,\text{obs}}$ partners the validation sample $\boldsymbol{x}_{\text{obs}}$ and $\boldsymbol{w}_{x,\text{unobs}}$ contains the surrogates for $\boldsymbol{x}_{\text{unobs}}$.

Table 6.2 Various missing data and measurement error models for partially observed predictor data.

Model for missingness or measurement error	Verbal description
$r_i \mid p \overset{\text{ind.}}{\sim} \text{Bernoulli}(p)$	Missing completely at random
$r_i \mid \phi_0, \phi_1 \overset{\text{ind.}}{\sim} \text{Bernoulli}\big(\text{logit}^{-1}(\phi_0 + \phi_1\, y_i)\big)$	Missing at random
$r_i \mid \phi_0, \phi_1 \overset{\text{ind.}}{\sim} \text{Bernoulli}\big(\text{logit}^{-1}(\phi_0 + \phi_1\, x_i)\big)$	Missing not at random
$w_i \overset{\text{ind.}}{\sim} N(x_i, \sigma_w^2), \quad x_i \overset{\text{ind.}}{\sim} N(\mu_x, \sigma_x^2)$	Classical measurement error
$x_i \overset{\text{ind.}}{\sim} N(w_i, \sigma_x^2)$	Berkson measurement error

Table 6.2 lists various models for the unobserved predictor vector $\boldsymbol{x}_{\text{unobs}}$. Appropriate priors should be placed on the missingness and measurement error parameters. The verbal descriptions match common usage in the literature and standard texts on the topics such as Little and Rubin (2002) and Carroll et al. (2006). The distinction between "missing at random" and "missing not at random" is subtle. In the former case, the missingness depends on completely observed data and graph theoretic results (e.g. Lauritzen 1996) can be used to show that

$$\{\boldsymbol{\beta}, \boldsymbol{u}, \sigma_u, \sigma_\varepsilon, \mu_x, \sigma_x\} \text{ is independent of } \{\phi_0, \phi_1\} \,\Big|\, \{\boldsymbol{x}, \boldsymbol{y}\}.$$

Such is not the case for data missing not at random, and the missingness mechanism impinges on inference for the regression parameters.

Figure 6.32 shows directed acyclic graphs corresponding to the models listed in Table 6.2. We see that nonparametric regression with partially observed predictor data corresponds to making inference for a more elaborate graphical model. According to this approach, the unobserved predictor data vector $\boldsymbol{x}_{\text{unobs}}$ has a similar status to the regression parameters: another *hidden node* for which inference is sought. Similar comments apply to the auxiliary parameters such as $\boldsymbol{\phi} \equiv (\phi_0, \phi_1)$.

The `HRW` package contains R scripts for demonstrating fitting of each of the models in Table 6.2 via Stan. The scripts are named:

```
npReg.R                  npRegMCAR.R
npRegMAR.R               npRegMNAR.R
npRegClassicMeaErr.R     npRegBerksonMeaErr.R
```

Simulated data are used in each of these scripts, which allows comparison of the estimates with the true model parameters and functions. They can be modified for real data and related models. Sections 6.8.3 and 6.8.4 provide illustrations.

There is insufficient space here to describe each of the aforementioned scripts and the output that they produce. Instead we encourage the reader to study and run those scripts that are of interest. We will provide details on just one of

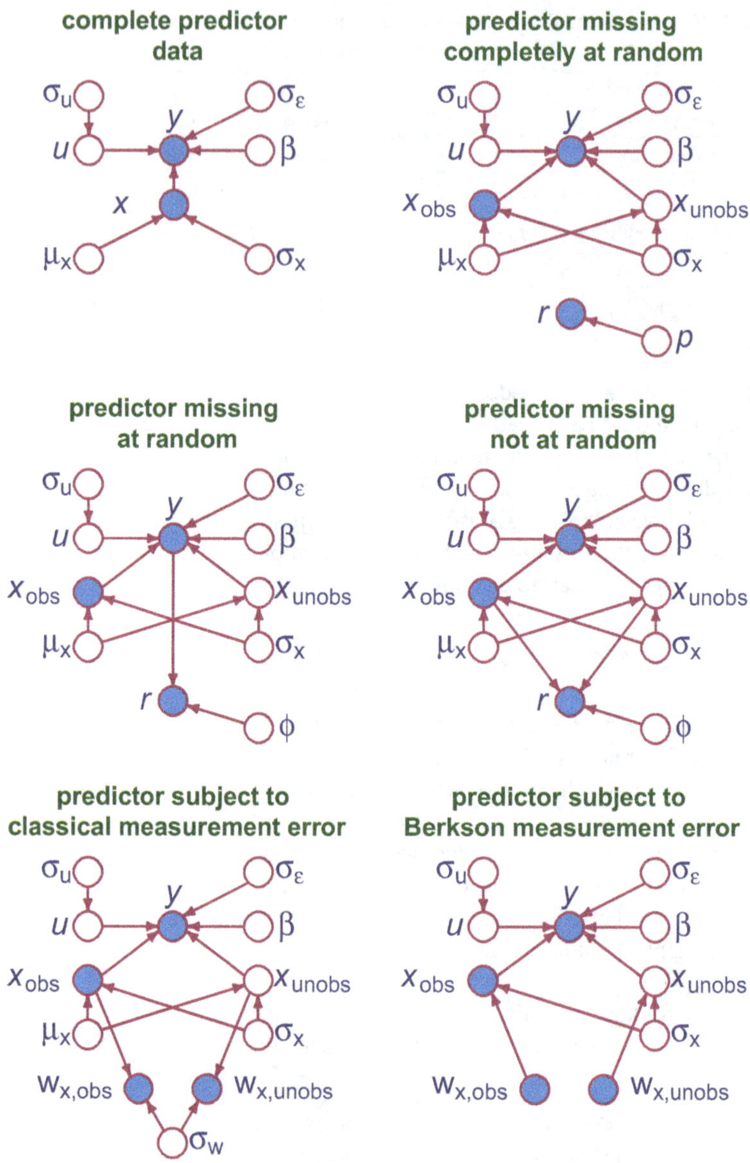

Fig. 6.32 Directed acyclic graphs for nonparametric regression with complete predictor data and five partially observed Gaussian predictor data models.

them, `npRegClassicMeaErr.R`, for nonparametric regression subject to classical measurement error.

The data in `npRegClassicMeaErr.R` are simulated according to the model

$$y_i \sim N(f(x_i), 0.35^2), \qquad w_i \overset{\text{ind.}}{\sim} N(x_i, 0.1^2), \qquad x_i \overset{\text{ind.}}{\sim} N(\tfrac{1}{2}, \tfrac{1}{36}), \quad 1 \le i \le n,$$

where

$$f(x) = 3\exp\{-78(x-0.38)^2\} + \exp\{-200(x-0.75)^2\} - x$$

and $n = 1000$. The w_i are fully observed surrogates for the x_i. Only 10% of the x_is are observed, giving a validation sample of size $n_{\text{obs}} = 100$.

The full Bayesian model in `npRegClassicMeaErr.R` is

$$\begin{gathered}
y_i \mid \beta_0, \beta_1, u_1, \ldots, u_k, \sigma_\varepsilon, x_i \overset{\text{ind.}}{\sim} N\big(\beta_0 + \beta_1 x_i + \textstyle\sum_{k=1}^{k} u_k z_k(x_i), \sigma_\varepsilon^2\big), \\
x_i \mid \mu_x, \sigma_x \overset{\text{ind.}}{\sim} N(\mu_x, \sigma_x^2), \quad w_i \mid x_i, \sigma_w \overset{\text{ind.}}{\sim} N(x_i, \sigma_w^2), \\
\mu_x \sim N(0, \sigma_\mu^2), \quad \sigma_x \sim \text{Half-Cauchy}(A_x), \\
\beta_0, \beta_1 \overset{\text{ind.}}{\sim} N(0, \sigma_\beta^2), \quad u_1, \ldots, u_k \mid \sigma_u \overset{\text{ind.}}{\sim} N(0, \sigma_u^2), \\
\sigma_\varepsilon \sim \text{Half-Cauchy}(A_\varepsilon), \quad \sigma_u \sim \text{Half-Cauchy}(A_u), \\
\sigma_w \sim \text{Half-Cauchy}(A_w)
\end{gathered} \tag{6.30}$$

where $\sigma_u, \sigma_\beta, A_\varepsilon, A_u, A_x, A_w > 0$ are hyperparameters which are all set to 10^5. Note that `npRegClassicMeaErr.R` uses the simple truncated line spline basis with

$$z_k(x) = (x - \kappa_k)_+, \quad 1 \le k \le K,$$

where the κ_k are knots placed at the quantiles of the x_is. This is because the spline basis functions have to be constructed inside the Bayesian inference machine code. This creates difficulties if using O'Sullivan splines. The input data are

$$\boldsymbol{x}_{\text{obs}}, \quad \boldsymbol{w} = \begin{bmatrix} \boldsymbol{w}_{x,\text{obs}} \\ \boldsymbol{w}_{x,\text{unobs}} \end{bmatrix} \quad \text{and} \quad \boldsymbol{y} = \begin{bmatrix} \boldsymbol{y}_{x,\text{obs}} \\ \boldsymbol{y}_{x,\text{unobs}} \end{bmatrix}$$

where, for example, $\boldsymbol{y}_{x,\text{obs}}$ is the 100×1 vector of y_is that are partnered by an exactly observed x_i and $\boldsymbol{y}_{x,\text{unobs}}$ is the 900×1 vector of y_is with no such x_i partner. The Stan code for the model specification is:

```
model
{
   yxObs  ~ normal(XxObs*beta+ZxObs*u,sigmaEps);
   xObs   ~ normal(muX,sigmaX);
   wxObs  ~ normal(xObs,sigmaW);

   yxUnobs  ~ normal(XxUnobs*beta+ZxUnobs*u,sigmaEps);
   xUnobs   ~ normal(muX,sigmaX);
   wxUnobs  ~ normal(xUnobs,sigmaW);

   u ~ normal(0,sigmaU) ;   beta ~ normal(0,sigmaBeta);
```

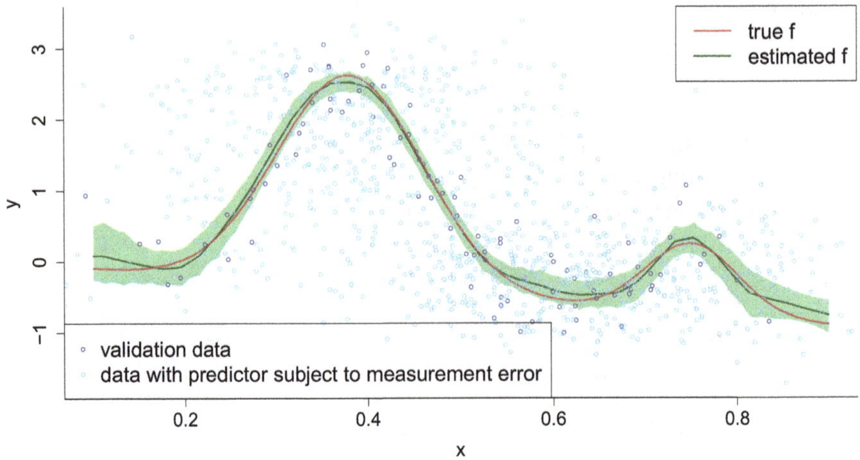

Fig. 6.33 Bayes estimate of the regression function for data simulated according to the nonparametric regression model (6.30) with partially observed predictor data, and the remaining predictor data subject to classical measurement error. The shaded region corresponds to pointwise 95% credible sets. The true mean function from which the data were generated is also shown.

```
   muX ~ normal(0,sigmaMu); sigmaX ~ cauchy(0,Ax);
   sigmaW ~ cauchy(0,Aw);   sigmaEps ~ cauchy(0,Aeps);
   sigmaU ~ cauchy(0,Au);
}
```

where the design matrices for the unobserved predictor data `XxUnobs` and `ZxUnobs` are constructed according to the Stan code:

```
for (i in 1:nUnobs)
{
   XxUnobs[i,1] = 1  ; XxUnobs[i,2] = xUnobs[i];
   for (k in 1:ncZ)
      ZxUnobs[i,k] = (xUnobs[i]-knots[k])
                            *step(xUnobs[i]-knots[k]);
}
```

Figure 6.33 shows the estimated function and pointwise 95% credible set based on a burn-in of size 2000 and a kept sample of size 1000. In this figure notice that only the 100 dark-colored points are scattered about the true regression function. For the other 900 light-colored points no such visual alignment occurs due to measurement error affecting their horizontal axis values. Despite this mismatch between much of the observed data and the signal, the MCMC-based Bayesian fitting procedure is able to estimate the signal very well.

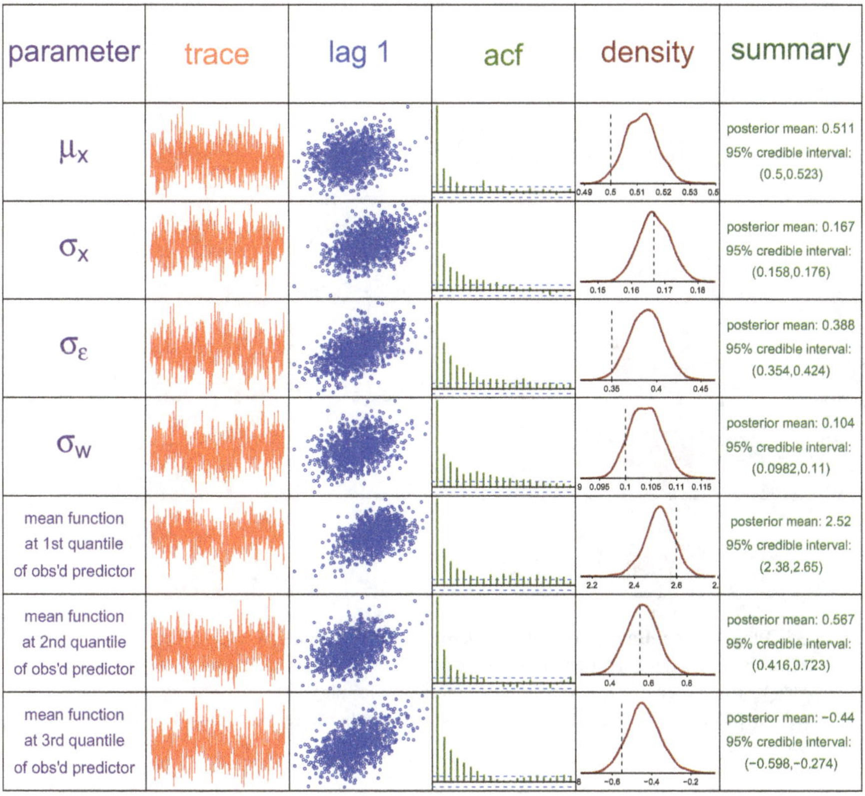

Fig. 6.34 Plots and summaries of MCMC samples for parameters of interest from the classical measurement error example: trace plot, lag-1 plot, estimated autocorrelation function, estimated posterior density function, and numerical summary.

Figure 6.34 shows the MCMC samples, lag-1 plots, autocorrelation function, estimated posterior density function, and numerical summaries for several key parameters, including the regression function estimates at quartiles of the observed predictor data. The true values are also shown as dashed vertical lines on the posterior density plots.

Figure 6.35 is a similar graphic for the first five unobserved x_i values. One interesting aspect of the posterior density functions of some of these $x_{\text{unobs},i}$ is that they are bimodal. This is because of two-peaked nature of the regression curve. A particular (w_i, y_i) pair could correspond to two distinct regions for its unobserved corresponding x_i.

To run and locate `npRegClassicMeaErr.R` issue the commands:

```
> library(HRW) ; demo(npRegClassicMeaErr,package = "HRW")
> system.file("demo","npRegClassicMeaErr.R",package = "HRW")
```

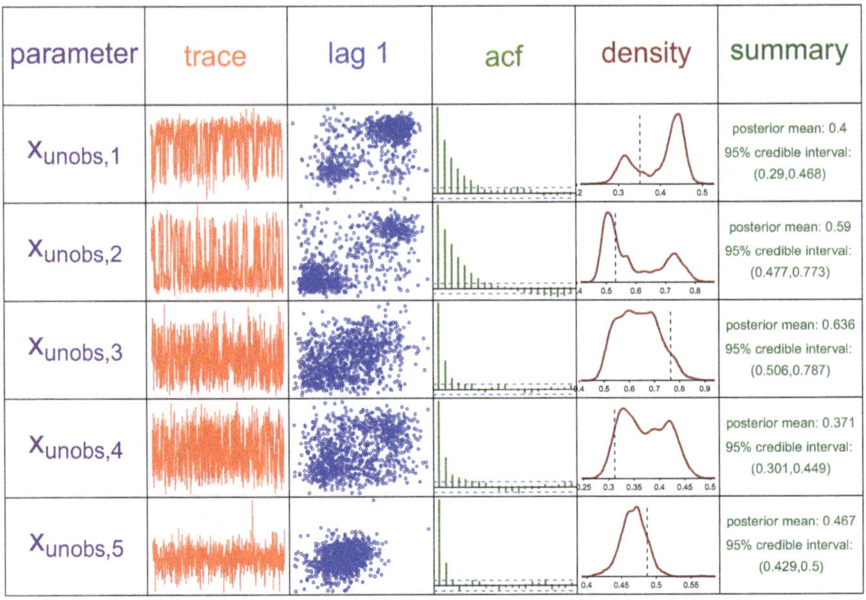

Fig. 6.35 Plots and summaries of MCMC samples for the first five unobserved predictors from the classical measurement error example: trace plot, lag-1 plot, estimated autocorrelation function, estimated posterior density function, and numerical summary. The dashed vertical lines in the density plots indicate true values of the unobserved predictor values (known due to data being simulated).

6.8.3 Example: Pima Indians Diabetes Study

This example involves a well-known dataset from a diabetes study on Pima Indians, available on the University of California at Irvine, USA, Machine Learning Repository, and also available as `PimaIndiansDiabetes` in the `mlbench` (Leisch and Dimitriadou 2010). The original source of the data is the U.S. National Institute of Diabetes and Digestive and Kidney Diseases. The response variable is presence or absence of diabetes, and eight predictors are available. The repository website mentions that it seems very likely that zero values encode missing data even though the dataset donors did not indicate this. We make that assumption here. Specifically, 11 of the body mass index records are coded as zero and we take those to be missing.

Let

$$\begin{aligned} x_i &\equiv \text{body mass index (kilogram/meter}^2) \\ \text{and } y_i &\equiv \text{indicator of having diabetes}, \quad 1 \le i \le n, \end{aligned} \tag{6.31}$$

where $n = 786$. A logistic nonparametric regression model with predictors not missing at random is:

$$
\begin{aligned}
y_i \,|\, \beta_0, \beta_1, x_i &\overset{\text{ind.}}{\sim} \text{Bernoulli}\Big(\text{logit}^{-1}\big(\beta_0 + \beta_1\, x_i + \textstyle\sum_{k=1}^{k} u_k\, z_k(x_i)\big)\Big),\\
x_i \,|\, \mu_x, \sigma_x &\overset{\text{ind.}}{\sim} N(\mu_x, \sigma_x^2), \quad r_i \equiv I(x_i \text{ observed}),\\
r_i \,|\, \phi_0, \phi_1 &\overset{\text{ind.}}{\sim} \text{Bernoulli}\big(\text{logit}^{-1}(\phi_0 + \phi_1\, x_i)\big)\\
\mu_x \sim N(0, \sigma_\mu^2), \quad &\sigma_x \sim \text{Half-Cauchy}(A_x), \quad \beta_0, \beta_1 \overset{\text{ind.}}{\sim} N(0, \sigma_\beta^2),\\
u_1, \ldots, u_k \,|\, \sigma_u &\overset{\text{ind.}}{\sim} N(0, \sigma_u^2), \quad \phi_0, \phi_1 \overset{\text{ind.}}{\sim} N(0, \sigma_\phi^2),\\
\sigma_u &\sim \text{Half-Cauchy}(A_u).
\end{aligned}
\tag{6.32}
$$

The R script `PIDana.R`, in the `HRW` package, facilitates fitting of (6.32) via `rstan`. The script stores the data in arrays `xObs` (the observed x_is), `yxObs` (the y_is corresponding to observed x_is) and `yxUnobs` (the y_is corresponding to unobserved x_is). The array `xUnobs`, corresponding to the unobserved x_is, is a hidden node in the model's directed acyclic graph and is a parameter with the same status as regression model parameters such as β_0, β_1, and σ_u. The Stan code for the model specification is:

```
transformed parameters
{
   matrix[n,2] X;        matrix[n,ncZ] Z;
   for (i in 1:nObs)
   {
      X[i,1] = 1   ;   X[i,2] = xObs[i] ;
      Z[i] = ZxObs[i] ;
   }
   for (i in 1:nUnobs)
   {
      X[i+nObs,1] = 1  ; X[i+nObs,2] = xUnobs[i];
      for (k in 1:ncZ)
         Z[i+nObs,k] = (xUnobs[i]-knots[k])
                       *step(xUnobs[i]-knots[k]);
   }
}
model
{
   y ~ bernoulli_logit(X*beta+Z*u);
   r ~ bernoulli_logit(X*phi);
   col(X,2) ~ normal(muX,sigmaX);
   u ~ normal(0,sigmaU) ; beta ~ normal(0,sigmaBeta);
   muX ~ normal(0,sigmaMu); phi ~ normal(0,sigmaPhi);
   sigmaX ~ cauchy(0,Ax);   sigmaU ~ cauchy(0,Au);
}
```

Figure 6.36 summarizes the MCMC output, based on a burn-in of size 1000 with 5000 kept samples and a thinning factor of 5. Convergence is seen to be excellent in

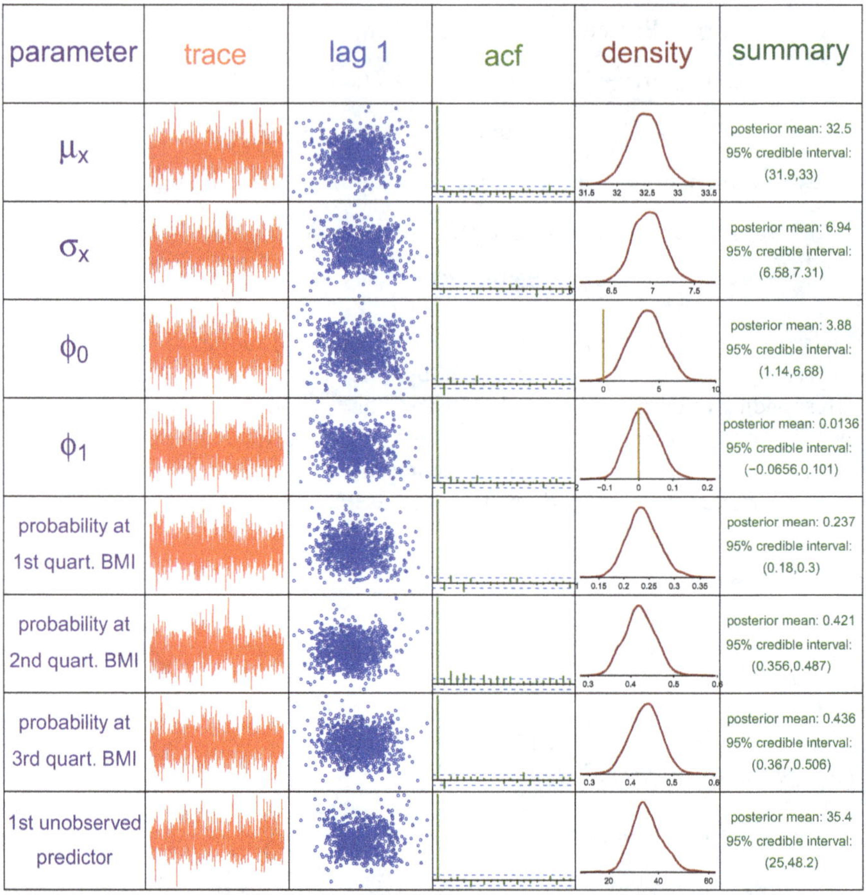

Fig. 6.36 Plots and summaries of MCMC samples for parameters of interest from the example involving the nonparametric logistic regression, with predictor missing not at random, model (6.32) applied to data from the diabetes study on Pima Indians: lag-1 plot, estimated autocorrelation function, estimated posterior density function, and numerical summary.

this case. The 95% credible set for ϕ_1 is $(-0.0646, 0.116)$ which indicates that the missingness is not significantly related to the missing predictor.

Figure 6.37 shows the fitted probability curve, corresponding to the pointwise posterior means. As expected, prevalence of diabetes is seen to increase with body mass index. However, the curve has an interesting plateau corresponding to body mass index values ranging from about 33 to 39.

The following commands can be used to run `PIDana.R` and obtain its location on the computer on which `HRW` is installed:

```
> library(HRW) ; demo(PIDana,package = "HRW")
> system.file("demo","PIDana.R",package = "HRW")
```

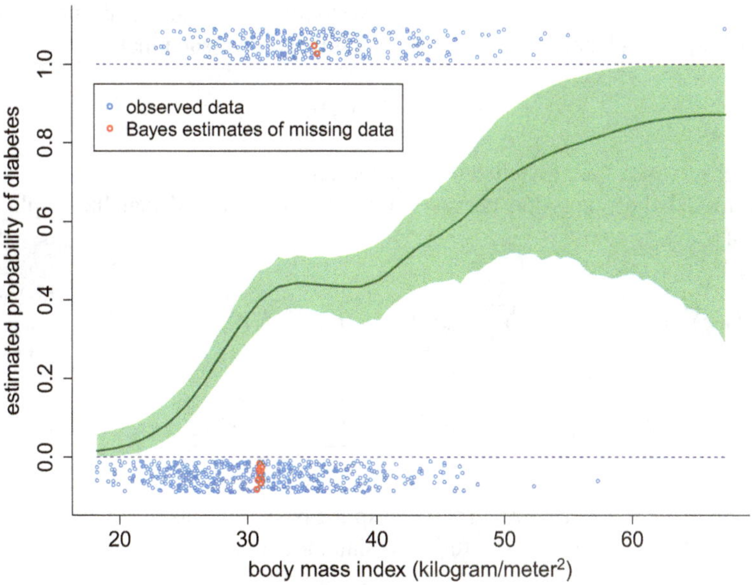

Fig. 6.37 Posterior mean and pointwise 95% credible set for the probability of diabetes as function of body mass index for the Pima Indians dataset, based on the missing not at random nonparametric logistic regression model (6.32). The observed data and Bayes estimates of the missing data are shown with some vertical jittering.

6.8.4 Example: Mental Health Clinical Trial

We now illustrate `rstan` fitting of a semiparametric regression model where the data are subject to measurement error. These data have been analyzed previously by Berry et al. (2002) and are from a 6-week clinical trial of a drug versus a placebo. The response is a physician-assessed score of the patient's mental health, and the predictor is the same score at baseline. A transformed and rescaled form of the data is in the `BCR` data frame of the `HRW` package.

Of central interest is the contrast function

$$c(x) \equiv f_1(x) - f_0(x)$$

where, respectively, f_0 and f_1 are the mean response scores, conditional on the baseline scores, for the placebo and drug populations. To capture c we work with the model

$$y_i \sim \begin{cases} N(f_0(x_i), \sigma_\varepsilon^2) \text{ for the placebo group} \\ N(f_1(x_i), \sigma_\varepsilon^2) \text{ for the drug group} \end{cases}, \quad 1 \leq i \leq n,$$

where (x_i, y_i) is the ith baseline and end of treatment mental health score pair. The measurement error in the y_is is subsumed into the variability about the $f_\ell(x_i)$,

$\ell = 0, 1$. Following Berry et al. (2002) we assume that the x_is are not observed exactly but, instead, are only available via a surrogate w_i such that

$$w_i \mid x_i \sim N(x_i, 0.35).$$

The unobserved x_is are assumed to be Gaussian.

We model the regression curves via mixed model-based penalized splines as follows:

$$f_0(x) = \beta_0 + \beta_1 x + \sum_{k=1}^{K} u_{0k} z_k(x), \quad u_{0k} \overset{\text{ind.}}{\sim} N(0, \sigma_{u0}^2) \quad \text{and}$$

$$f_1(x) = \beta_0 + \beta_0^{\text{drug}} + (\beta_1 + \beta_1^{\text{drug}})\, x + \sum_{k=1}^{K} u_{1k} z_k(x), \quad u_{1k} \overset{\text{ind.}}{\sim} N(0, \sigma_{u1}^2)$$

where the z_k, $1 \le k \le K$, form a suitable spline basis, such as the O'Sullivan spline basis described in Sect. 2.2. The full Bayesian model is

$$\begin{gathered}
\boldsymbol{y} | \boldsymbol{\beta}, \boldsymbol{u}_0, \boldsymbol{u}_1 \sim N(\boldsymbol{X\beta} + \boldsymbol{Z}_0 \boldsymbol{u}_0 + \boldsymbol{Z}_1 \boldsymbol{u}_1, \sigma_\varepsilon^2), \\
\boldsymbol{u}_0 | \sigma_0^2 \sim N(\boldsymbol{0}, \sigma_{u0}^2), \qquad \boldsymbol{u}_1 | \sigma_1^2 \sim N(\boldsymbol{0}, \sigma_{u1}^2), \\
w_i \mid x_i \overset{\text{ind.}}{\sim} N(x_i, 0.35), \quad x_i \overset{\text{ind.}}{\sim} N(\mu_x, \sigma_x^2), \quad \mu_x \sim N(0, \sigma_{\mu_x}^2), \\
\sigma_x \sim \text{Half-Cauchy}(A_x), \quad \boldsymbol{\beta} \sim N(\boldsymbol{0}, \sigma_{\boldsymbol{\beta}}^2 \boldsymbol{I}), \\
\sigma_{u0} \sim \text{Half-Cauchy}(A_u), \quad \sigma_{u1} \sim \text{Half-Cauchy}(A_u), \\
\sigma_\varepsilon \sim \text{Half-Cauchy}(A_\varepsilon).
\end{gathered} \tag{6.33}$$

The σ_{μ_x}, A_x, $\sigma_{\boldsymbol{\beta}}$, A_u, and A_ε are positive-valued hyperparameters. In the analyses described below, each of these are set to 10^5, corresponding to noninformativity. In (6.33), the coefficient vectors are

$$\boldsymbol{\beta} \equiv \begin{bmatrix} \beta_0 \\ \beta_1 \\ \beta_0^{\text{drug}} \\ \beta_1^{\text{drug}} \end{bmatrix}, \quad \boldsymbol{u}_0 \equiv \begin{bmatrix} u_{01} \\ \vdots \\ u_{0K} \end{bmatrix} \quad \text{and} \quad \boldsymbol{u}_1 \equiv \begin{bmatrix} u_{11} \\ \vdots \\ u_{1K} \end{bmatrix}.$$

The design matrices $\boldsymbol{X}$ and $\boldsymbol{Z}_0$ are

$$\boldsymbol{X} \equiv \begin{bmatrix} 1 & x_1 & 1 - I_1 & (1 - I_1)x_1 \\ \vdots & \vdots & \vdots & \vdots \\ 1 & x_n & 1 - I_n & (1 - I_n)x_n \end{bmatrix} \quad \text{and} \quad \boldsymbol{Z}_0 \equiv \begin{bmatrix} I_1\, z_1(x_1) & \cdots & I_1\, z_K(x_1) \\ \vdots & \ddots & \vdots \\ I_n\, z_1(x_n) & \cdots & I_n\, z_K(x_n) \end{bmatrix}$$

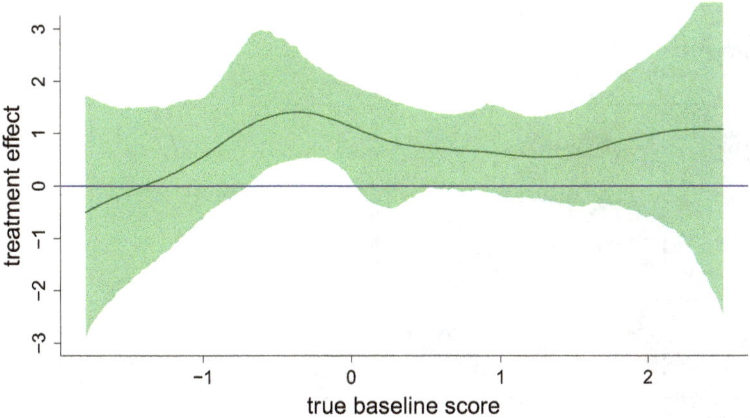

Fig. 6.38 Estimated contrast function and pointwise 90% credible sets, according to the Bayesian model (6.33), for the mental health clinical trial data used in Berry et al. (2002).

where

$$I_i \equiv \begin{cases} 1 \text{ if } (x_i, y_i) \text{ is from the placebo group,} \\ 0 \text{ if } (x_i, y_i) \text{ is from the drug group.} \end{cases}$$

The design matrix $\mathbf{Z}_1$ is defined analogously to $\mathbf{Z}_0$, but with I_i replaced by $1 - I_i$. According to this model, the contrast function is

$$c(x) = \beta_0^{\text{drug}} + \beta_1^{\text{drug}} x + \sum_{k=1}^{K} (u_{1k} - u_{0k})\, z_k(x).$$

The R script BCRana.R fits (6.33) using Stan via the rstan package. The results displayed in Figs. 6.38 and 6.39 are based on a burn-in of size 5000, kept samples of size 5000, and a thinning factor of 5. Figure 6.38 shows the fitted contrast function with 90% pointwise credible sets. Here we use credible sets at the 90% level, rather than at the 95% level, since the former level is used in Fig. 5 of Berry et al. (2002). Inspection of that figure shows good correspondence between the two contrast function estimates, although note that Berry et al. (2002) work with the negative of our contrast function. The drug is seen to exhibit borderline efficacy for higher values of the true baseline score. The script BCRana.R can be run via:

```
> library(HRW) ; demo(BCRana,package = "HRW")
```

and located by running the code:

```
> system.file("demo","BCRana.R",package = "HRW")
```

In Fig. 6.39 we display some MCMC diagnostics for key parameters such as the contrast function estimates at the quartiles of the true baseline scores. The chains are seen to be well behaved.

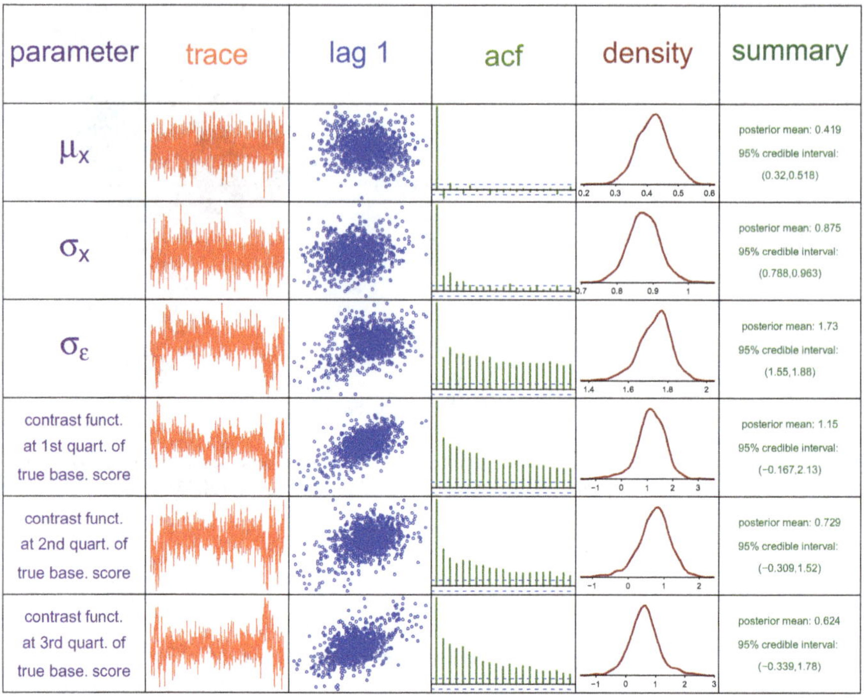

Fig. 6.39 Various plots and summaries for the contrast function MCMC samples at the quartiles of the predictor variable: trace plot, lag-1 plot, estimated autocorrelation function, estimated posterior density function, and numerical summary. The MCMC samples are produced by the R script `BCRana.R`, which uses Stan to fit model (6.33) to the mental health clinical trial data from Berry et al. (2002).

Figure 6.40 summarizes the MCMC output of 5 of the true baseline scores that are observed with measurement error in model (6.33). It provides, for example, a Bayes estimate of 0.95 for the 44th true baseline score, and corresponding 95% credible set of (0.0184, 1.89).

6.8.5 Extension to Finite Mixture Models

Partially observed Gaussian predictor models such as (6.30) and those depicted in Fig. 6.32 can be useful for handling missingness and measurement error, but also suffer from the limitation that the partially observed predictor is assumed to be Gaussian. Carroll et al. (1999) and Richardson et al. (2002) describe extensions of (6.30) in which the partially observed predictor has a finite mixture distribution such as the partially observed predictors x_i, $1 \leq i \leq n$, being independent with density function

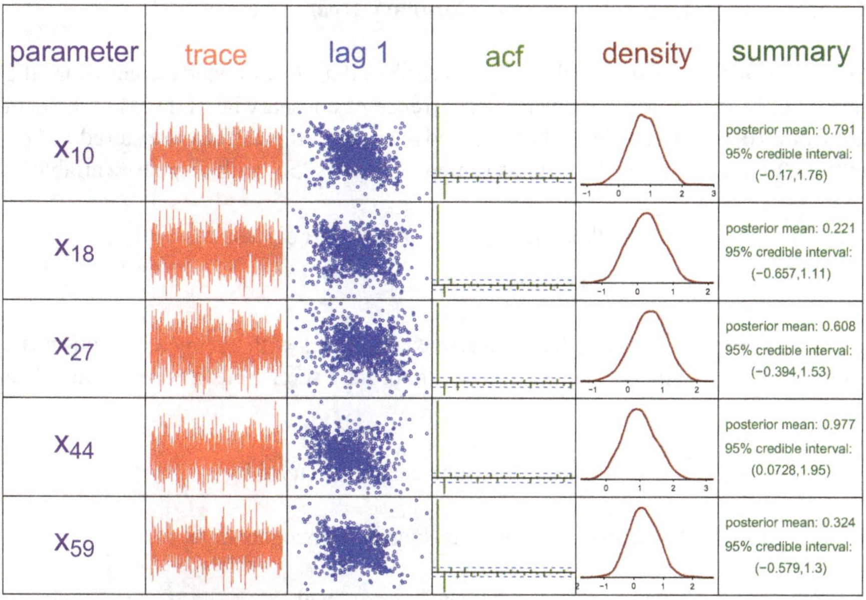

Fig. 6.40 Various plots and summaries for MCMC samples of five true baseline scores that are observed with measurement error according to model (6.33): trace plot, lag-1 plot, estimated autocorrelation function, estimated posterior density function, and numerical summary. The MCMC samples are produced by the R script `BCRana.R`, which uses Stan to fit model (6.33) to the mental health clinical trial data from Berry et al. (2002).

$$p(x_i|\omega_{x1},\ldots,\omega_{xK},\mu_{x1},\ldots,\mu_{xK},\sigma^2_{x1},\ldots,\sigma^2_{xK})$$
$$=\sum_{k=1}^{K}\omega_{xk}\,(2\pi\sigma^2_{xk})^{-1/2}\exp\{-(x_i-\mu_{xk})^2/(2\sigma^2_{xk})\}.$$

Here $(\omega_{xk},\mu_{xk},\sigma^2_{xk})$, $1\le k\le K$, are mixture parameters such that $\sum_{k=1}^{K}\omega_{xk}=1$ and $\sigma^2_{xk}>0$. Another extension covered in Richardson et al. (2002) is the linear regression classical measurement error model

$$w_i\overset{\text{ind.}}{\sim}N(\alpha_0+\alpha_1\,x_i,\sigma^2_w)$$

rather than

$$w_i\overset{\text{ind.}}{\sim}N(x_i,\sigma^2_w)$$

as treated in (6.30). This model can be used because validation data are available. We now provide an illustration of these extensions via an example from a coronary heart disease study.

6.8.5.1 Example: Cholesterol and Coronary Heart Disease

Section 6 of Richardson et al. (2002) describes data from a study concerning the impact of blood cholesterol level on occurrence of coronary heart disease. The data are from Roeder et al. (1996). Fuller details are given there. They are stored in `CHD` in the `HRW` package. The dataset has sample size $n = 256$, with response variable

$$y_i \equiv \begin{cases} 1 & i\text{th subject is a coronary heart disease case,} \\ 0 & \text{otherwise,} \end{cases}$$

$1 \le i \le n$, where a coronary heart disease case equates with having a previous heart attack, a history of *angina pectoris*, or an abnormal exercise electrocardiogram. The predictor of primary interest is

$$x \equiv \text{low density lipoprotein cholesterol level}$$

which is difficult to measure. Its easier-to-measure surrogate is

$$w \equiv \text{total cholesterol level}$$

of which x_i is a component. A secondary predictor is $v \equiv$ age in years. Let (x_i, w_i, v_i), $1 \le i \le 256$, denote the full set of predictor data.

Following Richardson et al. (2002) we formed a dataset for which most of the x_is, 184 out of 256, are not observed but all of the w_is are observed. A validation set, for which both x_i and w_i are observed, was obtained by selecting 32 coronary heart disease cases and 40 controls at random. As in Richardson et al. (2002) and Roeder et al. (1996) the data are analyzed as if they had come from a cohort study.

The model is

$$\begin{gathered} y_i \mid \beta_0, \beta_1, \beta_2, x_i \overset{\text{ind.}}{\sim} \text{Bernoulli}\big(\text{logit}^{-1}(\beta_0 + \beta_1\, x_i + \beta_2\, v_i)\big), \\ x_i \,|\omega, \mu_{x1}, \mu_{x2}, \sigma_{x1}, \sigma_{x2} \overset{\text{ind.}}{\sim} \omega\, N(\mu_{x1}, \sigma_{x1}^2) + (1-\omega)\, N(\mu_{x2}, \sigma_{x2}^2), \\ \omega \sim \text{Uniform}(0,1), \\ \mu_{x1}, \mu_{x2} \overset{\text{ind.}}{\sim} N(0, \sigma_\mu^2), \quad \sigma_{x1}, \sigma_{x2} \overset{\text{ind.}}{\sim} \text{Half-Cauchy}(A_x), \\ w_i \overset{\text{ind.}}{\sim} N(\alpha_0 + \alpha_1\, x_i, \sigma_w^2), \quad \sigma_w \sim \text{Half-Cauchy}(A_w), \\ \beta_0, \beta_1 \overset{\text{ind.}}{\sim} N(0, \sigma_\beta^2), \quad \alpha_0, \alpha_1 \overset{\text{ind.}}{\sim} N(0, \sigma_\alpha^2). \end{gathered} \tag{6.34}$$

Note that the Finite Normal Mixture component of the model

$$x_i \,|\omega, \mu_{x1}, \mu_{x2}, \sigma_{x1}, \sigma_{x2} \overset{\text{ind.}}{\sim} \omega\, N(\mu_{x1}, \sigma_{x1}^2) + (1-\omega)\, N(\mu_{x2}, \sigma_{x2}^2)$$

is equivalent to

$$x_i|\mu_{x1}, \mu_{x2}, \sigma_{x1}, \sigma_{x2}, a_i \sim \begin{cases} N(\mu_{x_1}, \sigma_{x_1}^2), & a_i = 1 \\ N(\mu_{x_2}, \sigma_{x_2}^2), & a_i = 0 \end{cases} \tag{6.35}$$

where $a_i|\omega \sim \text{Bernoulli}(\omega)$, $1 \leq i \leq n$, are auxiliary variables. Representation (6.35) allows easier implementation in a Bayesian inference engines such as BUGS or Stan. Only the former allows for discrete unobserved random variables, so we used it for fitting (6.34) via the script `RLJGana.R`, available in the `HRW` package. This script uses the package `BRugs` (Ligges et al. 2017) to access BUGS from within R. However, not all operating system that support R also support `BRugs`. Stan fitting of (6.34) is possible using `target +=` applied to the logarithm of the Finite Normal Mixture density function. The hyperparameters were set to be $\sigma_{\beta} = A_x = A_w = 10^5$. All predictor data were standardized before being fed into BUGS. The MCMC output was then back-transformed to correspond to the original data. The script `RLJGana.R` can be run and located using:

```
> library(HRW) ; demo(RLJGana,package = "HRW")
> system.file("demo","RLJGana.R",package = "HRW")
```

Figure 6.41 shows the MCMC samples for the regression coefficients in (6.34). To match Richardson et al. (2002) we work with the cholesterol variables divided by 100. Low density lipoprotein cholesterol level is seen to have marginal statistical significance, while age is very significant.

In Fig. 6.42 we plot the Bayes estimate of the low density lipoprotein cholesterol level density function based on the two-component Finite Normal Mixture model. A slight departure from normality is apparent, which is consistent with the analysis in Richardson et al. (2002).

Finally, we note that Richardson et al. (2002) worked with a more elaborate model in which the number of components in the Finite Normal Mixture is a model parameter. R implementation of such models is more challenging since Bayesian inference engines such as BUGS and Stan do not yet have post-development versions that can handle models of this type.

6.9 Arbitrarily Complicated Bayesian Semiparametric Regression

As witnessed in the previous section, the graphical models approach to semiparametric regression in combination with Bayesian inference engines such as Stan is an effective means by which missing data and measurement error can be accounted for in R-based analyses. However, this paradigm is even more powerful and, at least in theory, allows *arbitrarily complicated* Bayesian semiparametric regression models to be entertained. In practice, the viability of effective fitting and inference for particular model is limited by factors such as the versatility and speed of the Bayesian inference engine. Fortunately, there are auspicious developments on the Bayesian inference engine front. For example the Stan engine, and its interface

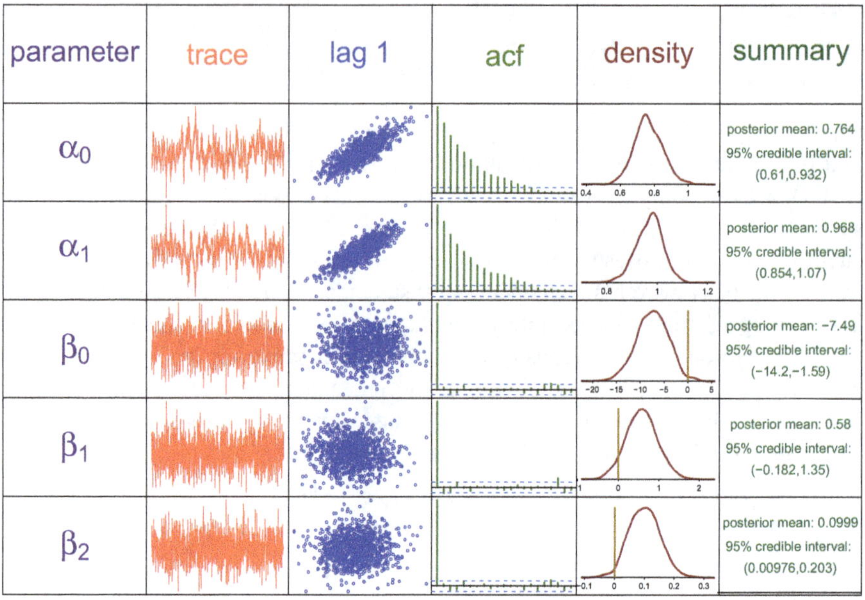

Fig. 6.41 Various plots and summaries for the contrast function MCMC samples at the quartiles of the predictor variable: trace plot, lag-1 plot, estimated autocorrelation function, estimated posterior density function, and numerical summary. The MCMC samples are produced by the R script `RLJGana.R`, which uses Stan to fit model (6.34) to the coronary heart disease data from Roeder et al. (1996).

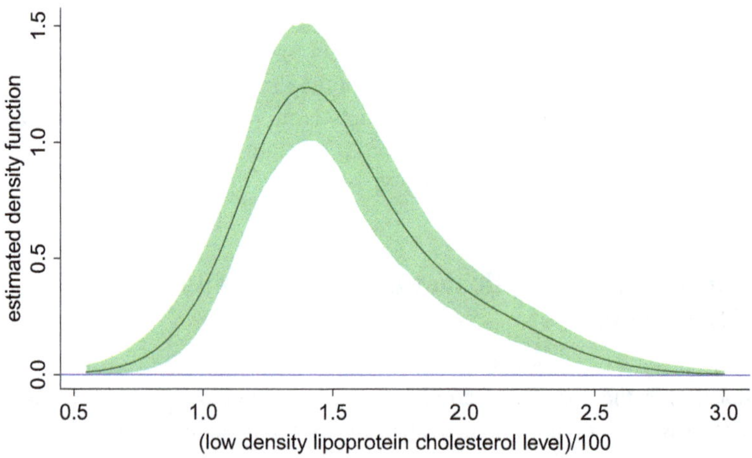

Fig. 6.42 Bayes estimate of the density function of low density lipoprotein cholesterol level based on the MCMC samples produced by the R script `RLJGana.R`, which uses Stan to fit model (6.34) to the coronary heart disease data from Roeder et al. (1996). The shaded region corresponds to pointwise 95% credible sets.

with R via the `rstan` package, emerged just five years before the completion of this book and generally is more flexible and faster than previous such engines. BUGS, Stan, and future Bayesian inference engines are putting more and more complicated semiparametric regression models within reach of R users. In this final section, we provide some illustrations of complicated models that are currently feasible in Stan. However, the underlying principles extend to numerous other models, some of which are explored in the exercises.

6.9.1 Binary Response Group-Specific Curves Model

The data in Fig. 6.43 show the alive/dead status for corals of two different taxa, *Pocillopora* and *Porites*, as a function of initial size at 25 different study sites. The source of the data is Kayal et al. (2015), which contains a fuller description

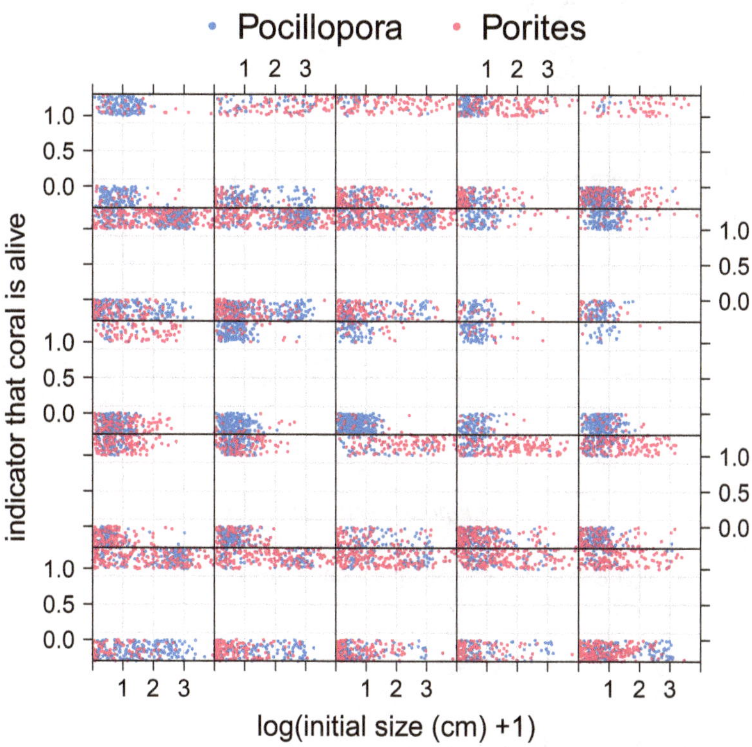

Fig. 6.43 Raw coral data. Alive/dead status versus log(initial size (cm) + 1) for corals of taxa *Pocillipora* and *Porites* at 25 study sites. The ordinates of the data are jittered to aid visualization.

of the study. The data frame `coral` in the `HRW` package contains these data. There is ecological interest in comparing the survival probabilities of the two coral taxa.

An appropriate semiparametric regression model, which benefits from the graphical models approach and Bayesian inference engines such as Stan, is one that caters for group-specific curves when the response is binary. In this case the group corresponds to study site. The Gaussian response case was treated in Sect. 4.3 but in that case the `lme()` function was seen to handle the problem well. The binary response case is more challenging due to the intractability of the likelihood. The R function `glmmPQL()`, from the package `MASS` (Ripley et al. 2015), can accommodate this extension to some extent but the standard error approximations can be quite inaccurate.

The generic form of the binary response group-specific curves model is

$$y_{ij} \sim \text{Bernoulli}(\text{logit}^{-1}(f(x_{ij}) + g_i(x_{ij}))), \quad 1 \le i \le m, \ 1 \le j \le n_i$$

where (x_{ij}, y_{ij}) is the jth predictor/response measurement on the ith group. As in Sect. 4.3, the global mean function is

$$f(x) = \beta_0 + \beta_1\, x + \sum_{k=1}^{K_{\text{gbl}}} u_{\text{gbl},k} z_{\text{gbl},k}(x), \quad u_{\text{gbl},k} | \sigma_{\text{gbl}} \overset{\text{ind.}}{\sim} N(0, \sigma^2_{\text{gbl}}) \tag{6.36}$$

where β_0 and β_1 are fixed effects. The deviation from f for the ith group is the function

$$g_i(x) = U_{0i} + U_{1i}\, x + \sum_{k=1}^{K_{\text{grp}}} u_{\text{grp},ik}\, z_{\text{grp},k}(x),$$
$$\text{where} \begin{bmatrix} U_{0i} \\ U_{1i} \end{bmatrix} \overset{\text{ind.}}{\sim} N(\mathbf{0}, \boldsymbol{\Sigma}) \quad \text{and} \quad u_{\text{grp},ik} | \sigma_{\text{grp}} \overset{\text{ind.}}{\sim} N(0, \sigma^2_{\text{grp}}).$$

The coral data can be divided into two categories:

$$A \equiv \textit{Pocillopora} \quad \text{and} \quad B \equiv \textit{Porites},$$

so for contrasting the two types we should extend (6.36) to

$$\left.\begin{aligned} f^A(x) &= \beta_0^{\text{A}} + \beta_1^{\text{A}}\, x + \sum_{k=1}^{K_{\text{gbl}}} u^A_{\text{gbl},k} z_{\text{gbl},k}(x) \\ g_i^A(x) &= U_{0i}^{\text{A}} + U_{1i}^{\text{A}}\, x + \sum_{k=1}^{K_{\text{grp}}} u^A_{\text{grp},ik} z_{\text{grp},k}(x) \end{aligned}\right\} \text{for type A}$$

and

$$\left.\begin{aligned} f^B(x) &= \beta_0^{\mathrm{A}} + \beta_0^{\mathrm{BvsA}} + (\beta_1^{\mathrm{A}} + \beta_1^{\mathrm{BvsA}})\,x + \sum_{k=1}^{K_{\mathrm{gbl}}} u^B_{\mathrm{gbl},k} z_{\mathrm{gbl},k}(x) \\ g_i^B(x) &= U_{0i}^{\mathrm{B}} + U_{1i}^{\mathrm{B}}\,x + \sum_{k=1}^{K_{\mathrm{grp}}} u^B_{\mathrm{grp},ik} z_{\mathrm{grp},k}(x) \end{aligned}\right\} \text{ for type B.}$$

This allows us to estimate the contrast function

$$c(x) \equiv f_B(x) - f_A(x) = \beta_0^{\mathrm{BvsA}} + \beta_1^{\mathrm{BvsA}}\,x + \sum_{k=1}^{K_{\mathrm{gbl}}} (u^B_{\mathrm{gbl},k} - u^A_{\mathrm{gbl},k}) z_{\mathrm{gbl},k}(x). \tag{6.37}$$

In this binary response situation with a logit link c corresponds to the *log odds ratio* of B compared with A. Define

$$I_{ij}^A \equiv \begin{cases} 1 \text{ if } (x_{ij}, y_{ij}) \text{ is of type } A, \\ 0 \text{ if } (x_{ij}, y_{ij}) \text{ is of type } B. \end{cases}$$

The full Bayesian model is

$$\begin{aligned} &\boldsymbol{y}|\,\boldsymbol{\beta}, \boldsymbol{u}^A_{\mathrm{gbl}}, \boldsymbol{u}^B_{\mathrm{gbl}}, \boldsymbol{U}, \boldsymbol{u}_{\mathrm{grp}} \sim \text{Bernoulli}(\text{logit}^{-1}(\boldsymbol{X\beta} + \boldsymbol{Z}^A_{\mathrm{gbl}}\boldsymbol{u}^A_{\mathrm{gbl}} \\ &\qquad + \boldsymbol{Z}^B_{\mathrm{gbl}}\boldsymbol{u}^B_{\mathrm{gbl}} + \boldsymbol{Z}_U\boldsymbol{U} + \boldsymbol{Z}^A_{\mathrm{grp}}\boldsymbol{u}^A_{\mathrm{grp}} + \boldsymbol{Z}^B_{\mathrm{grp}}\boldsymbol{u}^B_{\mathrm{grp}})), \\ &\boldsymbol{u}^A_{\mathrm{gbl}}|\,\sigma^A_{\mathrm{gbl}} \sim N(\boldsymbol{0}, (\sigma^A_{\mathrm{gbl}})^2), \qquad \boldsymbol{u}^B_{\mathrm{gbl}}|\,\sigma^B_{\mathrm{gbl}} \sim N(\boldsymbol{0}, (\sigma^B_{\mathrm{gbl}})^2), \\ &\boldsymbol{U}|\,\boldsymbol{\Sigma} \sim N(\boldsymbol{0}, \boldsymbol{I}_m \otimes \boldsymbol{\Sigma}), \quad \boldsymbol{\beta} \sim N(\boldsymbol{0}, \sigma^2_{\boldsymbol{\beta}}\boldsymbol{I}), \\ &\boldsymbol{u}^A_{\mathrm{grp}}|\,\sigma_{\mathrm{grp}} \sim N(\boldsymbol{0}, \sigma^2_{\mathrm{grp}}), \qquad \boldsymbol{u}^B_{\mathrm{grp}}|\,\sigma_{\mathrm{grp}} \sim N(\boldsymbol{0}, \sigma^2_{\mathrm{grp}}) \\ &\sigma^A_{\mathrm{gbl}} \sim \text{Half-Cauchy}(A_{\mathrm{gbl}}), \quad \sigma^B_{\mathrm{gbl}} \sim \text{Half-Cauchy}(A_{\mathrm{gbl}}), \\ &\sigma_{\mathrm{grp}} \sim \text{Half-Cauchy}(A_{\mathrm{grp}}), \quad \boldsymbol{\Sigma} \text{ is } 4 \times 4 \text{ with} \\ &\boldsymbol{\Sigma}|\,a_1, a_2, a_3, a_4 \sim \text{Inverse-Wishart}\big(5, 4\,\text{diag}(a_1, a_2, a_3, a_4)^{-1}\big), \\ &a_1, a_2, a_3, a_4 \overset{\text{ind.}}{\sim} \text{Inverse-Gamma}(\tfrac{1}{2}, 1/A_U^2) \end{aligned} \tag{6.38}$$

where the coefficient vectors are

$$\boldsymbol{\beta} \equiv \begin{bmatrix} \beta_0^A \\ \beta_1^A \\ \beta_0^{\text{BvsA}} \\ \beta_1^{\text{BvsA}} \end{bmatrix}, \quad \boldsymbol{u}_{\text{gbl}}^A \equiv \begin{bmatrix} u_{\text{gbl},1}^A \\ \vdots \\ u_{\text{gbl},K_{\text{gbl}}}^A \end{bmatrix}, \quad \boldsymbol{U} \equiv \begin{bmatrix} U_{01}^A \\ U_{11}^A \\ U_{01}^B \\ U_{11}^B \\ \vdots \\ U_{0m}^A \\ U_{1m}^A \\ U_{0m}^B \\ U_{1m}^B \end{bmatrix} \quad \text{and} \quad \boldsymbol{u}_{\text{grp}}^A \equiv \begin{bmatrix} u_{\text{grp},11}^{\text{A}} \\ \vdots \\ u_{\text{grp},1K_{\text{grp}}}^{\text{A}} \\ \vdots \\ u_{\text{grp},m1}^{\text{A}} \\ \vdots \\ u_{\text{grp},mK_{\text{grp}}}^{\text{A}} \end{bmatrix}$$

with $\boldsymbol{u}_{\text{gbl}}^B$ defined analogously to $\boldsymbol{u}_{\text{gbl}}^A$ and $\boldsymbol{u}_{\text{grp}}^B$ defined analogously to $\boldsymbol{u}_{\text{grp}}^A$. The design matrices $\boldsymbol{X}$, $\boldsymbol{Z}_{\text{gbl}}^A$ and $\boldsymbol{Z}_U$ are

$$\boldsymbol{X} \equiv \begin{bmatrix} \mathbf{1} & \boldsymbol{x}_1 & \mathbf{1} - \boldsymbol{I}_1^A & (\mathbf{1} - \boldsymbol{I}_1^A) \odot \boldsymbol{x}_1 \\ \vdots & \vdots & \vdots & \vdots \\ \mathbf{1} & \boldsymbol{x}_m & \mathbf{1} - \boldsymbol{I}_m^A & (\mathbf{1} - \boldsymbol{I}_m^A) \odot \boldsymbol{x}_m \end{bmatrix},$$

$$\boldsymbol{Z}_{\text{gbl}}^A \equiv \begin{bmatrix} \boldsymbol{I}_1^A \odot z_{\text{gbl},1}(\boldsymbol{x}_1) & \cdots & \boldsymbol{I}_1^A \odot z_{\text{gbl},K_{\text{gbl}}}(\boldsymbol{x}_1) \\ \vdots & \ddots & \vdots \\ \boldsymbol{I}_m^A \odot z_{\text{gbl},1}(\boldsymbol{x}_m) & \cdots & \boldsymbol{I}_m^A \odot z_{\text{gbl},K_{\text{gbl}}}(\boldsymbol{x}_m) \end{bmatrix}$$

$$\text{and} \quad \boldsymbol{Z}_U \equiv \underset{1 \le i \le m}{\text{blockdiag}} \left[\boldsymbol{I}_i^A \;\; \boldsymbol{I}_i^A \odot \boldsymbol{x}_i \;\; (\mathbf{1} - \boldsymbol{I}_i^A) \;\; (\mathbf{1} - \boldsymbol{I}_i^A) \odot \boldsymbol{x}_i \right]$$

with $\boldsymbol{x}_i$ equaling the $n_i \times 1$ vector containing the x_{ij}, $1 \le j \le n_i$, and $\boldsymbol{I}_i^A$ equaling the $n_i \times 1$ vector containing the I_{ij}^A, $1 \le j \le n_i$. The matrix $\boldsymbol{Z}_{\text{gbl}}^B$ is defined in a similar manner to $\boldsymbol{Z}_{\text{gbl}}^A$, but with $\boldsymbol{I}_i^A$ replaced by $\mathbf{1} - \boldsymbol{I}_i^A$. The design matrices $\boldsymbol{Z}_{\text{grp}}^A$ and $\boldsymbol{Z}_{\text{grp}}^B$ have block diagonal structure similar to $\boldsymbol{Z}_U$ with blocks analogous to $\boldsymbol{Z}_{\text{gbl}}^A$ and $\boldsymbol{Z}_{\text{gbl}}^B$ but there is allowance for a different, typically smaller, spline basis of size K_{grp}. The prior on $\boldsymbol{\Sigma}$, in terms of the auxiliary vector $\boldsymbol{a} \equiv (a_1, a_2, a_3, a_4)$, has entries that are marginally noninformative, as explained in Huang and Wand (2013). A directed acyclic graph representation of (6.38) is given in Fig. 6.44.

We now describe R fitting of (6.38) with

$$x_{ij} = \log(\text{initial size (cm)} + 1), \qquad y_{ij} = \begin{cases} 0 & \text{if coral dead,} \\ 1 & \text{if coral alive,} \end{cases}$$

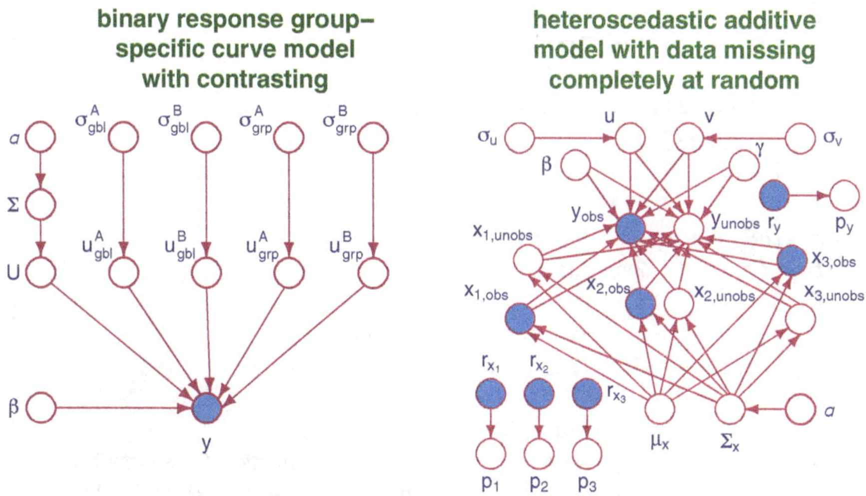

Fig. 6.44 Left panel: Directed acyclic graph corresponding to the binary response group-specific curves model with contrasting, given by (6.38). Right panel: Directed acyclic graph corresponding to the heteroscedastic additive with missing predictor data model (6.41).

and the I_{ij}^A being indicators of the coral have taxon *Pocillopora*. For these variables i ranges over $\{1, \ldots, 25\}$ corresponding to the sites and j indexes the measurement within the ith site. The R script `coralAna.R` fits (6.38) to the coral data of Fig. 6.43 using the `rstan` package. The main part of the Stan code is:

```
   vector[numObs] fmean;        vector[numObs] fullMean;
   fmean = (Xbase*beta + XB*betaBvsA + ZA*uGblA + ZB*uGblB);
   for (iAll in 1:numObs)
      fullMean[iAll] = (fmean[iAll]
             + U[idnum[iAll],3]*XA[iAll,1]
             + U[idnum[iAll],4]*XA[iAll,2]
             + U[idnum[iAll],1]*XB[iAll,1]
             + U[idnum[iAll],2]*XB[iAll,2]
             + dot_product(uGrpA[idnum[iAll]],ZgrpA[iAll])
             + dot_product(uGrpB[idnum[iAll]],ZgrpB[iAll]));
}
model
{
   matrix[2*ncXbase,2*ncXbase] scaleSigmaGrp;
   y ~ bernoulli_logit(fullMean);
   for (i in 1:numGrp)
     U[i] ~ multi_normal(zeroVec,SigmaGrp);
   uGblA ~ normal(0,siguGblA);  uGblB ~ normal(0,siguGblB);
   for (i in 1:numGrp)
```

```
{
   for (k in 1:ncZgrp)
   {
      uGrpA[i,k] ~ normal(0,siguGrp);
      uGrpB[i,k] ~ normal(0,siguGrp);
   }
}
scaleSigmaGrp = rep_matrix(0,4,4);
for (k in 1:(2*ncXbase))
{
   a[k] ~ inv_gamma(0.5,pow(AU,-2));
   scaleSigmaGrp[k,k] = 4/a[k];
}
SigmaGrp ~ inv_wishart(5,scaleSigmaGrp);
beta ~ normal(0,sigmaBeta); betaBvsA ~ normal(0,sigmaBeta);
siguGblA ~ cauchy(0,AuGbl); siguGblB ~ cauchy(0,AuGbl);
siguGrp ~ cauchy(0,AuGrp);
```

The fitted group-specific curves are shown in Fig. 6.45. Specifically, we plot pointwise posterior means and 95% credible sets. The fitted curves are seen to vary quite markedly between sites.

Figure 6.46 shows the estimated contrast function defined by (6.37), corresponding to the log odds ratio of *Porites* as a function of log(initial size + 1), compared with coral of taxon *Pocillopora*. The pointwise 95% credible sets, shown by the shaded region, show that there is a significant difference between the two taxa at any one initial size.

Finally, we show that the convergence of MCMC is excellent for the main quantities of interest in Stan fitting of (6.38). Figure 6.47 shows various aspects of the MCMC samples for the contrast function at the quartiles of the predictor.

To run `coralAna.R` issue the commands:

```
> library(HRW) ; demo(coralAna,package = "HRW")
```

To locate, and possibly copy and modify, `coralAna.R` use:

```
> system.file("demo","coralAna.R",package = "HRW")
```

6.9.2 Heteroscedastic Additive Model with Missingness

The data shown in Fig. 6.48 is part of the data frame `Ozone` in the R package `mlbench` (source: Breiman and Friedman 1985). Subsets of these data have been analyzed previously using ordinary Gaussian additive models (e.g. Hastie and Tibshirani 2000). A version of these data appears in Exercise 3 of Chap. 3. However,

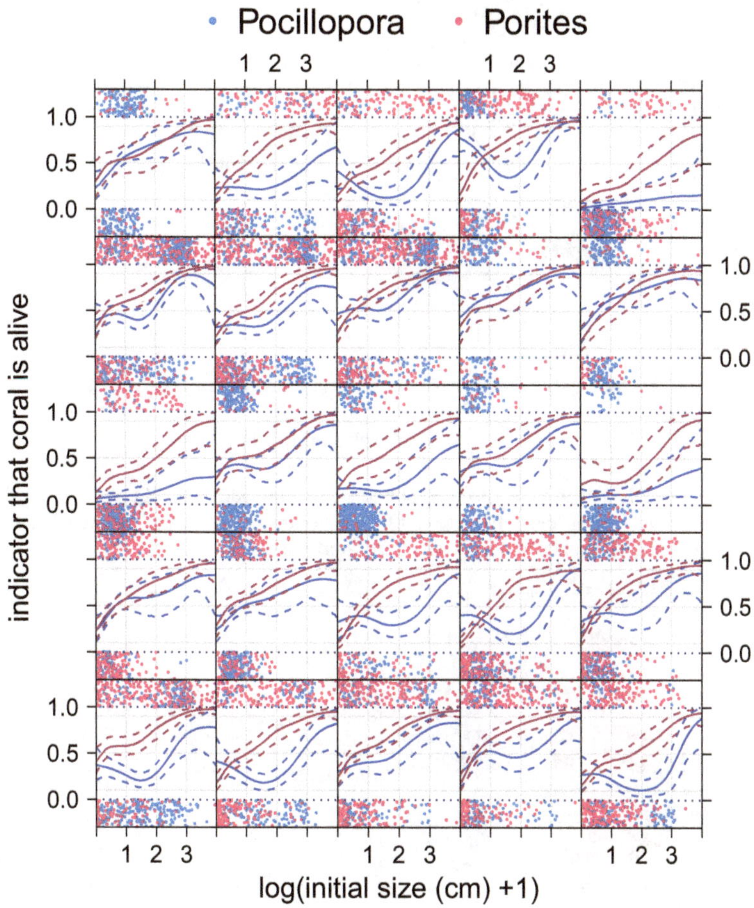

Fig. 6.45 Fitted group-specific probability function estimates, based on model (6.38), for the two coral taxa *Pocillopora* and *Porites* as a function of log(initial size (cm) + 1). The dashed curves represent pointwise 95% credible sets. Each panel corresponds to a different study site.

- there is pronounced heteroscedasticity in the data, implying that the constant variance assumption is questionable;
- as shown in Table 6.3, there is a substantial amount of missingness in the data. Most analyses ignore this aspect and perform a complete case analysis—which involves omitting each record of the data frame containing at least one missing value. For the data in Fig. 6.48 this involves omission of $151/366 \approx 41\%$ of the records.

We now describe the setting up a R-based fitting of a *heteroscedastic* additive model that includes *all of the observed data.*

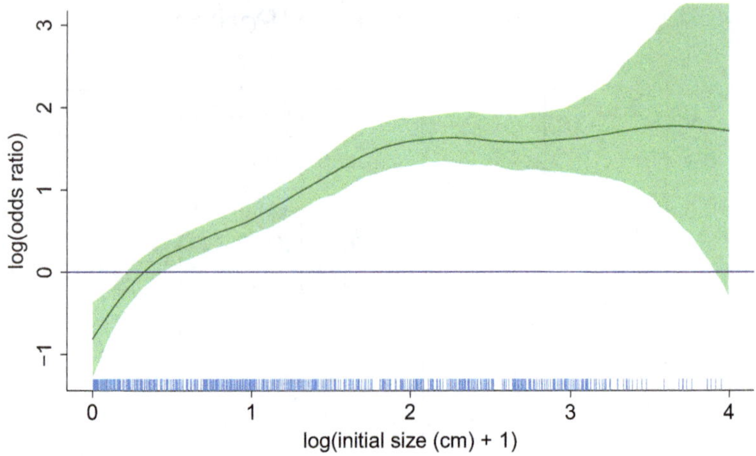

Fig. 6.46 Estimated contrast function corresponding to the log odds ratio of coral of taxon *Porites* being alive as a function of log(initial size (cm) + 1), compared with coral of taxon *Pocillopora*. The shaded region corresponds to pointwise 95% credible sets.

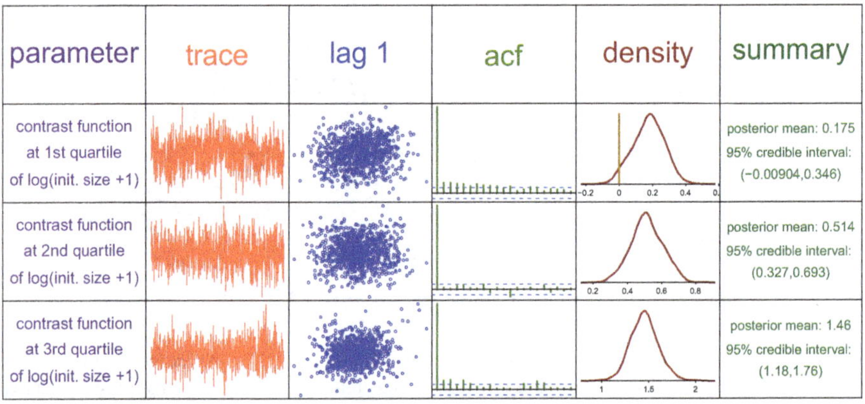

Fig. 6.47 Various plots and summaries for the contrast function MCMC samples at the quartiles of the predictor variable: trace plot, lag-1 plot, estimated autocorrelation function, estimated posterior density function, and numerical summary. The MCMC samples are produced by the R script `coralAna.R`, which uses Stan to fit model (6.38) to the coral data from Kayal et al. (2015).

A *heteroscedastic additive model* (e.g. Rigby and Stasinopoulos 2005) with three predictors is

$$y_i \sim N\Big(\beta_0 + f_1(x_{1i}) + f_2(x_{2i}) + f_3(x_{3i}),\\ \exp\big(\gamma_0 + h_1(x_{1i}) + h_2(x_{2i}) + h_3(x_{3i})\big)\Big), \quad 1 \le i \le n, \tag{6.39}$$

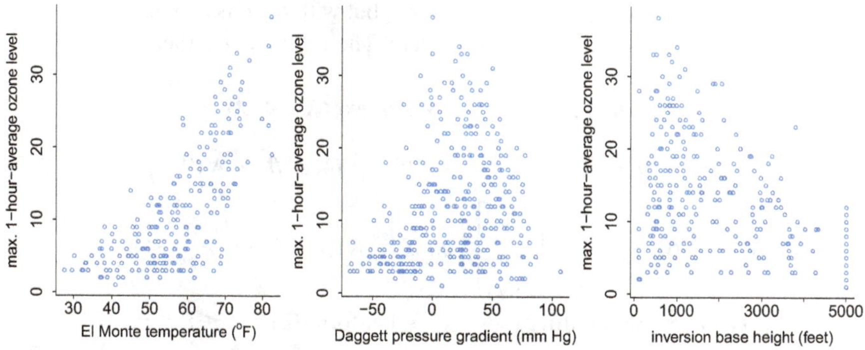

Fig. 6.48 Marginal effects scatterplots for the California ozone data example.

Table 6.3 Missingness percentages for the variables shown in Fig. 6.48, which are part of the `Ozone` data frame in the `R` package `mlbench`.

Variable	Missingness
Max. 1-hour-average ozone level	1.36%
El Monte temperature	37.95%
Daggett pressure gradient	0.27%
Inversion base height	4.10%

where f_j and h_j, $j = 1, 2, 3$, are smooth but otherwise arbitrary functions. Mixed model-based penalized spline models for these functions are

$$\begin{aligned} f_j(x) &= \beta_j\, x + \sum_{k=1}^{K_j} u_{jk}\, z_{jk}(x), \ u_{jk} \stackrel{\text{ind.}}{\sim} N(0, \sigma_{uj}^2) \\ \text{and} \quad h_j(x) &= \gamma_j\, x + \sum_{k=1}^{K_j} v_{jk}\, z_{jk}(x), \ v_{jk} \stackrel{\text{ind.}}{\sim} N(0, \sigma_{vj}^2). \end{aligned} \tag{6.40}$$

The $\{z_{jk} : 1 \le k \le K_j\}$, $j = 1, 2, 3$, are O'Sullivan cubic spline bases of sizes K_j respectively (Sect. 2.2).

Our aim is to fit the model given by (6.39) and (6.40) to the data depicted in Fig. 6.48 with

$$y_i = \text{maximum 1-hour average ozone level (parts per million) at Sandburg Air Force Base}$$

and the three predictors' variables

$$\begin{aligned} x_{1i} &= \text{temperature (degree Fahrenheit) in El Monte, California, USA} \\ x_{2i} &= \text{pressure gradient (millimeters of mercury) from Los Angeles International Airport to Daggett, California, USA} \\ \text{and} \quad x_{3i} &= \text{inversion base height (feet),} \end{aligned}$$

each on the ith day, $1 \le i \le n$, where $n = 366$, but with the missingness completely at random assumption for the unobserved data. The full model is then

$$
\begin{gathered}
\boldsymbol{y} \,|\, \boldsymbol{\beta}, \boldsymbol{u}, \boldsymbol{\gamma}, \boldsymbol{v} \sim N\big(\boldsymbol{X\beta} + \boldsymbol{Zu}, \exp(\boldsymbol{X\gamma} + \boldsymbol{Zv})\big),\\
\boldsymbol{u} \,|\, \sigma_u \sim N(\boldsymbol{0}, \sigma_u^2 \boldsymbol{I}), \quad \boldsymbol{v} \,|\, \sigma_v \sim N(\boldsymbol{0}, \sigma_v^2 \boldsymbol{I}), \quad \boldsymbol{\beta} \sim N(0, \sigma_{\boldsymbol{\beta}}^2 \boldsymbol{I}),\\
\boldsymbol{\gamma} \sim N(0, \sigma_{\boldsymbol{\gamma}}^2 \boldsymbol{I}), \quad \left.\begin{bmatrix} x_{1i} \\ x_{2i} \\ x_{3i} \end{bmatrix} \right| \boldsymbol{\mu}, \boldsymbol{\Sigma} \stackrel{\text{ind.}}{\sim} N(\boldsymbol{\mu}_x, \boldsymbol{\Sigma}_x), \ 1 \le i \le n,\\
r_{y_i} \,|\, p_y \stackrel{\text{ind.}}{\sim} \text{Bernoulli}(p_y), \quad p_y \sim \text{Uniform}(0,1), \ 1 \le i \le n,\\
r_{x_{ji}} \,|\, p_j \stackrel{\text{ind.}}{\sim} \text{Bernoulli}(p_j), \ p_{x_j} \sim \text{Uniform}(0,1), \ 1 \le i \le n, \ 1 \le j \le 3,\\
\sigma_{uj} \stackrel{\text{ind.}}{\sim} \text{Half-Cauchy}(A_u), \ \sigma_{vj} \stackrel{\text{ind.}}{\sim} \text{Half-Cauchy}(A_v), \ 1 \le j \le 3,\\
\boldsymbol{\mu}_x \sim N(\boldsymbol{0}, \sigma_x^2 \boldsymbol{I}), \quad \boldsymbol{\Sigma}_x \,|\, \boldsymbol{a} \sim \text{Inverse-Wishart}(4, 4\,\text{diag}(a_1, a_2, a_3)^{-1}),\\
\boldsymbol{a} \equiv (a_1, a_2, a_3), \quad a_1, a_2, a_3 \stackrel{\text{ind.}}{\sim} \text{Inverse-Gamma}(\tfrac{1}{2}, 1/A_x^2)
\end{gathered}
\tag{6.41}
$$

where

$$
r_{y_i} \equiv \begin{cases} 1 \text{ if } y_i \text{ is observed,} \\ 0 \text{ if } y_i \text{ is unobserved} \end{cases} \quad \text{and} \quad r_{x_{ji}} \equiv \begin{cases} 1 \text{ if } x_{ji} \text{ is observed,} \\ 0 \text{ if } x_{ji} \text{ is unobserved} \end{cases}
$$

for $1 \le i \le 366$, $1 \le j \le 3$.

The corresponding directed acyclic graph is the right panel of Fig. 6.44. Note that this figure uses vector notation such as $\boldsymbol{\sigma}_u \equiv (\sigma_{u1}, \sigma_{u2}, \sigma_{u3})$.

The R script `ozoneAna.R` in the `HRW` package facilitates fitting of (6.41) to the ozone data with the Stan inference engine. Figure 6.49 shows the resulting estimated mean and standard deviation functions with pointwise 95% credible sets. The first two predictors, El Monte temperature and Daggett pressure gradient, are shown to contribute nonlinear components to the overall standard deviation.

Figure 6.50 summarizes the MCMC output for the first unobserved response and predictor variables. It shows, for example, that the first unobserved El Monte temperature value has a 95% credible set of (46.1, 82.9) degree Fahrenheit.

To run and locate `ozoneAna.R` issue the commands:

```
> library(HRW) ; demo(ozoneAna,package = "HRW")
> system.file("demo","ozoneAna.R",package = "HRW")
```

6.9.3 *Practical Aspects of Graphical Models Approach to Bayesian Semiparametric Regression*

This final section has explored the notion that, with a graphical models approach to Bayesian semiparametric regression and a Bayesian inference engine accessible

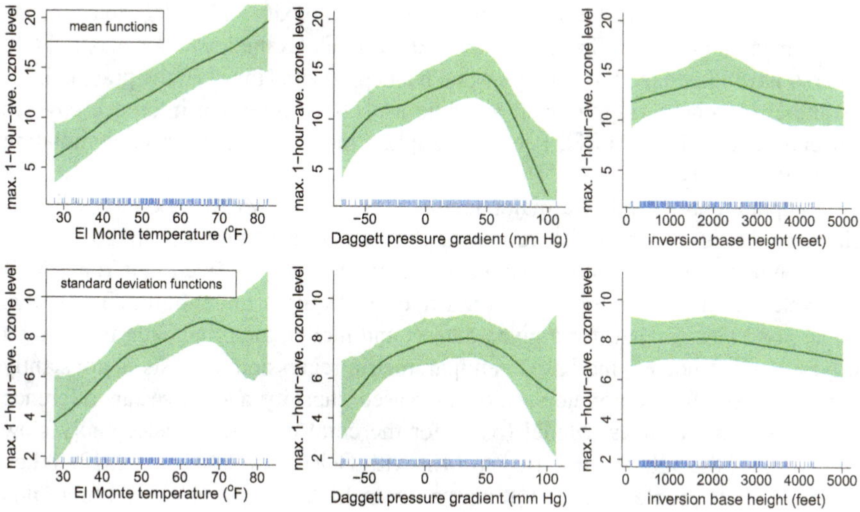

Fig. 6.49 Mean and standard deviation function estimates for the Bayesian heteroscedastic additive model with data missing completely at random model (6.41), applied to the data frame `Ozone` in the R package `mlbench`. The fits for each predictor are slices with the other two predictors set to their average values. The shaded regions correspond to pointwise 95% credible sets.

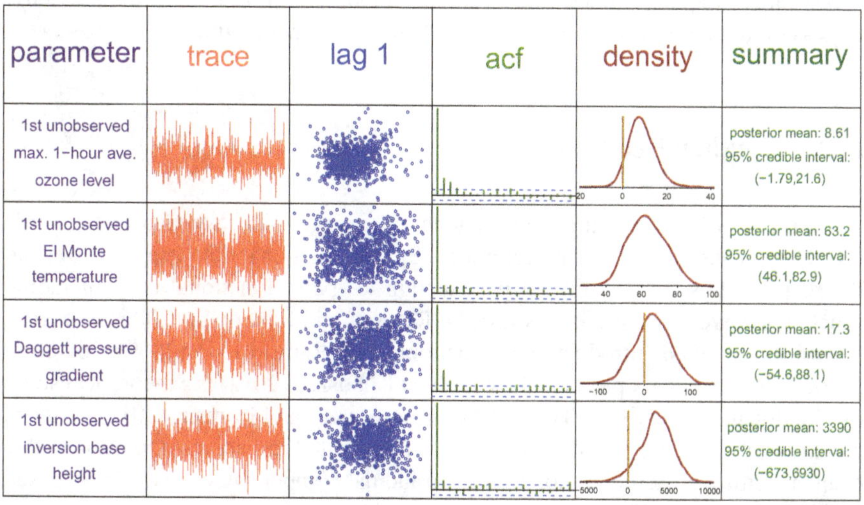

Fig. 6.50 Trace, lag-1 plots, estimated autocorrelation function, estimated posterior density function, and numerical summaries for the first unobserved response and predictors for the Stan fit of model (6.41) to the ozone data.

from R, the sky is the limit in terms of the complexity of models that can be entertained and analyses that can be carried out in R. Examples in Sects. 6.2.2, 6.8, and 6.9 have illustrated the benefits of such an approach. However, the practicalities of late-2010s computing, as well as some aspects of Bayesian inference engines currently accessible from R, need to be taken into account when contemplating analyses of this type.

Computational speed is a major factor. Some of the MCMC examples in this chapter take several hours to run on a common office computer in 2018. This makes model criticism and tweaking difficult. A series of overnight runs or, if available, the use of multiple machines may be required for a thorough analysis. Some good news is that, for a given dataset and model, the time taken for doing a Bayesian inference engine-based semiparametric regression analysis is constantly improving as office computers enjoy increased capacity and Bayesian inference engine software evolves. Model (6.38) for the coral data was unacceptably slow when one of the authors first tried to fit it with BUGS via `BRugs` in 2013. When he revisited the problem with Stan via `rstan` in 2016, the computational times had become reasonable and led to the analysis reported here. If you are reading this book during the year 2030, or later, then it may be that computing speed is no longer an issue for this example.

A last piece of advice is to regularly check the BUGS and Stan development websites for updates and to work with latest versions. This is particularly the case for Stan which, at the time of writing, is still in its early years.

Despite these downsides, the graphical models approach to semiparametric regression in R is a very advantageous and auspicious paradigm.

6.10 Further Reading

There is a very large literature on functional linear regression. See Cardot and Sarda (2011) for a survey and further reference. Sparse functional data is a topic beyond the scope of this chapter. For an introduction to sparse functional data, including functional regression, see James (2011). Ramsay and Silverman (2006) is a well-known and excellent introduction to functional data. Ramsay et al. (2009) provide an introduction to the `fda` package. Horváth and Kokoszka (2012) is a theoretical study of functional data analysis. The handbook by Ferraty and Romain (2011) contains recent survey articles, two of which have just been cited and are most relevant to this chapter. Fitting an additive model to the principal component scores was proposed by Müller and Yao (2008) and called the functional additive model. The functional generalized additive model was proposed by McLeanet al. (2013) and, under the name "continuous additive model" by Müller et al. (2013). Nonadditive models can be fit by a generalization of kernel regression to functional data; see Ferraty and Vieu (2006).

Survival analysis is a major topic having some overlap with semiparametric regression, but not covered in this book. The R package `survival` (Therneau and

Lumley 2015) supports semiparametric Cox proportional hazard regression models via the function `pspline()`. Similar comments apply to the function `hare()` in the package `polspline` (Kooperberg 2015). Other pointers to R software for semiparametric survival analysis models are provided in the *Comprehensive R Archive Network Task View: Survival Analysis*. At the time of this writing the relevant website is `cran.r-project/web/views/Survival.html`.

Time series analysis also has significant overlap with semiparametric regression, although the support within R is not very strong at the time of writing. The most relevant package is `tsDyn` (Di Narzo and Stigler 2016) which has additive autoregressive models as one capability. The function `stl()` uses nonparametric regression methodology to decompose a time series into seasonal, trend, and irregular components.

Wavelet regression is another major topic having strong connections with semiparametric regression. Nason (2008) is the definitive book on wavelet regression via R. As explained in Wand and Ormerod (2011), wavelets can be used instead of splines in semiparametric regression to handle jagged and jumpy effects. The R function `ZDaub()`, available within supplemental materials on the *Electronic Journal of Statistics* web-page for Wand and Ormerod (2011) is the wavelet equivalent of `ZOSull()` in the `HRW` package. Apart from the predictor data, its main argument is `numLevels` which is an integer L between 2 and 10 which leads to a wavelet basis of size $2^L - 1$. The underlying wavelet calculations are carried out in `wavethresh` (Nason 2016). The appropriate penalization for wavelets is different, with ℓ_1-type penalties being more suitable and is facilitated by R packages such as `lars` (Hastie and Efron 2013). This basis requires only a constant term as its unpenalized companion, as opposed to cubic O'Sullivan splines which have a linear companion. Apart from these differences, every model treated in this book is such that wavelets can be used instead of splines if it is believed that certain effects are jagged and/or jumpy (Wand and Ormerod 2011).

6.11 Exercises

1. Obtain quantile regression plots analogous to those shown in Fig. 6.10 for price (złoty) per square meter as a function of construction date for the Warsaw apartment data in the data frame `WarsawApts` of the `HRW` package. As in Fig. 6.10, the quantile function estimates can be obtained individually via the `rqss()` function in the package `quantreg` or using the LMS method via the functions `vgam()` and `lms.bcn()` in the package `VGAM`. Use four degrees of freedom for the λ and σ function fits and ten degrees of freedom for the μ function fit. For stability of `lms.bcn()`, the predictor data should be pre-transformed to the interval $[0, 1]$ and the response data be pre-transformed to the interval $[0.01, 1]$, since `lms.bcn()` requires strictly positive response data. The fitted quantile functions should then be transformed to the original

units. The scripts `MichIncMultQSS.R` and `MichIncLMS.R` in the `HRW` package, corresponding to Fig. 6.10, provide further guidance.

2. As explained in Sect. 6.2.2, the t distribution provides a mechanism by which semiparametric regression models can be made robust to response data outliers. A related issue concerns whether such an approach is resistant to gross outliers in the response data, and is explored in this exercise.

   a. Write an R script that simulates data according to the model

   $$y_i \,|\, x_i \overset{\text{ind.}}{\sim} t\big(f(x_i), 0.35, 1.5\big), \quad x_i \overset{\text{ind.}}{\sim} \text{Uniform}(0,1), \quad 1 \le i \le n$$

   with $f(x) \equiv \sin(4\pi\, x)$ and $n = 300$ and then fit the Bayesian t response penalized spline model given by the $d = 1$ special case of (6.2) using Stan, via the package `rstan`. The script should compare the fitted function with the true mean function $f(x)$.
   b. Select a y_i at random and replace its value by 10,000. Rerun the script and assess the effect of this gross outlier on the penalized spline fit.
   c. Repeat part b. with a single gross outlier set to 100,000 and then set to 1,000,000. Comment on the extent to which the Bayesian t response penalized spline model is resistant to gross outliers.

3. Section 6.2.2 contains a demonstration of use of the function `vgam()`, within the package `VGAM`, for fitting t response additive models of the form

$$y_i \overset{\text{ind.}}{\sim} t\Big(\sum_{j=1}^{d} f_j(x_{ji}), \sigma_\varepsilon, \nu\Big), \quad 1 \le i \le n,$$

where we are using the t distribution notation defined just after (6.1). The R script `MichInctAddMod.R` in the `HRW` package contains code for this analysis. Commands for running and locating `MichInctAddMod.R` are:

```
> library(HRW) ; demo(MichInctAddMod,package = "HRW")
> system.file("demo","MichInctAddMod.R",package = "HRW")
```

The $d = 1$ special case is

$$y_i \overset{\text{ind.}}{\sim} t\Big(f(x_i), \sigma_\varepsilon, \nu\Big), \quad 1 \le i \le n, \tag{6.42}$$

based on the sample of predictor/response pairs (x_i, y_i), $1 \le i \le n$. Consider the following extension of (6.42):

$$y_i \overset{\text{ind.}}{\sim} t\Big(f(x_i), g(x_i), h(x_i)\Big), \quad 1 \le i \le n, \tag{6.43}$$

where both g and h are smooth positive-valued functions. Use the `vgam()` and `studentt3()` functions in `VGAM` to fit (6.43) to predictor/response data corresponding to x and y generated by:

```
> library(Ecdat)         ; data(Workinghours)
> x <- Workinghours$age ; y <- Workinghours$income/10
```

4. In Sect. 6.3 an additive model was fit to the principal component scores computed from the second derivatives of absorbances in the Tecator dataset. Repeat this analysis using the first derivative.
    a. If you use the first k scores, which value of k gives the smallest AIC?
    b. Compare using first and second derivative by AIC values. Which is best?
    c. Divide the dataset into training and test samples. Do first derivatives provide a better or inferior out-of-sample prediction performance compared to second derivatives?
5. The dataset `yields` used in Sect. 6.6 contains Japanese yield curves in columns 31–60.
    a. Repeat the analysis in that section except use Japanese instead of European yield curves to predict U.S. yield curves. As a predictor of USA yield curves, which seems better, Japanese or European yield curves?
    b. Use both European and Japanese yield curves to predict U.S. yield curves. Does this bivariate model provide better predictions compared to using either European or Japanese yield curves alone?
6. Starting with the data frame `PimaIndiansDiabetes` in the `mlbench` R package creates a training dataset consisting of a random subset of 668 observations on the 9 variables and save the remaining 100 observations as a test dataset. Ignore the fact that, for some variables, 0 is used to code missing observations. Using the functions `tune.svm()` and `svm()` in the package `e10701` builds a support vector machine for classifying subjects into presence or absence of diabetes. The parameter search should be conducted over

$$(\gamma, C) \in \{10^{-5}, 10^{-4}, \ldots, 10^{5}\} \times \{10^{-5}, 10^{-4}, \ldots, 10^{5}\}.$$

    Obtain the confusion matrix and estimated misclassification rate using the test dataset.
7. The data frame `plankton` in the `HRW` package contains 4000 flow cytometric measurements on 5 species of plankton `Dunaliella`, `Hemiselmis`, `Isochrysis`, `Pavlova`, and `Pyramimonas` (source: Boddy et al. 2001). The total number of records in the data frame is 4000. This exercise is concerned with building a *multi-class* support vector machine for classification of plankton species via the R function `svm()` in the `e1071`. To aid visualization of the classifier we will first work with just two features: *forward scatter* and *red fluorescence under red light.*

a. Start an R session and issue the following commands to make the plankton data available to the current session:

```
> library(HRW) ; data(plankton)
```

b. Issue the following commands to divide the row indices into a training sample of size 3000 and a test sample of size 1000:

```
> set.seed(1)
> indsTrain <- sample(1:4000,3000)
> indsTest <- setdiff(1:4000,indsTrain)
> planktonTrain <- plankton[indsTrain,]
> planktonTest <- plankton[indsTest,]
```

c. Next, plot the training data by issuing these commands:

```
> speciesNames <- as.character(unique(plankton$species))
> pointCols <- c("red","blue","green3","orange","purple")
> plot(plankton$forwScatt[indsTrain],
+       plankton$redFluorRedLight[indsTrain],
+       col = pointCols[plankton$species[indsTrain]],
+       cex = 0.5,xlab = "forward scatter",
+       ylab = "red fluorescence under red light")
> legend("topleft",legend = speciesNames,col = pointCols,
+        pch = rep(1,5),pt.cex = 0.5,cex = 1.5)
```

d. Fit a radial basis function support vector machine with $(\gamma, C) = (1, 1)$ via the commands:

```
> library(e1071)
> fitSVM <- svm(factor(species) ~ forwScatt
+              + redFluorRedLight,data = planktonTrain,
+              gamma = 1,cost = 1)
```

e. Display the support machine classifier as an image plot via the commands:

```
> meshSize <- 201
> forwScattGrid <- seq(min(plankton$forwScatt),
+                     max(plankton$forwScatt),
+                     length = meshSize)
> redFluorRedLightGrid <-
+         seq(min(plankton$redFluorRedLight),
+             max(plankton$redFluorRedLight),
+             length = meshSize)
> pixelMesh <- expand.grid(forwScattGrid,
+                          redFluorRedLightGrid)
> names(pixelMesh) <- c("forwScatt","redFluorRedLight")
> classMat <- matrix(as.numeric(predict(fitSVM,pixelMesh)),
+             meshSize,meshSize)
> imageCols <- c("tomato","cyan","palegreen",
+                "navajowhite","thistle")
> image(forwScattGrid,redFluorRedLightGrid,classMat,
+       col = imageCols,xlab = "forward scatter",
+       ylab = "red fluorescence under red light")
```

f. Add the test data to the image plot:

```
> points(plankton$forwScatt[indsTest],
+        plankton$redFluorRedLight[indsTest],cex = 0.5,
+        col = pointCols[plankton$species[indsTest]])
```

g. Compute the confusion matrix and estimated misclassification rate using the R commands:

```
> predsTest <- cbind(plankton$forwScatt[indsTest],
+                    plankton$redFluorRedLight[indsTest])
> names(predsTest) <- c("forwScatt","redFluorRedLight")
> predClasses <- predict(fitSVM,predsTest)
> trueClasses <- plankton$species[indsTest]
> confusMat <- table(predClasses,trueClasses)
> dimnames(confusMat)[[1]] <- speciesNames
> dimnames(confusMat)[[2]] <- speciesNames
> print(confusMat)
> estMisclassRate <- (sum(confusMat)
+                    - sum(diag(confusMat)))/1000
> print(paste(100*estMisclassRate,"%",sep = ""))
```

h. Fit a radial basis function support machine with $(\gamma, C) = (1, 1)$ to the same training data, but with all six predictors included. Obtain the confusion matrix and estimated misclassification rate.

i. Re-do h., but instead using `tune.svm()` to select (γ, C) from the set

$$\{10^{-4}, 10^{-3}, \ldots, 10^{4}\} \times \{10^{-4}, 10^{-3}, \ldots, 10^{4}\}.$$

Obtain the confusion matrix and estimated misclassification rate.

8. The data frame `carAuction` in the R package `HRW` contains data on 49 features of 72,983 cars sold at auction in the USA (source: "Don't Get Kicked" competition, www.kaggle.com). Also in the data frame are indicators of whether each car was considered a bad buy. The breakdown of bad buy and good buy cars is:

Bad buy	Good buy
8976	64,007

A subset of these data is used to illustrate support vector machine classification in Sect. 6.7.1.

a. Randomly select 2983 of the cars to hold back as a test dataset. The remaining 70,000 cars form a training dataset.

b. Using the R package `LowRankQP`, obtain some penalized spline support vector machine classifiers based on the training dataset obtained in a. In each case, obtain the estimated misclassification rate, based on the test data, and compare it with the estimated misclassification rate with naïvely classifying all cars as good buys: 8976/(8976 + 64,007) ≈ 12.3%. The R script `carAucPenSplSVM.R`, corresponding to the illustration given in Sect. 6.7.1.1, provides some guidance.

c. Investigate the feasibility of radial basis function support vector machine classifiers in R based on the training dataset obtained in a.

9. The Yeo–Johnson family of density functions is given by

$$p(x;\mu,\sigma,\lambda) \equiv (2\pi\sigma^2)^{-1/2}\exp\left\{\frac{-\{\psi(\lambda,x)-\mu\}^2}{2\sigma^2}\right\}(|x|+1)^{\text{sign}(x)(\lambda-1)},$$

$$\lambda,\mu \in \mathbb{R},\ \ \sigma > 0.$$

where

$$\psi(\lambda,x) \equiv \begin{cases} \{(x+1)^{\lambda}-1\}/\lambda, & x \geq 0,\ \lambda \neq 0, \\ \log(x+1) & x \geq 0,\ \lambda = 0, \\ \{1-(1-x)^{2-\lambda}\}/(2-\lambda), & x < 0,\ \lambda \neq 2, \\ -\log(1-x), & x < 0,\ \lambda = 2. \end{cases}$$

(Yeo and Johnson 2000).

a. Consider the variables `wifeAge` and `otherIncome`, defined by
```
> library(Ecdat) ; data(Workinghours)
> wifeAge <- Workinghours$age
> otherIncome <- Workinghours$income/10
```
and corresponding to wife's age in years and other household income in thousands of U.S. dollars from the 1987 cross-section of the Michigan Panel Study of Income Dynamics (e.g. Lee 1995). Out of the 3382 wife's age and other house income pairs, 1017 correspond to households for which the wife's age is in the range 20–29 years old. Let

$$s_i, \quad 1 \leq i \leq 1017,$$

be the subset of `otherIncome` values for the 20–29 wife's age bracket and consider the following Bayesian model in which the s_i are modeled as being a random sample from a member of the Yeo–Johnson family density functions:

$$\begin{array}{l} s_i\,|\mu,\sigma,\lambda \quad \text{are independent with density function } p(s_i;\mu,\sigma,\lambda) \\ \lambda \sim N(0,\sigma_{\lambda}^2), \quad \mu \sim N(0,\sigma_{\mu}^2), \quad \sigma \sim \text{Half-Cauchy}(A). \end{array} \tag{6.44}$$

Write an R script that fits (6.44) using Stan via the package `rstan`. Set the hyperparameters to be $\sigma_{\lambda} = \sigma_{\mu} = A = 10^5$. The `target +=` facility, illustrated in Sect. 6.2.2, is required for model specification since the Yeo–Johnson family is not supported in Stan.

b. Repeat a. but for `otherIncome` values corresponding to the `wifeAge` values in the age brackets 30–39, 40–49, and 50–59.
c. Now consider the mixed model-based penalized spline model applied to the predictor/response pairs (x_i, y_i), $1 \le i \le n$, as follows:

$$\begin{array}{l} y_i \mid \boldsymbol{\beta}, \boldsymbol{u}, \boldsymbol{\gamma}, \boldsymbol{v}, \boldsymbol{\delta}, \boldsymbol{w} \text{ are independent with} \\ \text{density function } p(x_i; \mu(x_i), \sigma(x_i), \lambda(x_i)) \end{array} \tag{6.45}$$

where

$$\begin{array}{rll} & \mu(x) \equiv \beta_0 + \beta_1 x + \sum_{k=1}^{K} u_k z_k(x), & u_k \sim N(0, \sigma_u^2), \\ & \sigma(x) \equiv \gamma_0 + \gamma_1 x + \sum_{k=1}^{K} v_k z_k(x), & v_k \sim N(0, \sigma_v^2) \\ \text{and} & \lambda(x) \equiv \delta_0 + \delta_1 x + \sum_{k=1}^{K} w_k z_k(x), & w_k \sim N(0, \sigma_w^2). \end{array} \tag{6.46}$$

In (6.46), z_k, $1 \le k \le K$, is an O'Sullivan cubic spline basis over the range of the x_is. The priors are:

$$\beta_0, \beta_1 \stackrel{\text{ind.}}{\sim} N(0, \sigma_{\boldsymbol{\beta}}^2),\ \ \gamma_0, \gamma_1 \stackrel{\text{ind.}}{\sim} N(0, \sigma_{\boldsymbol{\gamma}}^2),\ \ \delta_0, \delta_1 \stackrel{\text{ind.}}{\sim} N(0, \sigma_{\boldsymbol{\delta}}^2),$$

$$\sigma_u \sim \text{Half-Cauchy}(A_u),\ \sigma_v \sim \text{Half-Cauchy}(A_v) \text{ and } \sigma_w \sim \text{Half-Cauchy}(A_w).$$

Use Stan and `rstan` to fit model (6.45) and (6.46) to the $n = 3382$ predictor/response pairs stored in the arrays `wifeAge` and `otherIncome`. For the MCMC fitting in Stan, work with input predictor data x_i, $1 \le i \le n$, pre-transformed to the unit interval. Do the same for the response data y_i, $1 \le i \le n$. Put $\sigma_{\boldsymbol{\beta}} = \sigma_{\boldsymbol{\gamma}} = \sigma_{\boldsymbol{\delta}} = A_u = A_v = A_w = 10^5$ and set the number of spline basis functions to be $K = 17$.
d. Based on the fit in c. obtain the Bayes estimates of the $100\tau\%$ quantile curves for $\tau \in \{0.05, 0.25, 0.5, 0.75, 0.95\}$. For a fixed $\tau \in (0, 1)$ this is given by

$$\psi^{-1}(\lambda(x)\,\mu(x) + \sigma(x)\Phi^{-1}(\tau))$$

where

$$\psi^{-1}(\lambda, x) \equiv \begin{cases} (\lambda x + 1)^{1/\lambda} - 1, & x \ge 0,\ \lambda \ne 0, \\ e^x - 1, & x \ge 0,\ \lambda = 0, \\ 1 - \{1 - (2 - \lambda)\,x\}^{1/(2-\lambda)}, & x < 0,\ \lambda \ne 2, \\ 1 - e^{-x}, & x < 0,\ \lambda = 2. \end{cases}$$

Transform the quantile function estimates back to the original units and plot them with the original data.

10. The data frame `lidar` in the package `HRW` contains data from a light detection and ranging experiment (source: Sigrist 1994). The dataset consists of the predictor/response pairs (x_i, y_i), $1 \le i \le n$, where

$$x_i = \text{distance traveled before the light is reflected back to its source,}$$

$$y_i = \text{logarithm of the ratio of received light from two laser sources}$$

and $n = 221$. In the data frame, these variables are named `range` and `logratio`, respectively.

a. Use the R function `plot()` to create a scatterplot of the response data versus the predictor data.
b. The scatterplot shows a strong nonlinear signal, but is also strongly heteroscedastic. Write an R script that uses Stan, via the package `rstan`, to fit the Bayesian heteroscedastic nonparametric regression model

$$y_i \overset{\text{ind.}}{\sim} N(f(x_i), g(x_i)), \quad 1 \le i \le n,$$

where

$$f(x) = \beta_0 + \beta_1 x + \sum_{k=1}^{K} u_k z_k(x), \quad u_k|\sigma_u \overset{\text{ind.}}{\sim} N(0, \sigma_u^2)$$

and

$$g(x) = \exp\left(\gamma_0 + \gamma_1 x + \sum_{k=1}^{K} v_k z_k(x)\right), \quad v_k|\sigma_v \overset{\text{ind.}}{\sim} N(0, \sigma_v^2)$$

are penalized spline models for f and g, with

$$z_k(x) = (x - \kappa_k)_+, \qquad 1 \le k \le K,$$

being the truncated line basis with knots

$$\kappa_k = \{(k+1)/(K+2)\}\text{th sample quantile of the } x_i\text{s}, \quad 1 \le k \le K,$$

and $K = 35$. Suitable noninformative priors should be placed on β_0, β_1, γ_0, γ_1, σ_u, and σ_v. It is advisable to standardize the data for the Bayesian analysis and then backtransform the results to correspond to the original data. The script should plot the scatterplot of the input data with the Bayes estimate of f and corresponding 95% pointwise credible sets added. Another plot should show the Bayes estimate of g with corresponding 95% pointwise credible sets.

c. Randomly select 20 predictor observations to be fictitiously unobserved. Write another R script that is similar to the script from part b. in that it fits the Bayesian heteroscedastic model using Stan but now treats the unobserved data as missing completely at random with missingness probability p having a Uniform prior distribution on (0, 1). The script should also obtain Bayes estimates and 95% credible sets for the unobserved predictor data and compare them with their actual values.

d. Randomly select 200 predictor observations to be fictitiously observed with Gaussian measurement error. Specifically, for these x_is we instead observe $w_i \sim N(x_i, \sigma_w^2)$ for some $\sigma_w > 0$. Set the true value of σ_w to be 50. Obtain a scatterplot of the observed data with different colors used for the (x_i, y_i) that are observed exactly and the (w_i, y_i). Write another R script that is similar to the scripts from part b. and c. in that it fits the Bayesian heteroscedastic model using Stan but now accounts for the measurement error in all but 21 of the predictor values. The script should also obtain Bayes estimates and 95% credible sets for σ_W the unobserved exact predictor data and compare them with their actual values.

11. The data frame `ragweed` in the R package `HRW` contains data on daily ragweed pollen counts for Kalamazoo, Michigan, USA, during that pollen's seasons for the years 1991–1994. Also available are data on temperature in degree Fahrenheit, wind speed in knots, and indicator of occurrence of rain. An additional variable, named `temperatureResidual` in `ragweed`, are the residuals from fitting five effective degrees of freedom smoothing splines, via the base R function `smooth.spline()`, to temperature versus day number for each annual ragweed pollen season. Of interest is the following Negative Binomial factor-by-curve additive/interaction model:

$$
\begin{aligned}
\texttt{pollenCount}_i \overset{\text{ind.}}{\sim} \text{Negative-Binomial}\Big(&\exp\big\{f_{\texttt{year}}(\texttt{dayInSeason}_i)\\
&+g(\texttt{temperatureResidual}_i)\\
&+h(\texttt{windSpeed}_i)\\
&+\beta\,\texttt{rain}_i\big\}, \kappa\Big)
\end{aligned}
\tag{6.47}
$$

where $x \sim \text{Negative-Binomial}(\mu, \kappa)$ denotes that x has the Negative Binomial probability mass function

$$
p(x;\mu,\kappa) = \frac{\kappa^{\kappa}\,\Gamma(x+\kappa)\,\mu^{x}}{\Gamma(\kappa)(\kappa+\mu)\Gamma(x+1)}, \qquad x = 0, 1, \ldots, \qquad \mu, \kappa > 0,
\tag{6.48}
$$

the $f_{\texttt{year}}$ are smooth functions for $\texttt{year} \in \{1991, 1992, 1993, 1994\}$, representing a factor-by-curve interaction between year and day in season, and g and h are also smooth functions. Using Stan via the R package `rstan` fit a Bayesian mixed model-based penalized spline version of (6.47). Note that the built-in Stan distribution `negative_binomial_2` is appropriate for the parametrization of the Negative Binomial distribution given by (6.48). Use priors and hyperparameters similar to those used in the examples of this chapter. For κ, use a noninformative Half-Cauchy prior. Plot the Bayes estimates of all functions with corresponding pointwise 95% credible sets and obtain Bayes estimates and 95% credible sets for β and κ.

References

Adler, D., Murdoch, D. and others (2017). `rgl`: 3D visualization using OpenGL. R package version 0.98.22.https://r-forge.r-project.org/projects/rgl/.

Akaike, H. (1973). Maximum likelihood identification of Gaussian autoregressive moving average models. *Biometrika*, **60**, 255–265.

Albert, J. (2007). *Bayesian Computation with* R. New York: Springer.

Al Kadiri, M., Carroll, R.J. and Wand, M.P. (2010). Marginal longitudinal semiparametric regression via penalized splines. *Statistics and Probability Letters*, **80**, 1242–1252.

Bachrach, L.K., Hastie, T., Wang, M.-C., Narasimhan, B. and Marcus, R. (1999). Bone mineral acquisition in healthy Asian, Hispanic, Black and Caucasian youth. A longitudinal study. *Journal of Clinical Endocrinology and Metabolism*, **84**, 4702–12.

Baltagi, B.H. (2013). *Econometric Analysis of Panel Data, Fifth Edition*. Chichester, U.K.: John Wiley & Sons.

Bande, M.F. and de la Fuente, M.O. (2016). `fda.usc`: Functional data analysis and utilities for statistical computing. R package version 1.3http://www.jstatsoft.org/v51/i04/.

Berry, S.M., Carroll, R.J. and Ruppert, D. (2002). Bayesian smoothing and regression splines for measurement error problems. *Journal of the American Statistical Association*, **97**, 160–169.

Biecek, P. (2014). `PBImisc`: A set of datasets used in my classes or in the book, *Modele liniowe i mieszane w R, wraz z przykladami w analizie danych*. R package version 0.999.http://cran.r-project.org.

Bishop, C.M. (2006). *Pattern Recognition and Machine Learning*. New York: Springer.

Bivand, R.S., Pebesma, E.J. and Gómez-Rubio, V. (2008). *Applied Spatial Data Analysis with* R. New York: Springer.

Boddy, L., Wilkins, M.F. and Morris, C.W. (2001). Pattern recognition in flow cytometry. *Cytometry*, **44**, 195–209.

de Boor, C. (2001). *A Practical Guide to Splines, Revised Edition*. New York: Springer-Verlag.

Braglia, L. (2016). `aplore3`: Datasets from Hosmer, Lemeshow and Sturdivant, 'Applied Logistic Regression' (Third Edition, 2013). R package version 0.9. http://www.r-project.org.

Breiman, L. and Friedman, J.H. (1985). Estimating optimal transformations for multiple regression and correlation. *Journal of the American Statistical Association*, **80**, 580–598.

Breslow, N.E. and Clayton, D.G. (1993). Approximate inference in generalized linear mixed models. *Journal of the American Statistical Association*, **88**, 9–25.

Brinkman, N.D. (1981). Ethanol fuel-A single-cylinder engine study of efficiency and exhaust emissions. *SAE Transactions*, **90**, 1410–1424.

Brooks, S.P. and Gelman, A. (1998). General methods for monitoring convergence of iterative simulations. *Journal of Computational and Graphical Statistics*, **7**, 434–455.

J. Harezlak et al., *Semiparametric Regression with R*, Use R!,
https://doi.org/10.1007/978-1-4939-8853-2

Brumback, B.A. and Rice, J.A. (1998). Smoothing spline models for the analysis of nested and crossed samples of curves (with discussion). *Journal of the American Statistical Association*, **93**, 961–994.

Cardot, H. and Sarda, P. (2011). Functional linear regression. In *The Oxford Handbook of Functional Data Analysis*, editors Ferraty, F. and Romain, Y., Oxford: Oxford University Press, pp. 21–46.

Carlin, B.P. and Louis, T.A. (2009). *Bayesian Methods for Data Analysis, Third Edition*. Boca Raton, Florida: Chapman & Hall/CRC.

Carpenter, B., Gelman, A., Hoffman, M.D., Lee, D., Goodrich, B., Betancourt, M., Brubaker, M., Guo, J., Li, P. and Riddell, A. (2017). Stan: A probabilistic programming language. *Journal of Statistical Software*, **76**, Issue 1, 1–32.

Carroll, R.J., Roeder, K. and Wasserman, L. (1999). Flexible parametric measurement error models, *Biometrics*, **55**, 44–54.

Carroll, R.J. and Ruppert, D. (1988). *Transformation and Weighting in Regression*. New York: Chapman & Hall.

Carroll, R.J., Ruppert, D., Stefanski, L.A. and Crainiceanu, C.M. (2006). *Measurement Error in Nonlinear Models, Second Edition*. Boca Raton, Florida: Chapman & Hall/CRC.

Chang, C.-C. and Lin, C.-J. (2011). LIBSVM : a library for support vector machines. *Association for Computing Machinery Transactions on Intelligent Systems and Technology*, **2**, 27:1–27:27.

Chouldechova, A., and Hastie, T. (2015). Generalized additive model selection. Unpublished manuscript. https://arxiv.org/pdf/1506.03850

Chouldechova, A., Hastie, T., and Spinu, V. (2018). `gamsel`: Fit regularization path for generalized additive models. R package version 1.8.1 http://www.r-project.org.

Cole, T.J. and Green, P.J. (1992). Smoothing reference centile curves: the LMS method and penalized likelihood. *Statistics in Medicine*, **11**, 1305–1319.

Coull, B.A., Ruppert, D. and Wand, M.P. (2001). Simple incorporation of interactions into additive models. *Biometrics*, **57**, 539–545.

Crainiceanu, C. and Ruppert, D. (2004). Likelihood ratio tests in linear mixed models with one variance component. *Journal of the Royal Statistical Society, Series B*, **66**, 165–185.

Crainiceanu, C.M., Ruppert, D., Claeskens, G. and Wand, M.P. (2005). Exact likelihood ratio tests for penalized splines. *Biometrika*, **92**, 91–103.

Crainiceanu, C., Ruppert, D. and Wand, M.P. (2005). Bayesian analysis for penalized spline regression using WinBUGS. *Journal of Statistical Software*, **14**, Issue 14, 1–24.

Craven, P. and Wahba, G. (1979). Smoothing noisy data with spline functions: estimating the correct degree of smoothing by the method of generalized cross-validation. *Numerische Mathematik*, **31**, 377–403.

Cressie, N.A.C. (2015). *Statistics for Spatial Data, Revised Edition*. New York: Wiley.

Cressie, N. and Wikle, C. (2011). *Statistics for Spatio-Temporal Data*. Hoboken, New Jersey: John Wiley & Sons.

Creswell, M. (1991). A multilevel bivariate model. In *Data Analysis with ML3*, editors Prosser, R., Rasbash, J. and Goldstein, H., London: Institute of Education, London, pp. 76–108.

Cristianini, N. and Shawe-Taylor, J. (2000). *An Introduction to Support Vector Machines and Other Kernel-Based Learning Methods*. Cambridge, U.K.: Cambridge University Press.

Croissant, Y. (2016). `Ecdat`: Data sets for econometrics. R package version 0.3. http://www.r-project.org.

Daniels, M.J. and Hogan, J.W. (2008). *Missing Data in Longitudinal Studies*. Boca Raton, Florida: Chapman & Hall/CRC.

Deb, P. and Trivedi, P.K. (1997). Demand for medical care by the elderly: A finite mixture approach. *Journal of Applied Econometrics*, **12**, 313–336.

Diggle, P., Heagerty, P., Liang, K.-L. and Zeger, S. (2002). *Analysis of Longitudinal Data, Second Edition*. Oxford, U.K.: Oxford University Press.

Diggle, P.J. and P.J. Ribeiro Jr. (2007). *Model-based Geostatistics*. New York: Springer.

Di Narzo, A.F. and Stigler, M. (2016). `tsDyn`: Nonlinear time series models with regime switching. R package version 0.9. http://github.com/MatthieuStigler/tsDyn/wiki.

Donnelly, C.A., Laird, N.M. and Ware, J.H. (1995). Prediction and creation of smooth curves for temporally correlated longitudinal data. *Journal of the American Statistical Association*, **90**, 984–989.

Duan, N. (1983). Smearing estimate: a nonparametric retransformation method. *Journal of the American Statistical Association*, **78**, 605–610.

Dunn, P.K. and Smyth, G.K. (1996). Randomized quantile residuals. *Journal of Computational and Graphical Statistics*, **5**, 236–244.

Durbán, M., Harezlak, J., Wand, M.P. and Carroll, R.J. (2005). Simple fitting of subject-specific curves for longitudinal data. *Statistics in Medicine*, **24**, 1153–1167.

Eilers, P.H.C. and Marx, B.D. (1996). Flexible smoothing with B-splines and penalties (with discussion). *Statistical Science*, **11**, 89–121.

Fahrmeir, L. and Kneib, T. (2011). *Bayesian Smoothing and Regression for Longitudinal, Spatial, and Event History Data*. Oxford, U.K.: Oxford University Press.

Fahrmeir, L. and Tutz, G. (1994). *Multivariate Statistical Modelling Based on Generalized Linear Models*. New York: Springer-Verlag.

Faraway, J.J. (2006). *Extending the Linear Model with R*. Boca Raton, Florida: Chapman & Hall/CRC.

Ferraty, F. and Romain, Y. (2011). *The Oxford Handbook of Functional Data Analysis*. Oxford: Oxford University Press.

Ferraty, F. and Vieu, P. (2006). *Nonparametric Functional Data Analysis*. New York: Springer.

Fitzmaurice, G., Davidian, M., Verbeke, G. and Molenberghs, G. (Editors) (2008). *Longitudinal Data Analysis*. Boca Raton, Florida: Chapman & Hall/CRC.

Fitzmaurice, G.M., Laird, N.M. and Ware, J.H. (2004). *Applied Longitudinal Analysis*. Hoboken, New Jersey: John Wiley & Sons.

Frees, E.W. (2004). *Longitudinal and Panel Data: Analysis and Applications in the Social Sciences*. Cambridge, U.K.: Cambridge University Press.

Fuller, W.A. (1987). *Measurement Error Models*. New York: John Wiley & Sons.

Gałecki, A. and Burzykowski, T. (2013). *Linear and Mixed-Effects Models Using* R. New York: Springer.

Gelman, A. (2006). Prior distributions for variance parameters in hierarchical models (Comment on article by Browne and Draper). *Bayesian Analysis*, **3**, 515–534.

Gelman, A., Carlin, J. B., Stern, H. S., Dunson, D. B., Vehtari, A., and Rubin, D. B. (2014), *Bayesian Data Analysis, 3rd Ed.*, Chapman & Hall Ltd (London; New York)

Gelman, A. and Hill, J. (2007). *Data Analysis using Regression and Multilevel/Hierarchical Models*. New York: Cambridge University Press.

Gelman, A. and Rubin, D.B. (1992). Inference from iterative simulation using multiple sequences. *Statistical Science*, **7**, 457–511.

Gelman, A., Sturtz, S., Ligges, U., Gorjanc, G. and Kerman, J. (2015). `R2WinBUGS`: Running WinBUGS and OpenBUGS from R/S-PLUS. R package version 2.1. http://www.r-project.org.

Goldstein, H. (2010). *Multilevel Statistical Models, Fourth Edition*. Chichester, U.K.: John Wiley & Sons.

Goldsmith, J., Bobb, J., Crainiceanu, C., Caffo, B. and Reich, D. (2011). Penalized functional regression. *Journal of Computational and Graphical Statistics*, **20**, 830–851.

Goldsmith, J., Scheipl, F., Huang, L., Wrobel, J., Gellar, J., Harezlak, J., McLean, M.W., Swihart, B., Xiao, L., Crainiceanu, C., Reiss, P., Chen, (2016). `refund`: Regression with functional data. R package version 0.1. http://www.r-project.org.

Green, P.J. and Silverman, B.W. (1994). *Nonparametric Regression and Generalized Linear Models*. London: Chapman and Hall.

Gu, C. (2017). `gss`: General smoothing splines. R package version 2.1. http://www.r-project.org.

Guo, W. (2002). Functional mixed effects models. *Biometrics*, **58**, 121–128.

Guo, J., Gabry, J. and Goodrich, B. (2017). `rstan`: R interface to Stan. R package version 2.17.2.http://mc-stan.org.

Gurrin, L.C., Scurrah, K.J. and Hazelton, M.L. (2005). Tutorial in biostatistics: spline smoothing with linear mixed models. *Statistics in Medicine*, **24**, 3361–3381.

Härdle, W., Hall, P. and Marron, J.S. (1988). How far are automatically chosen regression smoothing parameters from their optimum? *Journal of the American Statistical Association*, **83**, 86–101.

Hastie, T. (1996). Pseudosplines. *Journal of the Royal Statistical Society, Series B*, **58**, 379–396.

Hastie, T. J. (1992) Generalized additive models. Chapter 7 of Statistical Models in S eds J. M. Chambers and T. J. Hastie, Wadsworth & Brooks/Cole.

Hastie, T. (2017a). `gam`: Generalized additive models. R package version 1.14. http://www.r-project.org.

Hastie, T. (2017b). `mda`: Mixture and flexible discriminant analysis. R package version 0.4. http://www.r-project.org.

Hastie, T. (2016). `svmpath`: The support vector machine path algorithm. R package version 0.955http://www.jmlr.org/papers/volume5/hastie04a/hastie04a.pdf.

Hastie, T. and Efron, E. (2013). `lars`: Least angle regression, lasso and forward stagewise. R package version 1.2. http://www.r-project.org.

Hastie, T.J. and Tibshirani, R.J. (1986). Generalized additive models. *Statistical Science*, **1**, 297–310.

Hastie, T.J. and Tibshirani, R.J. (1990). *Generalized Additive Models*. Boca Raton, Florida: Chapman & Hall/CRC.

Hastie, T. and Tibshirani, R. (1993). Varying-coefficient models. *Journal of the Royal Statistical Society, Series B*, **55**, 757–796.

Hastie, T. and Tibshirani, R. (2000). Bayesian backfitting. *Statistical Science*, **15**, 196–223.

Hastie, T., Tibshirani, R. and Friedman, J. (2009). *The Elements of Statistical Learning, Second Edition*. New York: Springer.

Hastie, T., Tibshirani, R. and Wainwright, M. (2015). *Statistical Learning with Sparsity*. Boca Raton, Florida: CRC Press.

Hastie, T. and Zhu, J. (2006). Comment on article by J.M. Moguerza and A. Muñoz. *Statistical Science*, **21**, 352–357.

Hodges, J.S. (2014). *Richly Parameterized Linear Models*. Boca Raton, Florida: Chapman & Hall/CRC.

Hoff, P.D. (2010). *A First Course in Bayesian Statistical Methods*. New York: Springer.

Hoffman, M.D. and Gelman, A. (2014). The No-U-turn Sampler: adaptively setting path lengths in Hamiltonian Monte Carlo. *Journal of Machine Learning Research*, **15**, 1593–1623.

Horváth, L. and Kokoszka, P. (2012). *Inference for Functional Data with Applications*. New York: Springer.

Huang, A. and Wand, M.P. (2013). Simple marginally noninformative prior distributions for covariance matrices. *Bayesian Analysis*, **8**, 439–452.

Huber, P.J. (1981). *Robust Statistics*. New York: Wiley.

Huq, N.M. and Cleland, J. (1990). *Bangladesh Fertility Survey 1989 (Main Report)*. Dhaka, Bangladesh: National Institute of Population Research and Training.

Hurvich, C. M., Simonoff, J. S. and Tsai, C. (1998). Smoothing parameter selection in nonparametric regression using an improved Akaike information criterion. *Journal of the Royal Statistical Society, Series B*, **60**, 271–293.

Ivanescu, A., Staicu, A-M., Scheipl, F. and Greven, S. (2015). Penalized function-on-function regression. *Computational Statistics*, **30**, 539–568.

James, G. (2011). Sparseness and functional data analysis. In *The Oxford Handbook of Functional Data Analysis*, editors Ferraty, F. and Romain, Y., Oxford: Oxford University Press, pp. 298–326.

Jones, M.C. and Faddy, M.J. (2003). A skew extension of the t-distribution, with applications. *Journal of the Royal Statistical Society, Series B*, **65**, 159–174.

Kammann, E.E. and Wand, M.P. (2003). Geoadditive models. *Journal of the Royal Statistical Society, Series C*, **52**, 1–18.

Karatzoglou, A., Smola, A. and Hornik, K. (2016). kernlab: Kernel-based machine learning lab. R package version 0.9.http://www.r-project.org.

Kauermann, G., Krivobokova, T. and Fahrmeir, L. (2009). Some asymptotic results on generalized penalized spline smoothing. *Journal of the Royal Statistical Society, Series B*, **71**, 487–503.

Kauermann, G. and Opsomer, J.D. (2011). Data-driven selection of the spline dimension in penalized spline regression. *Biometrika*, **98**, 225–230.

Kayal, M., Vercelloni, J., Wand, M.P. and Adjeroud, M. (2015). Searching for the best bet in life-strategy: a quantitative approach to individual performance and population dynamics in reef-building corals. *Ecological Complexity*, **23**, 73–84.

Kipnis, V., Subar, A.F., Midthune, D., Freedman, L.S., Ballard-Barbash, R., Troiano, R., Bingham, S., Schoeller, D.A., Schatzkin, A. and Carroll, R.J. (2003). The structure of dietary measurement error: results of the OPEN biomarker study. *American Journal of Epidemiology*, **158**, 14–21.

Kleiber, C. and Kleiber, A. (2017). AER: Applied econometrics with R. R package version 1.2. http://cran.r-project.org.

Kneib, T., Heinzl, F., Brezger, A., Sabanes, B. and Klein, N. (2014). BayesX: R utilities accompanying the software package BayesX. R package version 0.2. http://www.bayesx.org.

Koenker, R. (2005). *Quantile Regression*. New York: Cambridge University Press.

Koenker, R., Ng, P. and Portnoy, S. (1994). Quantile smoothing splines. *Biometrika*, **81**, 673–68.

Koenker, R. (2017). quantreg: Quantile regression. R package version 5.34. http://www.r-project.org.

Kooperberg, C. (2015). polspline: Polynomial spline routines. R package version 1.1. http://www.r-project.org.

Kou, S.C. and Efron, B. (2002). Smoothers and the C_p, generalized maximum likelihood, and extended exponential criteria. *Journal of the American Statistical Association*, **97**, 766–782.

Krivobokova, T. (2013). Smoothing parameter selection in two frameworks for penalized splines. *Journal of the Royal Statistical Society, Series B*, **75**, 725–741.

Landman, B.A., Huang, A.J., Gifford, A., Vikram, D.S., Lim, I.A.L, Farrell, J.A.D., Bogovic, J.A., Hua, J., Chen, M., Jarso, S., Smith, S.A., Joel, S., Mori, S., Pekar, J.J., Barker, P.B., Prince, J.L. and van Zijl, P.C.M. (2010). Multi-parametric neuroimaging reproducibility: A 3T resource study. *NeuroImage*, **54**, 2854–2866.

Lange, K.L., Little, R.J.A. and Taylor, J.M.G. (1989). Robust statistical modeling using the t-distribution. *Journal of the American Statistical Association*, **84**, 881–896.

Lauritzen, S.L. (1996). *Graphical Models*. Oxford, U.K.: Clarendon Press.

Lee, M.-J. (1995). Semi-parametric estimation of simultaneous equations with limited dependent variables: a case study of female labour supply. *Journal of Applied Econometrics*, **10**, 187–200.

Lee, P.M. (2012). *Bayesian Statistics*. Chichester, U.K.: John Wiley & Sons.

Leisch, F. and Dimitriadou, E. (2010). mlbench: Machine learning benchmark problems. R package version 2.1. http://www.r-project.org.

Li, Y. and Ruppert, D. (2008). On the asymptotics of penalized splines. *Biometrika*, **95**, 415–436.

Ligges, U., Sturtz, S., Gelman, A., Gorjanc, G. and Jackson, C. (2017). BRugs: Interface to the OpenBUGS Markov chain Monte Carlo software. R package version 0.9. http://www.r-project.org.

Lin, Y. and Zhang, H. (2006). Component selection and smoothing in multivariate nonparametric regression. *Annals of Statistics*, **34**, 2272–2297.

Little, R.J. and Rubin, D.B. (2002). *Statistical Analysis with Missing Data, Second Edition*. Hoboken, New Jersey: John Wiley & Sons.

Loader, C. (1999). *Local Regression and Likelihood*. New York: Springer.

Loader, C. (2013). locfit: Local regression, likelihood and density estimation. R package version 1.5. http://www.r-project.org.

Longford, N.T. (2005). *Missing Data and Small-Area Estimation*. New York: Springer.

Lunn, D., Jackson, C., Best, N., Thomas, A. and Spiegelhalter, D. (2013). *The* BUGS *Book*. Boca Raton, Florida: CRC Press.

McLean, M., Hooker, G., Staicu, A.-M., Scheipl, F. and Ruppert, D. (2013). Functional generalized additive models. *Journal of Computational and Graphical Statistics*, **23**, 249–269.

Marley, J.K. and Wand, M.P. (2010). Non-standard semiparametric regression via BRugs. *Journal of Statistical Software*, **37**, Issue 5, 1–30.

Maronna, R.A., Martin, D.R. and Yohai, V.J. (2006). *Robust Statistics: Theory and Methods*. London: Wiley.

Marra, G. and Wood, S.N. (2011). Practical variable selection for generalized additive models. *Computational Statistics and Data Analysis*, **55**, 2372–2387.

McCullagh, P. and Nelder, J.A. (1989). *Generalized Linear Models, Second edition*. London: Chapman and Hall.

McCulloch, C.E., Searle, S.R. and Neuhaus, J.M. (2008). *Generalized, Linear, and Mixed Models, Second Edition*. New York: John Wiley & Sons.

Meyer, D., Dimitriadou, E., Hornik, K., Weingessel, A., Leisch, F., Chang, C.-C., Lin, C.-C. (2017). `e1071`: Miscellaneous functions of the department of statistics, probability theory group (formerly: E1071), TU Wien. R package version 1.6. http://www.r-project.org.

Minka, T., Winn, J., Guiver, J., Webster, S., Zaykov, Y., Yangel, B., Spengler, A. and Bronskill, J. (2014). Infer.NET 2.6, Microsoft Research Cambridge.http://research.microsoft.com/infernet

Moguerza, J.M. and Muñoz, A. (2006). Support vector machines with applications (with discussion). *Statistical Science*, **21**, 322–362.

Müller, H.G., Wu, Y. and Yao, F. (2013). Continuously additive models for nonlinear functional regression. *Biometrika*, **100**, 607–622.

Müller, H.G. and Yao, F. (2008). Functional additive models. *Journal of the American Statistical Association*, **103**, 1534–1544.

Munnell, A.H., Tootell, G.M.B., Browne, L.E. and McEneaney, J. (1996). Mortgage lending in Boston: Interpreting HDMA data. *American Economic Review*, 25–53.

Nason, G.P. (2008). *Wavelet Methods in Statistics with* R. New York: Springer.

Nason, G. (2016). `wavethresh`: Wavelets statistics and transforms. R package version 4.6.8. http://www.r-project.org.

Nychka, D., Furrer, R., Paige, J. and Sain, S. (2017). `fields`: Tools for spatial data. R package version 9.0. http://www.r-project.org.

O'Connell, M.A. and Wolfinger, R.D. (1997). Spatial regression models, response surfaces, and process optimization. *Journal of Computational and Graphical Statistics*, **6**, 224–241.

Ormerod, J.T. and Wand, M.P. (2018). `LowRankQP`: Low-rank quadratic programming. R package version 1.0.3. http://www.r-project.org.

Ormerod, J.T., Wand, M.P. and Koch, I. (2008). Penalised spline support vector classifiers: computational issues. *Computational Statistics*, **23**, 623–641.

O'Sullivan, F. (1986). A statistical perspective on ill-posed inverse problems (with discussion). *Statistical Science*, **1**, 505–527.

Parker, R.L. and Rice, J.A. (1985). Comment on article by B.W. Silverman. *Journal of the Royal Statistical Society, Series B*, **47**, 40–42.

Pearce, N.D. and Wand, M.P. (2006). Penalized splines and reproducing kernel methods. *The American Statistician*, **60**, 233–240.

Pearce, N.D. and Wand, M.P. (2009). Explicit connections between longitudinal data analysis and kernel machines. *Electronic Journal of Statistics*, **3**, 797–823.

Pearl, J. (1988). *Probabilistic Reasoning in Intelligent Systems*. San Francisco: Morgan Kaufmann Publishers.

Pinheiro, J.C. and Bates, D.M. (2000). *Mixed-Effects Models in* S *and* S-PLUS. New York: Springer-Verlag.

Pinheiro, J., Bates, D., DebRoy, S., Sarkar, D., EISPACK authors and R Core Team. (2017). `nlme`: Linear and nonlinear mixed effects models. R package version 3.1. http://www.r-project.org.

Pope, C.A., Dockery, D.W., Spengler, J.D. and Raizenne, M.E. (1991). Respiratory health and PM_{10} pollution: a daily time series analysis. *American Review of Respiratory Disease*, **144**, 668–674.

Pratt, J.H., Jones, J.J., Miller, J.Z., Wagner, M.A. and Fineberg, N.S. (1989). Racial differences in aldosterone excretion and plasma aldosterone concentrations in children. *New England Journal of Medicine*, **321**, 1152–1157.

Pya, N. (2017). scam: Shape constrained additive models. R package version 1.2-2.http://cran.r-project.org.

Pya, N. and Wood, S.N. (2016). A note on basis dimension selection in generalized additive modelling. Unpublished manuscript.http://arxiv.org/abs/1602.06696

Ramsay, J.O., Hooker, G. and Graves, S. (2009). *Functional Data Analysis with* R *and* MATLAB. New York: Springer.

Ramsay, J.O. and Silverman, B.W. (2006). *Functional Data Analysis, Second Edition*. New York: Springer.

Ramsay, J.O., Wickham, H., Graves, S. and Hooker, G. (2017). fda: Functional data analysis. R package version 2.4.7. http://www.functionaldata.org.

Rasmussen, C.E. and Williams, C.K.I. (2006). *Gaussian Processes for Machine Learning*. Cambridge, Massachusetts: MIT Press.

Rao, J.N.K. and Molina, I. (2015). *Small Area Estimation, Second Edition*. Hoboken, New Jersey: John Wiley & Sons.

Ravikumar, P., Lafferty, J., Liu, H. and Wasserman, L. (2009). Sparse additive models. *Journal of the Royal Statistical Society, Series B*, **71**, 1009–1030.

Reiss, P.T. and Ogden, R.T. (2009). Smoothing parameter selection for a class of semiparametric linear models. *Journal of the Royal Statistical Society, Series B*, **71**, 505–523.

Ribeiro, P.J. and Diggle, P.J. (2016). geoR: Analysis of geostatistical data. R package version 1.7. http://www.leg.ufpr.br/geoR.

Richardson, S., Leblond, L., Jaussent, I. and Green, P.J. (2002). Mixture models in measurement error problems with reference to epidemiological studies. *Journal of the Royal Statistical Society, Series A*, **165**, 549–566.

Rigby, R.A. and Stasinopoulos, D.M. (2005). Generalized additive models for location, scale and shape. *Journal of the Royal Statistical Society, Series C*, **54**, 507–554.

Ripley, B., Venables, B., Bates, D.M., Hornik, K., Gebhardt, A. and Firth, D. (2015). MASS: Support functions and datasets Venables and Ripley's 'Modern Applied Statistics with S'. R package version 7.3.http://www.stats.ox.ac.uk/pub/MASS4/.

Robinson, G.K. (1991). That BLUP is a good thing: the estimation of random effects. *Statistical Science*, **6**, 15–51.

Roeder, K., Carroll, R.J. and Lindsay, B.G. (1996). A semiparametric mixture approach to case-control studies with errors in covariables. **91**, 722–732.

Ruppert, D. (2002). Selecting the number of knots for penalized splines, *Journal of Computational and Graphical Statistics*, **11**, 735–757.

Ruppert, D. and Matteson, D.S. (2015). *Statistics and Data Analysis for Financial Engineering, Second Edition*. New York: Springer.

Ruppert, D., Wand, M.P. and Carroll, R.J. (2003). *Semiparametric Regression*. Cambridge, U.K.: Cambridge University Press.

Ruppert, D., Wand, M.P. and Carroll, R.J. (2009). Semiparametric regression during 2003–2007. *Electronic Journal of Statistics*, **3**, 1193–1256

Salibian-Barrera, M., Cubranic, D. and Alimadad, A. (2014). rgam: Robust generalized additive model. R package version 0.6. http://www.stat.ubc.ca/~matias/rgam/.

Sarkar, D. (2008). *Lattice*. New York: Springer.

Sarkar, D. (2017). lattice: Trellis graphics for R. R package version 0.20. http://lattice.r-forge.r-project.org/.

Scheipl, F. and Bolker, B. (2016). RLRsim: Exact (restricted) likelihood ratio tests for mixed and additive models. R package version 3.1.https://github.com/fabian-s/RLRsim.

Schölkopf, B. and Smola, A.J. (2002). *Learning with Kernels*. Cambridge, Massachusetts: MIT Press.

Sigrist, M.W. (Editor) (1994). *Air Monitoring by Spectroscopic Techniques*. New York: Wiley.

Sommer, A. (1982). *Nutritional Blindness*. New York: Oxford University Press.

Stan Development Team. (2017). *Stan Modeling Language User's Guide and Reference Manual, Stan Version 2.17.0.* http://mc-stan.org.

Stasinopoulos, D.M. and Rigby, R.A. (2008). Generalized additive models for location scale and shape (GAMLSS) in R. *Journal of Statistical Software*, **23**, Issue 7, 1–46.

Stasinopoulos, M. and Rigby, B. (2017). `gamlss`: Generalised additive models for location scale and shape. R package version 5.0.http://www.gamlss.org/.

Svensson, L.E.O. (1994). Estimating and interpreting forward interest rates: Sweden 1992–1994. International Monetary Fund Working Paper WP/94/114. Washington D.C.: International Monetary Fund Publication Services.

Taylor, J. and Tibshirani, R.J. (2015). Statistical learning and selective inference. *Proceedings of the National Academy of Sciences of the United States of America*, **112**, 7629–7634.

Therneau, T.M. and Lumley, T. (2015). `survival`: Survival analysis. R package version 2.38. http://www.r-project.org.

Tibshirani, R. (1996). Regression shrinkage and selection via the lasso. *Journal of the Royal Statistical Society, Series B, Methodological*, **58**, 267–288.

Tibshirani, R., Tibshirani, R., Taylor, J., Loftus, J. and Reid, S. (2017). `selectiveInference`: Tools for post-selection inference. R package version 1.2.4. http://cran.r-project.org.

Umlauf, N., Adler, D., Kneib, T., Lang, S. and Zeileis, A. (2015). Structured additive regression models: an R interface to BayesX. *Journal of Statistical Software*, **63**, Issue 21, 1–47.

Umlauf, N., Kneib, T., Lang, S. and Zeileis, A. (2016). `R2BayesX`: Estimate structured additive regression models with BayesX. R package version 1.1. http://cran.r-project.org.

Verbeke, G. and Molenberghs, G. (2000). *Linear Mixed Models for Longitudinal Data*. New York: Springer-Verlag.

Verbyla, A.P., Cullis, B.R., Kenward, M.G. and Welham, S.J. (1999). The analysis of designed experiments and longitudinal data by using smoothing splines (with discussion). *Journal of the Royal Statistical Society, Series C*, **48**, 269–312.

Verdinelli, I. and Wasserman, L. (1991). Bayesian analysis of outlier problems using the Gibbs sampler. *Statistics and Computing*, **1**, 105–117.

Wahba, G. (1990). *Spline Models for Observational Data.* Philadelphia: SIAM.

Wand, M.P. (2009). Semiparametric regression and graphical models. *Australian and New Zealand Journal of Statistics*, **51**, 9–41.

Wand, M.P. and Ripley, B.D. (2015). `KernSmooth`: Functions for kernel smoothing supporting Wand and Jones (1995). R package version 2.23. http://cran.r-project.org

Wand, M.P. and Jones, M.C. (1995). *Kernel Smoothing*. London: Chapman and Hall.

Wand, M.P. and Ormerod, J.T. (2008). On semiparametric regression with O'Sullivan penalized splines. *Australian and New Zealand Journal of Statistics*, **50**, 179–198.

Wand, M.P. and Ormerod, J.T. (2011). Penalized wavelets: embedding wavelets into semiparametric regression. *Electronic Journal of Statistics*, **5**, 1654–1717.

Wang, Y. (1998). Mixed effects smoothing spline analysis of variance. *Journal of the Royal Statistical Society, Series B*, **60**, 159–174.

Wickham, H. and Chang, W. (2016). `ggplot2`: Create elegant data visualisations using the grammar of graphics. R package version 2.2.1 https://ggplot2.tidyverse.org.

Wong, R.K.W., Yao, F. and Lee, T.C.M. (2013). `robustgam`: Robust estimation for generalized additive models. R package version 0.1.7. http://www.r-project.org.

Wood, S.N. (2003). Thin plate regression splines. *Journal of the Royal Statistical Society, Series B*, **65**, 95–114.

Wood, S.N. (2006a). *Generalized Additive Models*. Boca Raton, Florida: Chapman & Hall/CRC.

Wood, S.N. (2006b). Low rank scale invariant tensor product smooths for generalized additive mixed models. *Biometrics*, **62**, 1025–1036.

Wood, S.N. (2017). `mgcv`: Mixed GAM computation vehicle with GCV/AIC/REML smoothness estimation. R package version 1.8.http://cran.r-project.org.

Wood, S.N., Li, Z., Shaddick, G. and Augustin, N.H. (2017). Generalized additive models for gigadata: modelling the U.K. black smoke network daily data. *Journal of the American Statistical Association*, **112**, 1199–1210.

Wu, H. and Zhang, J.-T. (2006). *Nonparametric Regression Methods for Longitudinal Data Analysis*. Hoboken, New Jersey: John Wiley & Sons.

Xiao, L., Li, Y. and Ruppert, D. (2013). Fast bivariate P-splines: the sandwich smoother. *Journal of the Royal Statistical Society, Series B*, **75**, 577–599.

Xiao, L., Ruppert, D., Zipunnikov, V. and Crainiceanu, C. (2016). Fast covariance estimation for high-dimensional function data. *Statistics and Computing*, **26**, 409–421.

Yee, T.W. (2004). Quantile regression via vector generalized additive models. *Statistics in Medicine*, **23**, 2295–2315.

Yee, T.W. (2010). The `VGAM` package for categorical data analysis. *Journal of Statistical Software*, **32**, Issue 10, 1–34.

Yee, T.W. (2017). `VGAM`: Vector generalized linear and additive models. R package version 1.0. http://www.stat.auckland.ac.nz/~yee/VGAM.

Yeo, I.-K. and Johnson, R.A. (2000). A new family of power transformations to improve normality or symmetry. *Biometrika*, **87**, 954–959.

Zeger, S. and Diggle, P.J. (1994). Semiparametric models for longitudinal data with application to CD4 cell numbers in HIV seroconverters. *Biometrics*, **50**, 689–699.

Zeger, S., Liang, K.–Y. and Albert, P.S. (1988). Models for longitudinal data: a generalized estimating equation approach. *Biometrics*, **44**, 1049–1060.

Zhang, C., Shin, S.J., Wang, J., Wu, Y., Zhang, H.H. and Liu, Y. (2013). `probsvm`: Class probability estimation for support vector machines. R package version 1.00. http://www.r-project.org.

Zhang, D., Lin, X., Raz, J. and Sowers, M. (1998). Semiparametric stochastic mixed models for longitudinal data. *Journal of the American Statistical Association*, **93**, 710–719.

Zhang, H.H. and Lin, C-Y. (2013). `cosso`: Fit regularized nonparametric regression models using COSSO. R package version 2.1. http://www.r-project.org.

Zhao, T., Li, X., Liu, H. and Roeder, K. (2014) `SAM`: Sparse additive modeling. R package version 1.0.5. http://www.r-project.org.

Zhao, Y., Staudenmayer, J., Coull, B.A. and Wand, M.P. (2006). General design Bayesian generalized linear mixed models. *Statistical Science*, **21**, 35–51.

Index

J. Harezlak et al., *Semiparametric Regression with R*, Use R!,
https://doi.org/10.1007/978-1-4939-8853-2

E

F

G

H

Printed in the USA
CPSIA information can be obtained
at www.ICGtesting.com
CBHW071046280624
10813CB00003B/244